Birkhäuser

Trends in Mathematics

Trends in Mathematics is a series devoted to the publication of volumes arising from conferences and lecture series focusing on a particular topic from any area of mathematics. Its aim is to make current developments available to the community as rapidly as possible without compromise to quality and to archive these for reference.

Proposals for volumes can be submitted using the Online Book Project Submission Form at our website www.birkhauser-science.com.

Material submitted for publication must be screened and prepared as follows:

All contributions should undergo a reviewing process similar to that carried out by journals and be checked for correct use of language which, as a rule, is English. Articles without proofs, or which do not contain any significantly new results, should be rejected. High quality survey papers, however, are welcome.

We expect the organizers to deliver manuscripts in a form that is essentially ready for direct reproduction. Any version of TeX is acceptable, but the entire collection of files must be in one particular dialect of TeX and unified according to simple instructions available from Birkhäuser.

Furthermore, in order to guarantee the timely appearance of the proceedings it is essential that the final version of the entire material be submitted no later than one year after the conference.

More information about this series at http://www.springer.com/series/4961

Piotr Kielanowski • Anatol Odzijewicz
Emma Previato
Editors

Geometric Methods in Physics XXXV

Workshop and Summer School, Białowieża, Poland, June 26 – July 2, 2016

Birkhäuser

Editors
Piotr Kielanowski
Departamento de Física
CINVESTAV
Ciudad de México, Mexico

Anatol Odzijewicz
Institute of Mathematics
University of Białystok
Białystok, Poland

Emma Previato
Department of Mathematics and Statistics
Boston University
Boston, MA, USA

ISSN 2297-0215 ISSN 2297-024X (electronic)
Trends in Mathematics
ISBN 978-3-319-63593-4 ISBN 978-3-319-63594-1 (eBook)
https://doi.org/10.1007/978-3-319-63594-1

Library of Congress Control Number: 2017962618

Mathmatics Subject Classification (2010): 01-06, 01A70, 20N99, 58A50, 58Z05, 81P16, 33D80, 51P05

Printed on acid-free paper

This book is published under the trade name Birkhäuser, www.birkhauser-science.com
The registered company is Springer International Publishing AG
The registered company address is: Gewerbestrasse 11, 6330 Cham, Switzerland

Contents

Part III: Abstracts of the Lectures at "School on Geometry and Physics"

Trends in Mathematics, ix–x

Preface

This book contains a selection of papers presented during the Thirty-Fifth "Workshop on Geometric Methods in Physics" (WGMPXXXV) and abstracts of lectures given during the Fifth "School on Geometry and Physics", which both took place in Białowieża, Poland, in the summer 2016. These two coordinated activities are an annual event. Information on the previous and the upcoming occurrences and related materials can be found at the URL: `http://wgmp.uwb.edu.pl`.

The volume is divided into four parts. It opens with a paper dedicated to the memory of S. Twareque Ali – for many years an active member of the Organizing Committee of our workshop who died suddenly in 2016. The second part, "Geometry and Physics", includes papers based on talks delivered during the workshop. The third part, "Integrability and Geometry", is based on the eponymous special session, organized by G.A. Goldin, A. Odesskii, E. Previato, E. Shemyakova and Th. Voronov. The final part contains extended abstracts of the lecture-series given during the Fifth "School on Geometry and Physics".

The WGMP is an international conference organized each year by the Department of Mathematical Physics in the Faculty of Mathematics and Computer Science of the University of Białystok, Poland. The main theme of the workshops, consistent with the title, is the application of geometric methods in mathematical physics and it includes a study of non-commutative systems, Poisson geometry, completely integrable systems, quantization, infinite-dimensional groups, supergroups and supersymmetry, quantum groups, Lie groupoids and algebroids as well as related topics. Participation in the workshops is open; the typical audience consists of physicists and mathematicians from many countries in several continents with a wide spectrum of interests.

Workshop and School are held in Białowieża, a village located in the east of Poland near the border with Belarus. Białowieża is situated on the edge of the Białowieża Forest, shared between Poland and Belarus, which is one of the last remnants of the primeval forest that covered the European Plain before human settlement and was designated a UNESCO World Heritage Site. The peaceful atmosphere of a small village, combined with natural beauty, yields a unique environment for learning and cooperating: as a result, the core audience of the WGMPs has become a strong scientific community, documented by this series of Proceedings.

The Organizing Committee of the 2016 WGMP gratefully acknowledges the financial support of the University of Białystok and the Belgian Science Policy Office (BELSPO), IAP Grant P7/18 DYGEST. Thanks also go to the U.S. National Science Foundation for providing support to participants in the "Integrability and Geometry" session of the event, Grant DMS 1609812. Last but not least, credit is due to early-career scholars and students from the University of Białystok, who contributed limitless time and effort to setting up and hosting the event, aside from being active participants in the scientific activities.

The Editors

Participants of the XXXV WGMP
(Photo by Tomasz Goliński)

Geometric Methods in Physics. XXXV Workshop 2016
Trends in Mathematics, xi–xvi

In Memory of S. Twareque Ali

Gerald A. Goldin

Abstract. We remember a valued colleague and dear friend, S. Twareque Ali, who passed away unexpectedly in January 2016.

S. Twareque Ali in Białowieża.

1. Remembering Twareque

Syed Twareque Ali, whom we all knew as Twareque, was born in 1942, and died in January 2016. This brief tribute is the second one I have prepared for him in a short period of time. With each sentence I reflect again on his extraordinary personality, his remarkable career – and, of course, on the profound influence he had in my life. Twareque was more than a colleague – he was a close friend, a confidant, and a teacher in the deepest sense.

When I remember Twareque, the first thing that comes to mind is his laughter. He found humor in his early changes of nationality: born in the British Empire, a subject of George VI, Emperor of India, he lived in pre-independence India, became a citizen of Pakistan, and then of Bangladesh – all without moving from home. Eventually he became a Canadian citizen, residing with his family in Montreal for many years.

Twareque's laughter was a balm. In times of sadness or disappointment, he was a source of optimism to all around him. His positive view of life was rooted in deep, almost unconsciously-held wisdom. Although he personally experienced profound nostalgia for those lost to him, he knew how to live with joy. He could laugh at himself, never taking difficulties too seriously.

And he loved to tell silly, inappropriate jokes – which, of course, cannot be repeated publicly. He introduced me to the clever novels by David Lodge, *Changing Places*, and *Small World*, which satirize the academic world mercilessly. In Lodge's characters, Twareque and I saw plenty of similarities to academic researchers we both knew in real life – especially, to ourselves.

Twareque was fluent in several languages, a true "citizen of the world." He loved poetry, reciting lengthy passages from memory in English, German, Italian, or Bengali. In Omar Khayyam's *Rubaiyat*, translated by Edward Fitzgerald, he found verses that spoke to him. These are among them:

> ...
>
> Come, fill the Cup, and in the Fire of Spring
> The Winter Garment of Repentance fling:
>   The Bird of Time has but a little way
> To fly – and Lo! the Bird is on the Wing.
>
> ...
>
> A Book of Verses underneath the Bough,
> A Jug of Wine, a Loaf of Bread – and Thou
>   Beside me singing in the Wilderness
> Oh, Wilderness were Paradise enow!
>
> ...
>
> The Moving Finger writes, and, having writ,
> Moves on; nor all your Piety nor Wit
>   Shall lure it back to cancel half a Line,
> Nor all your Tears wash out a Word of it.

2. A short scientific biography

Twareque obtained his M.Sc. in 1966 in Dhaka (which is now in Bangladesh). He received his Ph.D. from the University of Rochester, New York, USA, in 1973, where he studied with Gérard Emch. Professor Emch remained an inspiration to him for the rest of his life, and Twareque expressed his continuing gratitude. In 2007, together with Kalyan Sinha, he edited a volume in honor of Emch's 70th birthday [1]; and in 2015, he organized a memorial session for Emch at the 34th Workshop on Geometric Methods in Physics in Białowieża.

After earning his doctorate, Twareque held several research positions: at the International Centre for Theoretical Physics (ICTP) in Trieste, Italy; at the University of Toronto and at the University of Prince Edward Island in Canada; and at the Technical University of Clausthal, Germany in the Arnold Sommerfeld Institute for Mathematical Physics with H.-D. Doebner. He joined the mathematics faculty of Concordia University in Montreal as an assistant professor in 1981, becoming an associate professor in 1983 and a full professor in 1990.

During his career as a mathematical physicist, Twareque achieved wide recognition for his scientific achievements. He was known for his studies of quantization methods, coherent states and symmetries, and wavelet analysis. A short account cannot do justice to his accomplishments; the reader is referred for more detail to two published obituaries from which I have drawn [2, 3], and asked to forgive the many omissions. I cannot do better than to quote the summary in another tribute I wrote [4]:

"During the 1980s, Twareque worked on measurement problems in phase space, and on stochastic, Galilean, and Einsteinian quantum mechanics [5,6] Then he began to study coherent states for the Galilei and Poincaré groups, and collaborated with Stephan de Bièvre on quantization on homogeneous spaces for semidirect product groups.

"There followed his extensive, long-term, and indeed famous collaboration with Jean-Pierre Antoine and Jean Pierre Gazeau, focusing on square integrable group representations, continuous frames in Hilbert space, coherent states, and wavelets. Their joint work culminated in publication of the second edition of their book in 2014 – a veritable treasure trove of mathematical and physical ideas [7–10].

"Twareque's work on quantization methods and their meaning is exemplified by the important review he wrote with M. Engliš [11], and his work on reproducing kernel methods with F. Bagarello and Gazeau [12]."

Twareque's contributions of time and effort helped bring a number of scientific conference series to international prominence. Foremost among these was the Workshop on Geometric Methods in Physics (WGMP) in Białowieża (organized by Anatol Odzijewicz). Twareque attended virtually every meeting from 1991 to 2015, where we would see each other each summer. He was a long-time member of the local organizing committee, and co-edited the *Proceedings* volumes. Other conference series to which he contributed generously of his energy included the University of Havana International Workshops in Cuba (organized by Reinaldo Rodriguez

Ramos), and the Contemporary Problems in Mathematical Physics (Copromaph) series in Cotonou, Benin (organized by M. Norbert Hounkonnou).

He was also an active member of the Standing Committee of the International Colloquium on Group Theoretical Methods in Physics (ICGTMP) series. Twareque and his wife Fauzia came together to the 29th meeting of ICGTMP in Tianjin, China in 2012. She attended the special session where Twareque (to his surprise) was honored on the occasion of his 70th birthday. Their son Nabeel, of whom he always spoke with great pride, practices pediatric medicine in Montreal.

Twareque was a deep thinker, who sought transcendence through ideas and imagination. The truths of science and the elegance of mathematics in the quantum domain were part of the mysterious beauty for which he longed – a longing shared by many great scientists, a longing that we, too, share.

S. Twareque Ali in thought at WGMP XXXIII,
July 2, 2014. Photograph by G.A. Goldin.

As profoundly as Twareque cared about understanding the meanings of scientific ideas, he cared equally about inspiring his students to succeed. He helped them with personal as well as professional issues. As Anna Krasowska and Renata Deptula, two of his more recent students who came from Poland to work with him, wrote [2], "If anything in our lives became too complicated it was a clear sign we needed to talk to Dr. Ali. Every meeting with him provided a big dose of encouragement and new energy, never accompanied with any criticism or judgment." This was Twareque's gift – to understand, to inspire, to give of himself.

Twareque died suddenly and unexpectedly January 24, 2016 in Malaysia, after participating actively in the 8th Expository Quantum Lecture Series (EqualS8) – indeed, doing the kind of thing he loved most.

3. Concluding thoughts

Twareque believed passionately in world peace, in service to humanity, and in international cooperation. He understood the broad sweep of history. His tradition was Islam, as mine is Judaism, and although neither of us adhered to all the rituals of our traditions, we shared an interest in their history, their commonalities, and their contributions to world culture. We even researched correspondences between the roots of words in Arabic and Hebrew. On a first visit to Israel for a conference in 1993, we visited Jerusalem together. Twareque did much to aid the less privileged and less fortunate – in the best of our traditions, often anonymously.

Often one closes a retrospective on someone's life with a sunset, marking the ending of day and the beginning of night. My choice for Twareque is different. He is someone who joined a scientific mind with a spiritual heart, and for Twareque, the park and the forest in Białowieża were at the center of his spirituality. So I imagine him looking at us, even now, and marveling at the beauty of heavenly clouds reflected in the water.

Reflection of the heavens in Białowieża Park, July 4, 2013.
Photograph by G.A. Goldin.

Acknowledgment
I am deeply indebted to Twareque's family, friends, students, and colleagues. Thanks to the organizers of the 35th Workshop on Geometric Methods in Physics for this opportunity to honor and remember him.

References

[1] S. Twareque Ali, Kalyan B. Sinha (eds.), *A Tribute to Gérard G. Emch.* Hindustan Book Agency, New Delhi, 2007.

[2] J.-P. Antoine, J.P. Gazeau, G. Goldin, J. Harnad, M. Ismail, A. Krasowska, R. Deptula, R.R. Ramos. In memoriam, S. Twareque Ali (1942–2016). Le Bulletin, Centre de Recherches Mathématiques **22** (1) (2016), 9–12.

[3] F. Schroeck, J.-P. Antoine, G. Goldin, R. Benaduci. Syed Twareque Ali: Obituary. Journal of Geometry and Symmetry in Physics **41** (2016), 105–111.

[4] G.A. Goldin, *In Memoriam: Syed Twareque Ali. Proceedings of the* 31*st International Colloquium on Group Theoretical Methods in Physics.* Springer, 2017 (to be published).

[5] S.T. Ali, Stochastic localization, quantum mechanics on phase space and quantum space-time. La Rivista del Nuovo Cimento **8** (11) (1985), 1.

[6] S.T. Ali, J.A. Brooke, P. Busch, R. Gagnon, F. Schroeck, Current conservation as a geometric property of space-time. Canadian J Phys **66** (3) (1988), 238–44.

[7] S.T. Ali, J.-P. Antoine, J.P. Gazeau, Square integrability of group representations on homogeneous spaces I. Reproducing triples and frames. Ann. Inst. H. Poincaré **55** (1991), 829–856.

[8] S.T. Ali, J.-P. Antoine, J.P. Gazeau, Continuous frames in Hilbert space. Annals of Physics **222**, 1–37; Relativistic quantum frames, Ann. of Phys. **222** (1993), 38–88.

[9] S.T. Ali, J.-P. Antoine, J.P. Gazeau, U.A. Müller, Coherent states and their generalizations: A mathematical overview. Reviews in Math. Phys. **7** (1995), 1013–1104.

[10] S.T. Ali, J.-P. Antoine, J.P. Gazeau *Coherent States, Wavelets and their Generalizations* (and references therein). Springer, 2000; 2nd edn., 2014.

[11] S.T. Ali, M. Engliš, Quantization Methods: A Guide for Physicists and Analysts. Rev. Math. Phys. **17** (2005), 391–490.

[12] S.T. Ali, F. Bagarello, J.P. Gazeau, Quantizations from reproducing kernel spaces. Ann. Phys. **332** (2013), 127–142.

Gerald A. Goldin
Rutgers University
New Brunswick, NJ 08901, USA
e-mail: geraldgoldin@dimacs.rutgers.edu

Part I

Geometry & Physics

Geometric Methods in Physics. XXXV Workshop 2016
Trends in Mathematics, 3–9

Quasi-periodic Algebras and Their Physical Automorphisms

A. Antonevich and A. Glaz

Abstract. An automorphism of a quasi-periodic algebra on $\mathbb{R}^m$ is said to be physical, if it is generated by a mappings of $\mathbb{R}^m$. The aim of this work is to give a description of the mappings, corresponding to such automorphisms.

Mathematics Subject Classification (2010). Primary 46J10; Secondary 42A75.

Keywords. Quasi-periodic function, maximal ideal space, automorphism, algebraic unit.

1. Introduction: Invariant subalgebras

Invariant algebra is an important object in different fields of analysis In this paper we consider quasi-periodic algebras on $\mathbb{R}^m$ invariant under mappings of $\mathbb{R}^m$. Quasi-periodic functions and algebras arise naturally in many fields of analysis. A list of their applications are given, for example, in [1, 2]. Among them let us single out integrating of Hamiltonian systems and nonlinear equations, the theory of conductivity and the theory of quasi-crystals.

Let $B(X)$ be Banach algebra of all bounded functions on X equipped with sup-norm. Any mapping $\alpha : X \to X$ generates the composition operator

$$Wa(x) = a(\alpha(x)), \tag{1}$$

acting on $B(X)$. The operator W is linear and multiplicative, i.e., it is an endomorphism of $B(X)$. If α is invertible, then W is an automorphism of $B(X)$.

A closed subalgebra $\mathcal{A} \subset B(X)$ is said to be *invariant* with respect to α (shortly α-*invariant*) if $W(\mathcal{A}) \subset \mathcal{A}$. Then the operator W is an *endomorphism* of $\mathcal{A}$. If W is invertible on $\mathcal{A}$, then it is an *automorphism*. In this case the algebra $\mathcal{A}$ is called *two-sided invariant.*

For any given subalgebra $\mathcal{A}_0 \subset B(X)$ there exists the smallest invariant closed algebra $\mathcal{A}^+$ containing $\mathcal{A}_0$ and there exists the smallest two-sided invariant closed algebra $\mathcal{A}$ containing $\mathcal{A}_0$.

Among the motivations to construct invariant algebras can be pointed out the following.

1.1. Investigation of weighted composition operators

A weighted composition operator on $B(X)$ is an operator of the form

$$Bu(x) = a(x)u(\alpha(x)), \tag{2}$$

where the coefficient $a \in B(X)$ is a given function.

According to Gelfand–Naimark theorem any commutative C^*-algebra $\mathcal{A}$ with unity element is isomorphic to the algebra $C(\mathcal{M}(\mathcal{A}))$ of all continuous functions on a compact space $\mathcal{M}(\mathcal{A})$. This space is called *maximal ideal space of the algebra* $\mathcal{A}$. The isomorphism

$$\mathcal{A} \ni a \to \widehat{a} \in C(\mathcal{M}(\mathcal{A}))$$

is called *Gelfand transform.*

If $\mathcal{A} \subset B(X)$ is an α-invariant C^*-subalgebra, then endomorphism W induces a continuous mapping $\widehat{\alpha} : \mathcal{M}(\mathcal{A}) \to \mathcal{M}(\mathcal{A})$.

Proposition 1 ([3]). *Let $\mathcal{A} \subset B(X)$ be an invariant C^*-subalgebra and $a \in \mathcal{A}$. For the spectral radius $R(B)$ of the weighted composition operator* (2) *the following variational principle holds*

$$R(B) = \max_{\nu \in \Lambda_{\widehat{\alpha}}} \exp\left(\int_{\mathcal{M}(\mathcal{A})} \ln|\widehat{a}| d\nu\right),$$

where $\Lambda_{\widehat{\alpha}}$ is the set of all $\widehat{\alpha}$-invariant normalized Borel measures on $\mathcal{M}(\mathcal{A})$.

Let us consider operator B of the form (2) such that $a \in \mathcal{A}_0$, where $\mathcal{A}_0 \subset B(X)$ is a C^*-subalgebra. In order to apply the variational principle we need to find an α-invariant algebra $\mathcal{A}$ containing $\mathcal{A}_0$.

Example. Let $X = \mathbb{R}$, $\alpha(x) = qx$, $q \in \mathbb{R}$ and

$$Bu(x) = a(x)u(qx), \tag{3}$$

where a_0 is a continuous periodic function with period 1. In this case we need to construct the smallest algebra, containing all periodic functions with period 1 and invariant with respect to $\alpha(x) = qx$.

1.2. Cross-product construction

Let $\mathcal{A}$ be a C^*-algebra and $\tau : \mathcal{A} \to \mathcal{A}$ be an automorphism. There exists a set of C^*-algebras $\mathcal{B}$ such that

1. $\mathcal{A} \subset \mathcal{B}$;
2. there exists a unitary element $T \in \mathcal{B}$, such that $\tau(a) = T^{-1}aT$;
3. the algebra $\mathcal{B}$ is generated by $\mathcal{A}$ and T.

The largest among such algebras, denoted by $\mathcal{A} \times_\tau \mathbb{Z}$, is called *cross-product* of $\mathcal{A}$ and its automorphism τ. A canonical construction of the cross-product was proposed by von Neumann.

There exist a number of generalizations of cross-product construction to the case of endomorphism $\tau : \mathcal{A}_0 \to \mathcal{A}_0$ [4,5]. One of them is based on the following. If we construct by given algebra $\mathcal{A}_0$ a larger algebra $\mathcal{A}$ such that τ can be extended to automorphism of $\mathcal{A}$, then cross-product construction is reduced to the classical case of automorphism.

2. Almost periodic algebras

2.1. Quasi-periodic algebras

Let $CB(\mathbb{R}^m)$ be the space of all bounded continuous functions on $\mathbb{R}^m$. The smallest closed subspace in $CB(\mathbb{R}^m)$ containing all functions

$$e^{i2\pi<h,x>}, x \in \mathbb{R}^m, h \in \mathbb{R}^m$$

is the algebra $CAP(\mathbb{R}^m)$ of *continuous almost periodic functions* [6].

Any C^*-subalgebra of $\mathcal{A} \subset CAP(\mathbb{R}^m)$ is called *almost periodic.* A closed subalgebra $\mathcal{A} \subset CB(\mathbb{R}^m)$ is called *quasi-periodic,* if it is generated by a finite number of functions

$$e^{i2\pi\langle \pm h_j, x\rangle}, \ h_j \in \mathbb{R}^m, \ j = 1, 2, \ldots, N.$$

To any almost periodic function a corresponds *formal Fourier series*

$$a(x) \sim \sum_{j=1}^{\infty} C_j e^{i2\pi\langle \xi_j, x\rangle}.$$

The vectors ξ_j are called *frequencies of the function,* the set $\{\xi_j\}$ is called the *spectrum of the function a.*

For a given almost periodic algebra $\mathcal{A}$ denote by $H(\mathcal{A})$ the union of spectra of all functions from $\mathcal{A}$. The set $H(\mathcal{A}) \subset \mathbb{R}^m$ is a subgroup in $\mathbb{R}^m$ and is called the *frequencies group of the algebra $\mathcal{A}$.*

The subgroup $\Gamma \in \mathbb{R}^m$ with a finite number of generators is called the *quasi-lattice.* As an abstract group, any quasi-lattice Γ is isomorphic to $\mathbb{Z}^N$, where N is the number of independent generators.

In this terminology a subalgebra $\mathcal{A}$ is *quasi-periodic,* if $H(\mathcal{A})$ is a *quasi-lattice.* If $H(\mathcal{A}) \approx \mathbb{Z}^N$ then $\mathcal{A}$ is called the algebra with N quasi-periods.

2.2. Gelfand transform of almost periodic algebras

Let G be a commutative locally compact group. Any continuous homomorphism f from G into the unite circle $\mathbb{S}^1 = \{z \in \mathbb{C} : |z| = 1\}$ is called *the character of group G.* The set of all characters forms the *dual group $\widehat{G}$,* which is also locally compact.

According to the Pontryagin duality [7], if G is discrete, then the dual group $\widehat{G}$ is compact.

Theorem 2. *Let $\mathcal{A}$ be a C^*-subalgebra of $CAP(\mathbb{R}^m)$. Then*

$$\mathcal{M}(\mathcal{A}) = \widehat{H(\mathcal{A})},$$

i.e., the space of maximal ideals is the dual group to the frequencies group.

Consider the following examples.

1. $\mathbb{R}^m$ can be considered as a discrete group. It is a group with an uncountable set of generators. The dual group $\widehat{\mathbb{R}^m}$ is called *Bohr compact* and does not have an explicit description.

   Group $\mathbb{R}^m$ is the frequencies group of the algebra $CAP(\mathbb{R}^m)$ of all almost periodic functions. Therefore the space of maximal ideals of the algebra $CAP(\mathbb{R}^m)$ is the Bohr compact.
2. If $\mathcal{A}$ is a quasi-periodic algebra, then $H(\mathcal{A}) = \mathbb{Z}^N$ for some N and the dual group is a N-dimensional torus: $\widehat{\mathbb{Z}^N} = \mathbb{T}^N$. It follows that the Gelfand transform gives an isomorphism

$$\mathcal{A} \to C(\mathbb{T}^N) \sim CP(\mathbb{R}^N),$$

   where $CP(\mathbb{R}^N)$ is the space of continuous function on $\mathbb{R}^N$ periodic with period 1 for each variable.

The isomorphism $CP(\mathbb{R}^N) \to \mathcal{A}$ (inverse to Gelfand transform) can be constructed as follows.

Let us consider a linear embedding $\mathbb{R}^m \to L \subset \mathbb{R}^N$, where L is an m-dimensional vector subspace. Then the restrictions of functions from $CP(\mathbb{R}^N)$ on the L form a quasi-periodic algebra $\mathcal{A}_L$ on $\mathbb{R}^m$ whose frequencies group is the orthogonal projection of the lattice $\mathbb{Z}^N$ onto L.

A subspace $L \subset \mathbb{R}^N$ is said to be *totally irrational*, if there are no vectors from $\mathbb{Z}^N$ that are orthogonal to L (except zero vector). If the subspace L is totally irrational, then $H(\mathcal{A}_L) \approx \mathbb{Z}^N$ and $\mathcal{A}_L$ is a quasi-periodic algebra with N generators.

Using different totally irrational embedding of $\mathbb{R}^m$ into $\mathbb{R}^N$ there can be obtained any quasi-periodic algebra on $\mathbb{R}^m$ with N quasi-periods. These algebras are isomorphic to each other as abstract algebras, but differently realized as subalgebras of $CAP(\mathbb{R}^m)$.

Like in the paper [1] the space $\mathbb{R}^m$ will be called as the *physical space* and the space $\mathbb{R}^N$ as the *super-space*.

3. Automorphisms of quasi-periodic algebra

3.1. Statement of the problem

Let $\mathcal{A}$ be a quasi-periodic algebra on $\mathbb{R}^m$ with N quasi-periods. Each automorphism $\tau : \mathcal{A} \to \mathcal{A}$ is generated by a homeomorphism $\widehat{\tau}$ of the torus $\mathbb{T}^N$ (and by the corresponding covering mapping $\widetilde{\tau}$ of the super-space $\mathbb{R}^N$).

An automorphism τ of $\mathcal{A}$ is called *physical*, if it is generated by a mapping $\alpha : \mathbb{R}^m \to \mathbb{R}^m$ of the physical space. A general problem is to give a description of all physical automorphisms of the quasi-periodic algebras. We remark that the symmetry group of quasi-crystal consists on such mapping and the problem under consideration is connected with investigation of quasi-crystallographic groups [2].

Here we consider the following case of general problem.

Let $\mathcal{A}_0$ be a given quasi-periodic algebra on $\mathbb{R}^m$ and $\alpha : \mathbb{R}^m \to \mathbb{R}^m$ be a given mapping. In general, algebra $\mathcal{A}_0$ may be not invariant under α and the smallest invariant (two-sided invariant) algebra $\mathcal{A}$ containing $\mathcal{A}_0$ can be not quasi-periodic.

The question is: *for which mapping α of the physical space $\mathbb{R}^m$ the smallest invariant (two-sided invariant) algebra containing $\mathcal{A}_0$ is quasi-periodic?*

3.2. Invariant almost periodic algebras on $\mathbb{R}$

Let us show that this problem is meaningful even for linear mapping of $\mathbb{R}$:

$$\alpha(x) = qx, \ \ q \in \mathbb{R}.$$

Let $\mathcal{A}_0$ be the algebra of continuous functions on $\mathbb{R}$, periodic with the period 1. We will construct the smallest almost periodic algebra $\mathcal{A}$, containing $\mathcal{A}_0$ and invariant with respect to this α. As we have already noted, these issues are related to the study of operator (3).

Example 1. Let $\alpha(x) = \pi x$. Under the action of operator $(Wa)(x) = a(\pi x)$ on the $\mathcal{A}_0$ the functions with frequencies $\pi, \pi^2, \ldots$ appear. Due to the fact that number π is transcendental, there are no relations between these frequencies, and the group of frequencies of the smallest invariant algebra $\mathcal{A}^+$ is a free group with a countable number of generators $\pi, \pi^2, \ldots, \pi^k, \ldots$:

$$H(\mathcal{A}^+) = \mathbb{Z}^{\mathbb{N}}, \ \ \mathcal{M}(\mathcal{A}^+) = \mathbb{T}^{\mathbb{N}}.$$

Therefore in this case the smallest invariant almost periodic algebra $\mathcal{A}^+$ (and $\mathcal{A}$) is not quasi-periodic.

Example 2. Let $\alpha(x) = 2x$. Then $W(\mathcal{A}_0)$ is the algebra of periodic functions with a period $\frac{1}{2}$. Since $W(\mathcal{A}_0) \subset \mathcal{A}_0$, here $\mathcal{A}_0$ is α-invariant and $\mathcal{A}^+ = \mathcal{A}_0$.

But $\mathcal{A}_0$ is not a two-sided invariant. Under the action of W^{-n} the algebra of periodic functions with period 2^n is obtained. Therefore the smallest two-sided invariant algebra $\mathcal{A}$ is generated by periodic functions with periods 2^n and it is not quasi-periodic. Here

$$H(\mathcal{A}) = \left\{ \frac{k}{2^n} : k \in \mathbb{Z}, n \in \mathbb{N} \right\} \subset \mathbb{R}.$$

$H(\mathcal{A})$ is a group with a countable number of generators, but it is not free. For example, the relations $2^n h_n = h_0 = 1$ hold for the "natural" generators $h_n = 2^{-n}, n = 0, 1, \ldots$.

The dual group $\widehat{H(\mathcal{A})}$ is called *solenoid.*

The solenoid appeared in many areas. The first it has been found by L. Vietoris in 1927 as an example for the cohomology theory. Van Dantzig (1930) analyzed solenoid as an example of a compact Abelian group with a non-trivial topological structure. It arises as an example of the strange attractor for the system of differential equations (V.V. Nemytskii, V.V. Stepanov, 1940). The role of solinoid in the theory of dynamical systems was detected by S. Smale and R.F. Wilson.

Solenoid can be constructed like the Möbius strip. Let $\mathcal{K}$ be the Cantor discontinuum. Solenoid as a topological space can be obtained from the product of $[0,1] \times \mathcal{K}$ by identifying $\{0\} \times \mathcal{K}$ and $\{1\} \times \mathcal{K}$ by means of an invertible map $\phi : \mathcal{K} \to \mathcal{K} : (0, \omega) \sim (1, \phi(\omega))$.

Example 3. Let $q = 3+2\sqrt{2}$. Then $W(\mathcal{A}_0)$ is an algebra with a period q. It is easy to check that algebra $\mathcal{A}$ with the frequencies group

$$H(\mathcal{A}) = \{n + k(3 + 2\sqrt{2}) : n, k \in \mathbb{Z}\} = \{n + k2\sqrt{2} : n, k \in \mathbb{Z}\}$$

is invariant with respect to the corresponding α and it is invariant with respect to $\alpha^{-1}(x) = [3 - 2\sqrt{2}]x$. We get here the first example of a quasi-periodic algebra, two-sided invariant under a linear mapping.

Note that if we consider a very similar quasi-periodic algebra with a group of frequencies

$$\{n + k\sqrt{2} : n, k \in \mathbb{Z}\},$$

then there is no linear map with respect to which the algebra is two-sided invariant, in other words, there are no non-trivial symmetries.

3.3. Physical automorphisms on $\mathbb{R}^m$

The following definitions are similar to the well-known definitions from the number theory. The matrix $Q \in \mathbb{C}^{m\times m}$ is called *algebraic* if there is a polynomial

$$P(t) = p_n t^n + p_{n-1}t^{n-1} + \cdots + p_0, \quad p_k \in \mathbb{Z},$$

such that $P(Q) = 0$. It is called the *integer algebraic* if $p_n = 1$. Integer algebraic matrix Q is called the *algebraic unit* if the inverse Q^{-1} is also an algebraic integer (which is equivalent to $p_n = 1$ and $p_0 = \pm 1$).

The different structures of the smallest invariant almost periodic algebras from the examined above examples are determined by different algebraic properties of the corresponding numbers q. Indeed, number π is not algebraic, numbers 2 is algebraic integer but not an algebraic unit, and $q = 3 + 2\sqrt{2}$ is an algebraic unit, since it is a root of the polynomial $t^2 - 6t + 1 = 0$.

The next theorem asserts that for arbitrary m the results are similar.

Algebra $\mathcal{A}_0$ is called *irreducible with respect to* Q if minimal vector subspace S of $\mathbb{R}^m$, containing $H(\mathcal{A}_0)$ and invariant with respect to the conjugate map $x \to Q^T x$ is the $\mathbb{R}^m$.

Theorem 3 **([8]).** *Let $\mathcal{A}_0$ be a quasi-periodic algebra on $(\mathbb{R}^m)$, $\alpha(x) = Qx$ and $\mathcal{A}_0$ is irreducible with respect to Q.*

The smallest closed two-sided invariant subalgebra $\mathcal{A}$, that includes $\mathcal{A}_0$, is quasi-periodic if and only if Q is an algebraic unit.

In this case $\mathcal{M}(\mathcal{A}) = \mathbb{T}^N$, and the induced homeomorphism $\widehat{\alpha} : \mathbb{T}^N \to \mathbb{T}^N$ is an algebraic automorphism of the torus: the covering mapping of $\mathbb{R}^N$ is given by a *matrix* $M_Q \in \mathbb{Z}^{N\times N}$ *with determinant* ± 1. *The algebra* $\mathcal{A}$ *is realized as restriction of* $CP(\mathbb{R}^N)$ *on an* m*-dimensional subspace* L *invariant with respect to* M_Q.

Theorem 4. *For a given invertible mapping* $\alpha : \mathbb{R}^m \to \mathbb{R}^m$ *there exists a two-sided* α*-invariant quasi-periodic algebra* $\mathcal{A}$ *if and only if* α *can be represented in the form* $\alpha(x) = Qx + \varphi(x)$, *where* Q *is an algebraic unit and the mapping* $\varphi(x)$ *is quasi-periodic (all components are quasi-periodic functions).*

References

[1] I.A. Dynnikov, S.P. Novikov, *Topology of quasi-periodic functions on plane.* Uspekhi matem. nauk **60:1(361)** (2005), 3–28.

[2] Le Ty Quok Thang, S.A. Piunikhin S.A., Sadov V.A., *Geometry of quasi-crystals.* Uspekhi matem. nauk, 1993, 48:1(289), 41–102.

[3] A.B. Antonevich, *Linear functional equations. Operator approach.* Birkhäuser, 1996.

[4] R. Exel, *A new look at the crossed product of a* C^**-algebra by an endomorphism,* Ergodic Theory Dynam. Systems 23(2003) 1733–1750.

[5] A.B. Antonevich, V.I. Bakhtin, A.V. Lebedev, *Crossed product by an endomorphism, algebras of coefficients and transfer-operators,* Math. Sbornik, 202:9(2011), 3–34.

[6] B.M. Levitan, V.V. Zhikov, *Almost periodic functions and differential equations,* MGU, Moscow, 1978.

[7] L.S. Pontryagin, *Continuous Groups,* "Nauka", Moscow,1973. English transl. Topological Groups, Gordon and Breach, New York, 1066.

[8] A.B. Antonevich, A.N. Glaz, *Quasi-periodic algebras invariant with respect to linear mapping,* Doklady NAN Belarusi. – 2014. – N. 5. – N. 30–35.

A. Antonevich
Institute of Mathematics
University of Białystok
Ciołkowskego 1M
15-245 Białystok, Poland
e-mail: antonevich@bsu.by

A. Glaz
Belorusian State University
Nezavisimosti 4
220030 Minsk, Belarus
e-mail: anna-glaz@yandex.ru

Geometric Methods in Physics. XXXV Workshop 2016
Trends in Mathematics, 11–17

Berezin Symbols on Lie Groups

Ingrid Beltiţă, Daniel Beltiţă and Benjamin Cahen

Abstract. In this paper we present a general framework for Berezin covariant symbols, and we discuss a few basic properties of the corresponding symbol map, with emphasis on its injectivity in connection with some problems in representation theory of nilpotent Lie groups.

Mathematics Subject Classification (2010). Primary 22E27; Secondary 22E25, 47L15.

Keywords. Coherent states, Berezin calculus, coadjoint orbit.

1. Introduction

Let $\mathcal{V}$ be a finite-dimensional complex Hilbert space and N be a second countable smooth manifold with a fixed Radon measure μ. We denote by $L^2(N, \mathcal{V}; \mu)$ the complex Hilbert space of (equivalence classes of) $\mathcal{V}$-valued functions μ-measurable on N that are absolutely square integrable with respect to μ. We also endow the space of smooth functions $\mathcal{C}^\infty(N, \mathcal{V})$ with the Fréchet topology of uniform convergence on compact sets together with their derivatives of arbitrarily high degree.

If $\mathcal{H} \subseteq L^2(N, \mathcal{V})$ is a closed linear subspace with $\mathcal{H} \subseteq \mathcal{C}^\infty(N, \mathcal{V})$, then the inclusion map $\mathcal{H} \hookrightarrow \mathcal{C}^\infty(N, \mathcal{V})$ is continuous, hence for every $x \in N$ the evaluation map $K_x\colon \mathcal{H} \to \mathcal{V}$, $f \mapsto f(x)$, is continuous. The map

$$K\colon N \times N \to \mathcal{B}(\mathcal{V}), \quad K(x, y) := K_x K_y^*$$

is called the *reproducing kernel* of the Hilbert space $\mathcal{H}$. Then for every linear operator $A \in \mathcal{B}(\mathcal{H})$ we define its *full symbol* as

$$K^A\colon N \times N \to \mathcal{B}(\mathcal{V}), \quad K^A(x, y) := K_x A K_y^*\colon \mathcal{V} \to \mathcal{V}$$

The research of the first two named authors has been partially supported by grant of the Romanian National Authority for Scientific Research and Innovation, CNCS–UEFISCDI, project number PN-II-RU-TE-2014-4-0370.

and $K^A \in \mathcal{C}^\infty(N \times N, \mathcal{B}(\mathcal{V}))$. See [12, §I.2] for a detailed discussion of this construction, which goes back to [6] and [7].

Main problem

In the above setting, the full symbol map

$$\mathcal{B}(\mathcal{H}) \to \mathcal{C}^\infty(N \times N, \mathcal{B}(\mathcal{V})), \quad A \mapsto K^A$$

is injective, as easily checked (see also Proposition 1(1) below). Therefore it is interesting to find sufficient conditions on a continuous map $\iota\colon \Gamma \to N \times N$, ensuring that the corresponding ι-restricted symbol map

$$S^\iota\colon \mathcal{B}(\mathcal{H}) \to \mathcal{C}(\Gamma, \mathcal{B}(\mathcal{V})), \quad A \mapsto K^A \circ \iota$$

is still injective. The case of the diagonal embedding $\iota\colon \Gamma = N \hookrightarrow N \times N$, $x \mapsto (x,x)$, is particularly important and in this case the ι-restricted symbol map is called the (non-normalized) *Berezin covariant symbol map* and is denoted simply by S, hence

$$S\colon \mathcal{B}(\mathcal{H}) \to \mathcal{C}^\infty(N, \mathcal{B}(\mathcal{V})), \quad (S(A))(x) := K_x A K_x^*\colon \mathcal{V} \to \mathcal{V}.$$

In the present paper we will discuss the above problem and we will briefly sketch an approach to that problem based on results from our forthcoming paper [4]. This approach blends some techniques of reproducing kernels and some basic ideas of linear partial differential equations, in order to address a problem motivated by representation theory of Lie groups (see [8–11]). This problem is also related to some representations of infinite-dimensional Lie groups that occur in the study of magnetic fields (see [1] and [3]). Let us also mention that linear differential operators associated to reproducing kernels have been earlier used in the literature (see, for instance, [5]).

2. Basic properties of the Berezin covariant symbol map

In the following we denote by $\mathfrak{S}_p(\bullet)$ the Schatten ideals of compact operators on Hilbert spaces for $1 \le p < \infty$.

Proposition 1. *In the above setting, if $A \in \mathcal{B}(\mathcal{H})$, then one has:*

1. *If $A \ge 0$, then $S(A) \ge 0$, and moreover $S(A) = 0$ if and only if $A = 0$.*
2. *For all $f \in \mathcal{H}$ and $x \in N$ one has*

$$(Af)(x) = \int_N K^A(x,y) f(y) \mathrm{d}\mu(y).$$

3. *If $\{e_j\}_{j\in J}$ is an orthonormal basis of $\mathcal{H}$, then for all $x, y \in N$ one has*

$$K^A(x,y) = \sum_{j\in J} K_x e_j \otimes \overline{K_y A^* e_j} = \sum_{j \in J} e_j(x) \otimes \overline{(A^* e_j)(y)} \in \mathcal{B}(\mathcal{V}),$$

where for any $v, w \in \mathcal{V}$ we define their corresponding rank-one operator $v \otimes \overline{w} := (\,\cdot \mid w) v \in \mathcal{B}(\mathcal{V})$.

4. *If* $A \in \mathfrak{S}_2(\mathcal{H})$, *then*

$$\|A\|^2_{\mathfrak{S}_2(\mathcal{H})} = \iint\limits_{N\times N} \|K^A(x,y)\|^2_{\mathfrak{S}_2(\mathcal{V})} \mathrm{d}\mu(x)\mathrm{d}\mu(y)$$

and if $A \in \mathfrak{S}_1(\mathcal{H})$, *then*

$$\operatorname{Tr} A = \int\limits_N \operatorname{Tr} K^A(x,x)\mathrm{d}\mu(x).$$

Proof. See [4] for more general versions of these assertions, in which in particular the Hilbert space $\mathcal{V}$ is infinite-dimensional. Assertion (1) is a generalization of [12, Ex. I.2.3(c)], Assertion (1) is a generalization of [12, Prop. I.1.8(b)], while Assertion (1) is a generalization of [12, Cor. A.I.12]. □

3. Examples of Berezin symbols and specific applications

Here we specialize to the following setting:

1. G is a connected, simply connected, nilpotent Lie group with its Lie algebra $\mathfrak{g}$, whose center is denoted by $\mathfrak{z}$, and $\mathfrak{g}^*$ is the linear dual space of $\mathfrak{g}$, with the corresponding duality pairing $\langle\cdot,\cdot\rangle\colon \mathfrak{g}^*\times\mathfrak{g}\to\mathbb{R}$.
2. $\pi\colon G\to\mathcal{B}(\mathcal{H})$ be a unitary irreducible representation associated with the coadjoint orbit $\mathcal{O}\subseteq\mathfrak{g}^*$.

The group G will be identified with $\mathfrak{g}$ via the exponential map, so that $G=(\mathfrak{g},\cdot_G)$, where $\cdot_G$ is the Baker–Campbell–Hausdorff multiplication.

We use the notation $\mathcal{H}_\infty=\mathcal{H}_\infty(\pi)$ for the nuclear Fréchet space of smooth vectors of π. Let then $\mathcal{H}_{-\infty}$ be the space of antilinear continuous functionals on $\mathcal{H}_\infty$, $\mathcal{B}(\mathcal{H}_\infty,\mathcal{H}_{-\infty})$ be the space of continuous linear operators between the above space (these operators are thought of as possibly unbounded linear operators in $\mathcal{H}$), and $\mathcal{S}(\bullet)$ and $\mathcal{S}'(\bullet)$ for the spaces of Schwartz functions and tempered distributions, respectively. Then we have that

$$\mathcal{H}_\infty\hookrightarrow\mathcal{H}\hookrightarrow\mathcal{H}_{-\infty}.$$

Let $X_1,\dots,X_m$ be a Jordan–Hölder basis in $\mathfrak{g}$ and $e\subseteq\{1,\dots,m\}$ be the set of jump indices of the coadjoint orbit $\mathcal{O}$. Select $\xi_0\in\mathcal{O}$ and let $\mathfrak{g}=\mathfrak{g}_{\xi_0}\dotplus\mathfrak{g}_e$ be its corresponding direct sum decomposition, where $\mathfrak{g}_e$ is the linear span of $\{X_j\mid j\in e\}$ and $\mathfrak{g}_{\xi_0}:=\{x\in\mathfrak{g}\mid[x,\mathfrak{g}]\subseteq\operatorname{Ker}\xi_0\}$.

We need the notation for the Fourier transform. For $a\in\mathcal{S}(\mathcal{O})$ we set

$$\widehat{a}(x)=\int\limits_{\mathcal{O}}\mathrm{e}^{-\mathrm{i}\langle\xi,x\rangle}a(\xi)\mathrm{d}\xi,$$

where on $\mathcal{O}$ we consider the Liouville measure normalized such that the Fourier transform is unitary when extended to $L^2(\mathcal{O})\to L^2(\mathfrak{g}_e)$. We denote by $\check{F}$ the inverse Fourier transform of $F\in L^2(\mathfrak{g}_0)$.

Definition 2. 1. For $f \in \mathcal{H}$ and $\phi \in \mathcal{H}$, or $f \in \mathcal{H}_{-\infty}$ and $\phi \in \mathcal{H}_{\infty}$, let $\mathcal{A} \in C(\mathfrak{g}_e) \cap \mathcal{S}'(\mathfrak{g}_e)$ be the *coefficient mapping* for π, defined by

$$\mathcal{A}_{\phi} f(x) = \mathcal{A}(f,\phi)(x) := (f \mid \pi(x)\phi), \ x \in \mathfrak{g}_e.$$

2. For $f \in \mathcal{H}$ and $\phi \in \mathcal{H}$, or $f \in \mathcal{H}_{-\infty}$ and $\phi \in \mathcal{H}_{\infty}$, the *cross-Wigner distribution* $\mathcal{W}(f,\phi) \in \mathcal{S}'(\mathcal{O})$ is defined by the formula

$$\widehat{\mathcal{W}(f,\phi)} = \mathcal{A}_{\phi} f.$$

Proposition 3. *For $f, \phi \in \mathcal{H}$ we have that $\mathcal{A}(f,\phi) \in L^2(\mathfrak{g}_0)$, $\mathcal{W}(f,\phi) \in L^2(\mathcal{O})$. Moreover*

$$(\mathcal{A}(f_1,\phi_1) \mid \mathcal{A}(f_2,\phi_2))_{L^2(\mathfrak{g}_0)} = (f_1 \mid f_2)\overline{(\phi_1 \mid \phi_2)}$$
$$(\mathcal{W}(f_1,\phi_1) \mid \mathcal{W}(f_2,\phi_2))_{L^2(\mathcal{O})} = (f_1 \mid f_2)\overline{(\phi_1 \mid \phi_2)}$$

for all $f_1, f_2, \phi_1, \phi_2 \in \mathcal{H}$.

Proof. This follows from [2, Prop. 2.8(i)]. □

From now on we assume that

$$\phi \in \mathcal{H}_{\infty} \quad \text{with} \quad \|\phi\| = 1 \quad \text{is fixed.}$$

We let $V \colon \mathcal{H} \to L^2(\mathfrak{g}_e)$ be the isometry defined by

$$(Vf)(x) := (f \mid \phi_x) \text{ for all } x \in \mathfrak{g}_e,$$

where $\phi_x := \pi(x)\phi$. We denote

$$\mathcal{K} := \operatorname{Ran} V \subset L^2(\mathfrak{g}_0).$$

Then $\mathcal{K}$ is a reproducing kernel Hilbert space of smooth functions, with inner product equal to the $L^2(\mathfrak{g}_0)$-inner product, so the present construction is a special instance of the general framework of Section 1 with $\mathcal{V} = \mathbb{C}$.

The reproducing kernel of $\mathcal{K}$ is given by

$$K(x,y) = (\pi(x)\phi \mid \pi(y)\phi) = (\phi_x \mid \phi_y),$$

and $K_y(\cdot) := K(\cdot, y) \in \operatorname{Ran} V$, for all $y \in \mathfrak{g}_0$. We also note that

$$(\forall x \in \mathfrak{g}_0) \quad K_x = V\phi_x.$$

The Berezin covariant symbol of an operator $T \in \mathcal{B}(\mathcal{K})$ is then the bounded continuous function

$$S(T) \colon \mathfrak{g}_e \to \mathbb{C}, \quad S(T)(x) = (TK_x \mid K_x)_{\mathcal{K}}.$$

One thus obtains a well-defined bounded linear operator

$$S \colon \mathcal{B}(\mathcal{K}) \to \mathcal{C}^{\infty}(\mathfrak{g}_e) \cap L^{\infty}(\mathfrak{g}_e)$$

which also gives by restriction a bounded linear operator

$$S \colon \mathfrak{S}_2(\mathcal{K}) \to L^2(\mathfrak{g}_0).$$

To find accurate descriptions of the kernels of the above operators is a very important problem for many reasons, as explained in [8–11] also for other classes of Lie groups than the nilpotent ones.

The case of flat coadjoint orbits of nilpotent Lie groups

We now assume that the coadjoint orbit $\mathcal{O}$ is flat, hence its corresponding representation π is square integrable modulo the center of G.

Remark 4. Consider the representation $\rho\colon G \to \mathcal{B}(\mathcal{K})$,

$$\rho(g) = V\pi(g)V^*,$$

that is a unitary representation of G equivalent to π, thus it corresponds to the same coadjoint orbit $\mathcal{O}$. We denote by Op_ρ the Weyl calculus corresponding to this representation. The following then holds:

1. For $a \in \mathcal{S}'(\mathcal{O})$ one has $\mathrm{Op}_\rho(a) = V\mathrm{Op}(a)V^* = T_a$.
2. For $T \in \mathcal{B}(\mathcal{K})$ and $X \in \mathfrak{g}_0$, one has

$$S(\rho(x)^{-1}T\rho(x))(z) = S(T)(x \cdot z), \qquad \text{for all } z \in \mathfrak{g}_0. \tag{1}$$

Theorem 5. *Assume that in the constructions above,*

$$\phi \in \mathcal{H}_\infty \quad \textit{is such that } \mathcal{W}(\phi, \phi) \textit{ is a cyclic vector for } \alpha. \tag{2}$$

Then $S\colon \mathfrak{S}_2(\mathcal{K}) \to L^2(\mathfrak{g}_0)$ *is injective.*

Proof. The method of proof is based on specific properties of the Weyl–Pedersen calculus from [2]. □

We refer to [4] for a more complete discussion and for proofs of the above assertions in a much more general setting. To conclude this paper we will just briefly discuss an important example.

The special case of the Heisenberg groups

Let G be the Heisenberg group of dimension $2n+1$ and H be the center of G. Let $\{X_1, \dots, X_n, Y_1, \dots, Y_n, Z\}$ be a basis of $\mathfrak{g}$ in which the only nontrivial brackets are $[X_k, Y_k] = Z$, $1 \le k \le n$ and let $\{X_1^*, \dots, X_n^*, Y_1^*, \dots, Y_n^*, Z^*\}$ be the corresponding dual basis of $\mathfrak{g}^*$.

For $a = (a_1, a_2, \dots, a_n) \in \mathbb{R}^n$, $b = (b_1, b_2, \dots, b_n) \in \mathbb{R}^n$ and $c \in \mathbb{R}$, we denote by $[a, b, c]$ the element $\exp_G(\sum_{k=1}^n a_k X_k + \sum_{k=1}^n b_k Y_k + cZ)$ of G. Then the multiplication of G is given by

$$[a, b, c][a', b', c'] = [a + a', b + b', c + c' + \frac{1}{2}(ab' - a'b)]$$

and H consists of all elements of the form $[0, 0, c]$ with $c \in \mathbb{R}$.

The coadjoint action of G is then given by

$$\begin{aligned}\mathrm{Ad}^*([a,b,c])&\left(\sum_{k=1}^n \alpha_k X_k^* + \sum_{k=1}^n \beta_k Y_k^* + \gamma Z^*\right)\\ &=\sum_{k=1}^n (\alpha_k + \gamma b_k) X_k^* + \sum_{k=1}^n (\beta_k - \gamma a_k) Y_k^* + \gamma Z^*.\end{aligned}$$

Fix a real number $\lambda > 0$. By the Stone–von Neumann theorem, there exists a unique (up to unitary equivalence) unitary irreducible representation π_0 of G whose restriction to H is the character $\chi : [0,0,c] \to e^{i\lambda c}$. This representation is realized on $\mathcal{H}_0 = L^2(\mathbb{R}^n)$ as

$$\pi_0([a,b,c])(f)(x) = e^{i\lambda(c - bx + \frac{1}{2}ab)} f(x-a).$$

Here we take ϕ to be the function $\phi(x) = \left(\frac{\lambda}{\pi}\right)^{1/4} e^{-\lambda x^2/2}$. Then we have $\|\phi\|_2 = 1$.

Theorem 5 gives a new proof of the following known fact:

Corollary 6. *The map S is a bounded linear operator from $\mathfrak{S}_2(\mathcal{H}_0)$ to $L^2(\mathbb{R}^{2n})$ which is one-to-one and has dense range.*

References

[1] I. Beltiţă, D. Beltiţă, *Continuity of magnetic Weyl calculus.* J. Funct. Anal. **260** (2011), no. 7, 1944–1968.

[2] I. Beltiţă, D. Beltiţă, *Modulation spaces of symbols for representations of nilpotent Lie groups.* J. Fourier Anal. Appl. **17** (2011), no. 2, 290–319.

[3] I. Beltiţă, D. Beltiţă, *Algebras of symbols associated with the Weyl calculus for Lie group representations.* Monatsh. Math. **167** (2012), no. 1, 13–33.

[4] I. Beltiţă, D. Beltiţă, B. Cahen, *Berezin quantization for representations of nilpotent Lie groups.* Preprint, 2016.

[5] D. Beltiţă, J. Galé, *Linear connections for reproducing kernels on vector bundles.* Math. Z. **277** (2014), no. 1–2, 29–62.

[6] F.A. Berezin, *Quantization* (Russian). Izv. Akad. Nauk SSSR Ser. Mat. 38 (1974), 1116–1175. Transl. in Math. USSR Izv. **8** (1974), no. 5, 1109–1165.

[7] F.A. Berezin, *Quantization in complex symmetric spaces.* (Russian) Izv. Akad. Nauk SSSR Ser. Mat. **39** (1975), no. 2, 363–402, 472. Transl. in Math. USSR Izv. **9** (1975), no. 2, 341–379.

[8] B. Cahen, *Berezin quantization on generalized flag manifolds.* Math. Scand. **105** (2009), 66–84.

[9] B. Cahen, *Stratonovich–Weyl correspondence for compact semisimple Lie groups.* Rend. Circ. Mat. Palermo **59** (2010), 331–354.

[10] B. Cahen, *Berezin quantization and holomorphic representations.* Rend. Sem. Mat. Univ. Padova **129** (2013), 277–297.

[11] B. Cahen, *Stratonovich–Weyl correspondence via Berezin quantization.* Rend. Istit. Mat. Univ. Trieste **46** (2014), 157–180.

[12] K.-H. Neeb, *Holomorphy and Convexity in Lie Theory.* de Gruyter Expositions in Mathematics, 28. Walter de Gruyter & Co., Berlin, 2000.

Ingrid Beltiţă and Daniel Beltiţă
Institute of Mathematics "Simion Stoilow" of the Romanian Academy
P.O. Box 1-764, Bucharest, Romania
e-mail: ingrid.beltita@gmail.com
beltita@gmail.com

Benjamin Cahen
Laboratoire de Mathématiques et Applications de Metz
UMR 7122, Université de Lorraine (campus de Metz) et CNRS
Bât. A, Ile du Saulcy
F-57045 Metz Cedex 1, France
e-mail: benjamin.cahen@univ-lorraine.fr

Geometric Methods in Physics. XXXV Workshop 2016
Trends in Mathematics, 19–26

A Curious Differential Calculus on the Quantum Disc and Cones

Tomasz Brzeziński and Ludwik Dąbrowski

Abstract. A non-classical differential calculus on the quantum disc and cones is constructed and the associated integral is calculated.

Mathematics Subject Classification (2010). Primary 58B32.

Keywords. Non-commutative geometry; differential forms; integral forms.

1. Introduction

The aim of this note is to present a two-dimensional differential calculus on the quantum disc algebra, which has no counterpart in the classical limit, but admits a well-defined (albeit different from the one in [2]) integral, and restricts properly to the quantum cone algebras. In this way the results of [3] are extended to other classes of non-commutative surfaces and to higher forms. The presented calculus is associated to an orthogonal pair of skew-derivations, which arise as a particular example of skew-derivations on generalized Weyl algebras constructed recently in [1]. It is also a fundamental ingredient in the construction of the Dirac operator on the quantum cone [6] that admits a twisted real structure in the sense of [5].

The reader unfamiliar with non-commutative differential geometry notions is referred to [4].

2. A differential calculus on the quantum disc

Let $0 < q < 1$. The coordinate algebra of the quantum disc, or the quantum disc algebra $\mathcal{O}(D_q)$ [8] is a complex $*$-algebra generated by z subject to

$$z^*z - q^2zz^* = 1 - q^2. \tag{1}$$

To describe the algebraic contents of $\mathcal{O}(D_q)$ it is convenient to introduce a self-adjoint element $x = 1 - zz^*$, which q^2-commutes with the generator of $\mathcal{O}(D_q)$, $xz = q^2zx$. A linear basis of $\mathcal{O}(D_q)$ is given by monomials x^kz^l, x^kz^{*l}. We view

$\mathcal{O}(D_q)$ as a $\mathbb{Z}$-graded algebra, setting $\deg(z) = 1$, $\deg(z^*) = -1$. Associated with this grading is the degree-counting automorphism $\sigma : \mathcal{O}(D_q) \to \mathcal{O}(D_q)$, defined on homogeneous $a \in \mathcal{O}(D_q)$ by $\sigma(a) = q^{2\deg(a)}a$. As explained in [1] there is an orthogonal pair of skew-derivations $\partial, \bar{\partial} : \mathcal{O}(D_q) \to \mathcal{O}(D_q)$ twisted by σ and given on the generators of $\mathcal{O}(D_q)$ by

$$\partial(z) = z^*, \quad \partial(z^*) = 0, \qquad \bar{\partial}(z) = 0, \quad \bar{\partial}(z^*) = q^2 z, \tag{2}$$

and extended to the whole of $\mathcal{O}(D_q)$ by the (right) σ-twisted Leibniz rule. Therefore, there is also a corresponding first-order differential calculus $\Omega^1(D_q)$ on $\mathcal{O}(D_q)$, defined as follows.

As a left $\mathcal{O}(D_q)$-module, $\Omega^1(D_q)$ is freely generated by one forms $\omega, \bar{\omega}$. The right $\mathcal{O}(D_q)$-module structure and the differential $d : \mathcal{O}(D_q) \to \Omega^1(D_q)$ are defined by

$$\omega a = \sigma(a)\omega, \quad \bar{\omega} a = \sigma(a)\bar{\omega}, \qquad d(a) = \partial(a)\omega + \bar{\partial}(a)\bar{\omega}. \tag{3}$$

In particular,

$$dz = z^*\omega = q^2\omega z^*, \qquad dz^* = q^2 z\bar{\omega} = \bar{\omega} z, \tag{4}$$

and so, by the commutation rules (3),

$$\omega = \frac{q^{-2}}{1-q^2}\left(dzz - q^4 zdz\right), \qquad \bar{\omega} = \frac{q^{-2}}{1-q^2}\left(z^*dz^* - q^2 dz^* z^*\right). \tag{5}$$

Hence $\Omega^1(D_q) = \{\sum_i a_i db_i \mid a_i, b_i \in \mathcal{O}(D_q)\}$, i.e., $(\Omega^1(D_q), d)$ is truly a first-order differential calculus not just a degree-one part of a differential graded algebra. The appearance of $q^2 - 1$ in the denominators in (5) indicates that this calculus has no classical (i.e., $q = 1$) counterpart.

The first-order calculus $(\Omega^1(D_q), d)$ is a $*$-calculus in the sense that the $*$-structure extends to the bimodule $\Omega^1(D_q)$ so that $(a\nu b)^* = b^*\nu^* a^*$ and $(da)^* = d(a^*)$, for all $a, b \in \mathcal{O}(D_q)$ and $\nu \in \Omega^1(D_q)$, provided $\omega^* = \bar{\omega}$ (this choice of the $*$-structure justifies the appearance of q^2 in the definition of $\bar{\partial}$ in equation (2)). From now on we view $(\Omega^1(D_q), d)$ as a $*$-calculus, which allows us to reduce by half the number of necessary checks.

Next we aim to show that the module of 2-forms $\Omega^2(D_q)$ obtained by the universal extension of $\Omega^1(D_q)$ is generated by the anti-self-adjoint 2-form[1]

$$\mathsf{v} = \frac{q^{-6}}{q^2-1}(\omega^*\omega + q^8\omega\omega^*), \qquad \mathsf{v}^* = -\mathsf{v} \tag{6}$$

and to describe the structure of $\Omega^2(D_q)$. By (3), for all $a \in \mathcal{O}(D_q)$,

$$\mathsf{v}a = \sigma^2(a)\mathsf{v}. \tag{7}$$

Combining commutation rules (3) with the relations (4) we obtain

$$z^*dz = q^2 dzz^*, \qquad dzz - q^4 zdz = q^2(1-q^2)\omega, \tag{8}$$

[1] One should remember that the $*$-conjugation takes into account the parity of the forms; see [9].

and their $*$-conjugates. The differentiation of the first of equations (8) together with (3) and (1) yield

$$\omega\omega^* = (1-x)\mathsf{v}, \qquad \omega^*\omega = q^6(q^2x-1)\mathsf{v}, \tag{9}$$

which means that $\omega\omega^*$ and $\omega^*\omega$ are in the module generated by v. Next, by differentiating $\omega z^* = q^{-2}z^*\omega$ and $\omega z = q^2 z\omega$ and using (4) and (3) one obtains

$$\begin{aligned} d\omega z^* &= q^{-2}z^*d\omega + z(\omega^*\omega + q^4\omega\omega^*), \\ d\omega z &= q^2 z d\omega + (q^2+q^{-2})z^*\omega^2. \end{aligned} \tag{10}$$

The differentiation of $dz = z^*\omega$ yields

$$z^*d\omega = -q^2 z\omega^*\omega. \tag{11}$$

Multiplying this relation by z from left and right, and using commutation rules (1) and (3) one finds that $(1-x)d\omega = q^{-4}z^*d\omega z$. Developing the right-hand side of this equality with the help of the second of equations (10) we find

$$d\omega = \frac{1+q^{-4}}{q^2-1}z^{*2}\omega^2. \tag{12}$$

Combining (10) with (12) we can derive

$$z^{*3}\omega^2 = -z\frac{q^8}{q^4+1}\left(\omega^*\omega + q^4\omega\omega^*\right). \tag{13}$$

The multiplication of (13) by z^3 from the left and right and the usage of (1), (3) give

$$(1-x)(1-q^{-2}x)(1-q^{-4}x)\omega^2 = -\frac{q^8}{q^4+1}z^4\left(\omega^*\omega + q^4\omega\omega^*\right), \tag{14a}$$

$$(1-q^2x)(1-q^4x)(1-q^6x)\omega^2 = -\frac{q^8}{q^4+1}z^4\left(\omega^*\omega + q^4\omega\omega^*\right). \tag{14b}$$

Comparing the left-hand sides of equations (14), we conclude that

$$x\omega^2 = 0 = \omega^2 x \quad \text{and, by } *\text{-conjugation,} \quad x\omega^{*2} = 0 = \omega^{*2}x, \tag{15}$$

and hence in view of either of (14)

$$\omega^2 = -\frac{q^8}{q^4+1}z^4\left(\omega^*\omega + q^4\omega\omega^*\right). \tag{16}$$

By (9), the right-hand side of (16) is in the module generated by v, and so is ω^2 and its adjoint ω^{*2}. Thus, the module $\Omega^2(D_q)$ spanned by all products of pairs of one-forms is indeed generated by v.

Multiplying (12) and (11) by x and using relations (15) we obtain

$$xz\omega^*\omega = 0 = \omega^*\omega xz. \tag{17}$$

Following the same steps but now starting with the differentiation of $dz^* = q^2 z\omega^*$ (see (4)), we obtain the complementary relation

$$xz\omega\omega^* = 0 = \omega\omega^* xz. \tag{18}$$

In view of the definition of v, (17) and (18) yield $xz\mathsf{v} = 0 = \mathsf{v}xz$. Next, the multiplication of, say, the first of these equations from the left and right by z^* and the use of (1) yield $x(1-x)\mathsf{v} = 0$ and $x(1-q^2x)\mathsf{v} = 0$. The subtraction of one of these equations from the suitable scalar multiple of the other produces the necessary relation

$$x\mathsf{v} = 0 = \mathsf{v}x, \tag{19}$$

which fully characterizes the structure of $\Omega^2(D_q)$ as an $\mathcal{O}(D_q)$-module generated by v. In the light of (19), the $\mathbb{C}$-basis of $\Omega^2(D_q)$ consists of elements $\mathsf{v}z^n$, $\mathsf{v}z^{*m}$, and hence, for all $w \in \Omega^2(D_q)$, $wx = xw = 0$, i.e., $\Omega^2(D_q)$ is a torsion (as a left and right $\mathcal{O}(D_q)$-module). Since $\mathcal{O}(D_q)$ is a domain and $\Omega^2(D_q)$ is a torsion, the dual of $\Omega^2(D_q)$ is the zero module, hence, in particular $\Omega^2(D_q)$ is not projective. Again by (19), the annihilator of $\Omega^2(D_q)$,

$$\mathrm{Ann}(\Omega^2(D_q)) := \{a \in \mathcal{O}(D_q) \mid \forall w \in \Omega^2(D_q),\ aw = wa = 0\},$$

is the ideal of $\mathcal{O}(D_q)$ generated by x. The quotient $\mathcal{O}(D_q)/\mathrm{Ann}(\Omega^2(D_q))$ is the Laurent polynomial ring in one variable, i.e., the algebra $\mathcal{O}(S^1)$ of coordinate functions on the circle. When viewed as a module over $\mathcal{O}(S^1)$, $\Omega^2(D_q)$ is free of rank one, generated by v. Thus, although the module of 2-forms over $\mathcal{O}(D_q)$ is neither free nor projective, it can be identified with sections of a trivial line bundle once pulled back to the (classical) boundary of the quantum disc.

With (19) at hand, equations (9), (16), (12) and their $*$-conjugates give the following relations in $\Omega^2(D_q)$

$$d\omega = q^8 z^2 \mathsf{v}, \quad d\omega^* = -z^{*2}\mathsf{v}, \quad \omega\omega^* = \mathsf{v}, \quad \omega^*\omega = -q^6\mathsf{v}, \tag{20a}$$

$$\omega^2 = q^{12}\frac{q^2-1}{q^4+1}z^4\mathsf{v}, \qquad \omega^{*2} = q^{-4}\frac{q^2-1}{q^4+1}z^{*4}\mathsf{v}. \tag{20b}$$

One can easily check that (20), (19) and (7) are consistent with (3) with no further restrictions on v. Setting $\Omega^n(D_q) = 0$, for all $n > 2$, we thus obtain a 2-dimensional calculus on the quantum disc.

3. Differential calculus on the quantum cone

The quantum cone algebra $\mathcal{O}(C_q^N)$ is a subalgebra of $\mathcal{O}(D_q)$ consisting of all elements of the $\mathbb{Z}$-degree congruent to 0 modulo a positive natural number N. Obviously $\mathcal{O}(C_q^1) = \mathcal{O}(D_q)$, the case we dealt with in the preceding section, so we may assume $N > 1$. $\mathcal{O}(C_q^N)$ is a $*$-algebra generated by the self-adjoint $x = 1 - zz^*$ and by $y = z^N$, which satisfy the following commutation rules

$$xy = q^{2N}yx, \qquad yy^* = \prod_{l=0}^{N-1}\left(1 - q^{-2l}x\right), \qquad y^*y = \prod_{l=1}^{N}\left(1 - q^{2l}x\right). \tag{21}$$

The calculus $\Omega(C_q^N)$ on $\mathcal{O}(C_q^N)$ is obtained by restricting of the calculus $\Omega(D_q)$, i.e., $\Omega^n(C_q^N) = \{\sum_i a_0^i d(a_1^i)\cdots d(a_n^i)a_{n+1}^i \mid a_k^i \in \mathcal{O}(C_q^N)\}$. Since d is a degree-zero map $\Omega(C_q^N)$ contains only these forms in $\Omega(D_q)$, whose $\mathbb{Z}$-degree is a multiple of N.

We will show that all such forms are in $\Omega(C_q^N)$. Since $\deg(\omega)=2$, $\deg(\omega^*)=-2$ and $\deg(\mathsf{v})=0$, this is equivalent to

$$\Omega^1(C_q^N)=\mathcal{O}(D_q)_{\overline{-2}}\,\omega\oplus\mathcal{O}(D_q)_{\overline{2}}\,\omega^*,\qquad \Omega^2(C_q^N)=\mathcal{O}(C_q^N)\mathsf{v},$$

where $\mathcal{O}(D_q)_{\overline{s}}=\{a\in\mathcal{O}(D_q)\mid \deg(a)\equiv s \bmod N\}$.

As an $\mathcal{O}(C_q^N)$-module, $\mathcal{O}(D_q)_{\overline{-2}}$ is generated by z^{N-2} and z^{*2}, hence to show that $\mathcal{O}(D_q)_{\overline{-2}}\,\omega\subseteq\Omega^1(C_q^N)$ suffices it to prove that $z^{N-2}\omega, z^{*2}\omega\in\Omega^1(C_q^N)$. Using the Leibniz rule one easily finds that

$$dy=\left([N;q^2]-q^{-2N+4}\,[N;q^4]\,x\right)z^{N-2}\omega,$$

where $[n;s]:=\frac{s^n-1}{s-1}$. Hence, in view of (1) and (3),

$$y^*dy=[N;q^2]\left(1-q^4\frac{[N;q^4]}{[N;q^2]}x\right)\prod_{l=3}^{N}\left(1-q^{2l}x\right)z^{*2}\omega, \tag{22a}$$

$$dyy^*=q^{-2N}\,[N;q^2]\left(1-q^{-2N+4}\frac{[N;q^4]}{[N;q^2]}x\right)\prod_{l=0}^{N-3}\left(1-q^{-2l}x\right)z^{*2}\omega. \tag{22b}$$

The polynomial in x on the right-hand side of (22a) has roots in common with the polynomial on the right-hand side of (22b) if and only if there exists an integer $k\in[-2N+2,-N-1]\cup[2,N-1]$ such that

$$q^{2k}(q^{2N}+1)=q^2+1. \tag{23}$$

Equation (23) is equivalent to $q^2\,[N+k-1;q^2]+[k;q^2]=0$, with the left-hand side strictly positive if $k>0$ and strictly negative if $k\le -N$. So, there are no solutions within the required range of values of k. Hence the polynomials (22a), (22b) are coprime, and so there exists a polynomial (in x) combination of the left-hand sides of equations (22) that gives $z^{*2}\omega$. This combination is an element of $\Omega^1(C_q^N)$ and so is $z^{*2}\omega$. Next,

$$z^{*2}\omega\, y=q^{2N}(1-q^2x)(1-q^4x)z^{N-2}\omega,$$
$$yz^{*2}\omega=(1-q^{-2N+4}x)(1-q^{-2N+2}x)z^{N-2}\omega,$$

so again there is an x-polynomial combination of the left-hand sides (which are already in $\Omega^1(C_q^N)$) giving $z^{N-2}\omega$. Therefore, $\mathcal{O}(D_q)_{\overline{-2}}\,\omega\subseteq\Omega^1(C_q^N)$. The case of $\mathcal{O}(D_q)_{\overline{2}}$ follows by the $*$-conjugation.

Since $z^2\omega^*$, $z^{*2}\omega$ are elements of $\Omega^1(C_q^N)$,

$$\Omega^2(C_q^N)\ni z^2\omega^*z^{*2}\omega=q^{-4}(1-x)(1-q^{-2}x)\omega^*\omega=-q^2\mathsf{v}, \tag{24}$$

by the quantum disc relations and (20) and (19). Consequently, $\mathsf{v}\in\Omega^2(C_q^N)$. Therefore, $\Omega(C_q^N)$ can be identified with the subspace of $\Omega(D_q)$, of all the elements whose $\mathbb{Z}$-degree is a multiple of N.

4. The integral

Here we construct an algebraic integral associated to the calculus constructed in Section 2. We start by observing that since σ preserves the $\mathbb{Z}$-degrees of elements of $\mathcal{O}(D_q)$ and ∂ and $\bar{\partial}$ satisfy the σ-twisted Leibniz rules, the definition (2) implies that ∂ lowers while $\bar{\partial}$ raises degrees by 2. Hence, one can equip $\Omega^1(D_q)$ with the $\mathbb{Z}$-grading so that d is the degree zero map, provided $\deg(\omega) = 2$, $\deg(\omega^*) = -2$. Furthermore, in view of the definition of σ, one easily finds that

$$\sigma^{-1} \circ \partial \circ \sigma = q^4 \partial, \qquad \sigma^{-1} \circ \bar{\partial} \circ \sigma = q^{-4} \bar{\partial}, \tag{25}$$

i.e., ∂ is a q^4-derivation and $\bar{\partial}$ is a q^{-4}-derivation. Therefore, by [7], $\Omega(D_q)$ admits a divergence, for all right $\mathcal{O}(D_q)$-linear maps $f : \Omega^1(D_q) \to \mathcal{O}(D_q)$, given by

$$\nabla_0(f) = q^4 \partial\left(f(\omega)\right) + q^{-4} \bar{\partial}\left(f(\omega^*)\right). \tag{26}$$

Since the $\mathcal{O}(D_q)$-module $\Omega^2(D_q)$ has a trivial dual, ∇_0 is flat. Recall that by the *integral* associated to ∇_0 we understand the cokernel map of ∇_0.

Theorem 1. *The integral associated to the divergence* (26) *is a map* $\Lambda : \mathcal{O}(D_q) \to \mathbb{C}$, *given by*

$$\Lambda(x^k z^l) = \lambda \frac{[k+1; q^2]}{[k+1; q^4]} \delta_{l,0}, \qquad \text{for all } k \in \mathbb{N},\ l \in \mathbb{Z}, \tag{27}$$

where, for $l < 0$, z^l *means* z^{*-l} *and* $\lambda \in \mathbb{C}$.

Proof. First we need to calculate the image of ∇_0. Using the twisted Leibniz rule and the quantum disc algebra commutation rules (1), one obtains

$$\partial(x^k) = -q^{-2} \left[k; q^4\right] x^{k-1} z^{*2}. \tag{28}$$

Since $\partial(z^*) = 0$, (28) means that all monomials $x^k z^{*l+2}$ are in the image of ∂ hence in the image of ∇_0. Using the $*$-conjugation we conclude the $x^k z^{l+2}$ are in the image of $\bar{\partial}$ hence in the image of ∇_0. So Λ vanishes on (linear combinations of) all such polynomials. Next note that

$$\partial(z^2) = (q^2+1) - (q^4+1)x, \tag{29}$$

hence

$$\partial(z^* z^2 - q^4 z^2 z^*) = (1-q^4) z^*, \quad \partial(z^* z^2 - q^2 z^2 z^*) = (1-q^2)(1+q^4) x z^*.$$

This means that z^* and xz^* are in the image of ∂, hence of ∇_0. In fact, all the $x^k z^*$ are in this image which can be shown inductively. Assume $x^k z^* \in \mathrm{Im}(\partial)$, for all $k \leq n$. Then using the twisted Leibniz rule, (28) and (29) one finds

$$\partial(x^n z^2) = -q^2 \left[N; q^4\right] x^{n-1} + (q^2+1) \left[n+1; q^4\right] x^n - \left[n+2; q^4\right] x^{n+1}. \tag{30}$$

Since $\partial(z^*) = 0$, equation (30) implies that $\partial(z^n z^2 z^*)$ is a linear combination of monomials $x^{n-1} z^*$, $x^n z^*$ and $x^{n+1} z^*$. Since the first two are in the image of ∂ by the inductive assumption, so is the third one. Therefore, all linear combinations of $x^k z^*$ and $x^k z$ (by the $*$-conjugation) are in the image of ∇_0.

Putting together all this means that Λ vanishes on all the polynomials

$$\sum_{k,l=1}^{n} (c_{kl}x^k z^l + c'_{kl}x^k z^{*l}).$$

The rest of the formula (27) can be proven by induction. Set $\lambda = \Lambda(1)$. Since Λ vanishes on all elements in the image of ∇_0, hence also in the image of ∂, the application of Λ to the right-hand side of (28) confirms (27) for $k = 1$. Now assume that (27) is true for all $k \leq n$. Then the application Λ to the right-hand side of (30) followed by the use of the inductive assumption yields

$$\begin{aligned}\left[n+2;q^4\right]\Lambda\left(x^{n+1}\right) &= q^2\left[N;q^4\right]\Lambda\left(x^{n-1}\right) - (q^2+1)\left[n+1;q^4\right]\Lambda\left(x^n\right)\\ &= \lambda\left((q^2+1)\left[n+1;q^2\right] - q^2\left[n;q^2\right]\right)\\ &= \lambda\left[n+2;q^2\right].\end{aligned}$$

Therefore, the formula (27) is true also for $n+1$, as required. □

The restriction of Λ to the elements of $\mathcal{O}(D_q)$, whose $\mathbb{Z}$-degree is a multiple of N gives an integral on the quantum cone $\mathcal{O}(C_q^N)$.

Acknowledgment

The work on this project began during the first author's visit to SISSA, supported by INdAM-GNFM. He would like to thank the members of SISSA for hospitality. The second author was supported in part by the Simons Foundation grant 346300 and the Polish Government MNiSW 2015-2019 matching fund.

References

1. A. Almulhem & T. Brzeziński, *Skew derivations on generalized Weyl algebras.* J. Algebra 493 (2018), 194–235.
2. E.J. Beggs & S. Majid, *Spectral triples from bimodule connections and Chern connections.* J. Noncommut. Geom. 11 (2017) 669–701.
3. T. Brzeziński, *Non-commutative differential geometry of generalized Weyl algebras.* SIGMA Symmetry Integrability Geom. Methods Appl. **12** (2016), 059.
4. T. Brzeziński, *Differential and integral forms on non-commutative algebras.*
5. T. Brzeziński, N. Ciccoli, L. Dabrowski & A. Sitarz, *Twisted reality condition for Dirac operators.* Math. Phys. Anal. Geom. **19** (2016), 19:16.
6. T. Brzeziński & L. Dąbrowski, *in preparation.*
7. T. Brzeziński, L. El Kaoutit & C. Lomp, *Non-commutative integral forms and twisted multi-derivations.* J. Noncommut. Geom. **4** (2010), 281–312.
8. S. Klimek & A. Lesniewski, *A two-parameter quantum deformation of the unit disc.* J. Funct. Anal. **115** (1993), 1–23.
9. S.L. Woronowicz, *Differential calculus on compact matrix pseudogroups (quantum groups).* Comm. Math. Phys. **122** (1989), 125–170.

Tomasz Brzeziński
Department of Mathematics
Swansea University
Swansea SA2 8PP, UK

and

Department of Mathematics
University of Białystok
K. Ciolkowskiego 1M
15-245 Białystok, Poland
e-mail: T.Brzezinski@swansea.ac.uk

Ludwik Dąbrowski
SISSA
Via Bonomea 265
I-34136 Trieste, Italy
e-mail: dabrow@sissa.it

Geometric Methods in Physics. XXXV Workshop 2016
Trends in Mathematics, 27–33

Nambu Mechanics: Symmetries and Conserved Quantities

Marián Fecko

Abstract. In Nambu mechanics, continuous symmetry leads to a *relative integral invariant*, a differential form which only upon integration over a cycle provides a conserved real number. This differs sharply from what is the case in Hamiltonian mechanics, where conserved quantities are *functions* on (extended) phase space, which are constant on trajectories. The origin of the difference may be traced back to a shift in degrees of relevant form present in *action integral* for Nambu mechanics.

Mathematics Subject Classification (2010). Primary 53Z05; Secondary 49S05.

Keywords. Hamiltonian mechanics, Nambu mechanics, symmetry, conserved quantity, integral invariant.

1. Introduction

From times when seminal paper of Emmy Noether [1] was published (see also nice account in [2]), we know that there is close correspondence between symmetries of action integral and conserved quantities for the dynamics given by the action.

In Hamiltonian mechanics, as an example, the conserved quantity is represented by a *function* on the phase space of the system, which is constant on trajectories (see, e.g., [3, 4] or [5]). In practical applications of Hamiltonian mechanics, valuable information may then be obtained by evaluating the function (say, energy, a component of linear or angular momentum, etc.) in two points of particular trajectory and using the fact that the two numbers are guaranteed to be the same.

In 1973, Nambu [6] proposed a different dynamics, which later became known as *Nambu mechanics*. It is governed, in its basic version, by *two* "Nambu Hamiltonians" H_1 and H_2, each of them being a function on "Nambu phase space". Now, one easily proves that *both* H_1 and H_2 are *conserved* in the sense described above. So, one could conjecture that there are two symmetries of the corresponding action integral which lead to these particular conserved quantities.

However, construction of action integral for Nambu mechanics turns out to be a delicate matter (see [7] and [8]). Namely, the action is given by a *surface* (rather than line) integral in spite of the fact that equations of motion describe motion of *points* along *trajectories* in phase space (along "world-lines" in extended phase space; exactly like it is the case for Hamiltonian mechanics). This peculiarity then leads to the fact, that standard machinery for obtaining conserved quantity from symmetry leads, in Nambu mechanics, to a strange result: conserved quantity that one obtains for a continuous symmetry turns out to be a *relative integral invariant* rather than a function on the phase space.

2. Nambu mechanics – equations and action integral

In its basic version, Nambu equations read

$$\dot{x}_i = \epsilon_{ijk} \frac{\partial H_1}{\partial x_j} \frac{\partial H_2}{\partial x_k} \qquad i = 1, 2, 3. \tag{1}$$

Here, H_1 and H_2 are, in general, functions of x_1, x_2, x_3 and t.

As was observed in [7] and [8], equations (1) may be rewritten as "vortex lines equations"

$$i_{\dot{\gamma}} d\hat{\sigma} = 0, \tag{2}$$

where

$$\dot{\gamma} = \dot{x}^1 \partial_1 + \dot{x}^2 \partial_2 + \dot{x}^3 \partial_3 + \partial_t \tag{3}$$

is the velocity vector to curve γ on extended Nambu phase space and

$$\hat{\sigma} := x^1 dx^2 \wedge dx^3 - H_1 dH_2 \wedge dt \tag{4}$$

(see also [9]). Formally, Eq. (2) looks exactly like geometrical version of *Hamilton* equations

$$\dot{q}^a = \frac{\partial H}{\partial p_a} \qquad \dot{p}_a = -\frac{\partial H}{\partial q^a} \tag{5}$$

except for the fact, that for Hamilton equations the role of $\hat{\sigma}$ is played by

$$\sigma = p_a dq^a - H dt. \tag{6}$$

The similarity suggests that one could construct action integral for Nambu mechanics simply repeating the way it is done in Hamilton mechanics. Namely, it is well known (see again [3, 4] or [5]) that the action integral for the Hamiltonian case reads

$$S[\gamma] = \int_\gamma \sigma = \int_{t_1}^{t_2} (p_a \dot{q}^a - H) dt. \tag{7}$$

Then, replacing σ by $\hat{\sigma}$ might probably lead to the action for the Nambu case. The idea, however, does *not* work since one can not integrate *two*-form over *one*-dimensional object (curve). Instead, one is forced to integrate $\hat{\sigma}$ over a *surface*. A problem then arises how a surface may be naturally associated with Nambu trajectories.

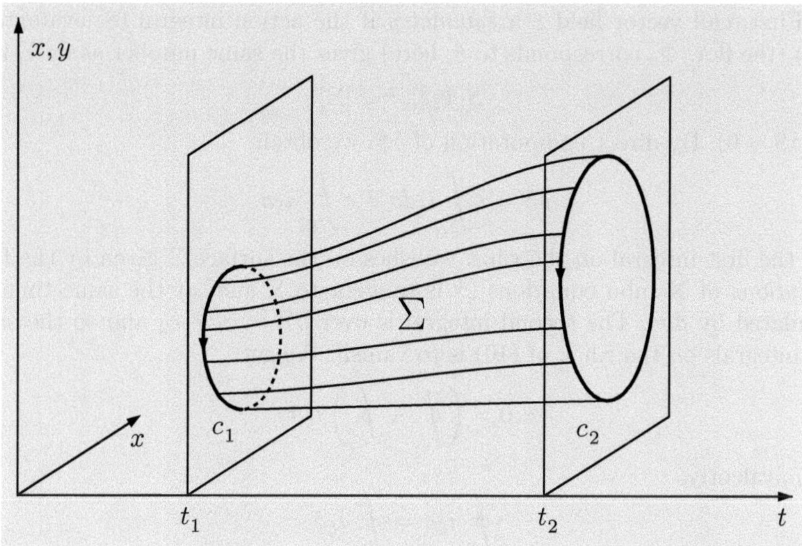

FIGURE 1. A two-chain Σ made up from a one-cycle c_1 using solutions of Nambu equations.

In Takhtajan's paper [8] it is done by the following trick: The value of action integral is associated with an appropriate one-parameter *family* of trajectories rather than with a single trajectory.

Namely, consider the family constructed as follows: Let, from each point p of a *one-cycle* (loop) c_1 at the time t_1, emanate the solution $\gamma(t)$ of Nambu equations (2), fulfilling initial condition $\gamma(t_1) = p$. At the time t_2, the points $\gamma(t_2)$ (for all $p \in c_1$) form a *one-cycle* (loop) c_2 again (image of c_1 w.r.t. the Nambu flow for $t_2 - t_1$) and the points $\gamma(t)$, for *all* $t \in \langle t_1, t_2 \rangle$ and *all* $p \in c_1$, form a *two-chain* (2-dimensional surface) Σ made of solutions (see Fig. 1; notice that $\partial\Sigma = c_1 - c_2$). The value of the action, assigned to the family, is defined to be

$$S[\Sigma] = \int_\Sigma \hat{\sigma}. \tag{8}$$

One then verifies [8, 10] that the surface given by the family of *solutions* of Nambu equations is indeed an *extremal* of the action integral (8).

3. Conserved quantity from a symmetry

Having introduced action integral for Nambu mechanics, we can mimic steps which lead from a symmetry of *Hamiltonian* action (7) to corresponding conserved quantity (function, there). And see what we get in this way in Nambu mechanics. (See more details in [10].)

First, call vector field ξ a *symmetry* if the action integral (8) evaluated on $\Phi_\epsilon(\Sigma)$ (the flow Φ_ϵ corresponds to ξ, here) gives the same number as on Σ itself

$$S[\Phi_\epsilon\Sigma] = S[\Sigma] \tag{9}$$

(i.e., $\delta S = 0$). By direct computation of δS, we obtain

$$\delta S = \epsilon \int_\Sigma i_\xi d\hat{\sigma} + \epsilon \oint_{\partial\Sigma} i_\xi \hat{\sigma}. \tag{10}$$

Now, the first integral on the r.h.s. vanishes on the surface Σ given by the family of *solutions* of Nambu equations ($\dot{\gamma}$ is tangent to Σ and, at the same time, it is annihilated by $d\hat{\sigma}$). The second integral is over $\partial\Sigma = c_1 - c_2$ and so the sum of both integrals on the r.h.s. of (10) is to vanish. We get

$$0 = \left(\oint_{c_1} - \oint_{c_2}\right) i_\xi\hat{\sigma} \tag{11}$$

or, equivalently,

$$\oint_{c_1} i_\xi\hat{\sigma} = \oint_{c_2} i_\xi\hat{\sigma}. \tag{12}$$

This is, however, nothing but a *conservation law*: for *solutions* of Nambu equations,

$$f_\xi(t_1; c_1) = f_\xi(t_2; c_2), \tag{13}$$

where f_ξ is given by the *integral*

$$f_\xi(t_a; c_a) := \oint_{c_a} i_\xi\hat{\sigma} \qquad a = 1, 2. \tag{14}$$

In full analogy with the Hamiltonian case, a more general definition of symmetry is possible. Rather than using differential version of (9), vanishing of the Lie derivative

$$\mathcal{L}_\xi\hat{\sigma} = 0\ , \tag{15}$$

we define *symmetry of Nambu system* as a vector field ξ obeying somewhat weaker condition,

$$\mathcal{L}_\xi\hat{\sigma} = d\chi_\xi \tag{16}$$

(*exactness* of the Lie derivative is enough). Or, by Cartan's formula,

$$i_\xi d\hat{\sigma} = -d(i_\xi\hat{\sigma} - \chi_\xi). \tag{17}$$

Upon integration over the surface Σ we get

$$\int_\Sigma i_\xi d\hat{\sigma} = -\oint_{\partial\Sigma} (i_\xi\hat{\sigma} - \chi_\xi). \tag{18}$$

Since the l.h.s. vanishes (on solutions), it holds

$$\oint_{c_1} (i_\xi\hat{\sigma} - \chi_\xi) = \oint_{c_2} (i_\xi\hat{\sigma} - \chi_\xi). \tag{19}$$

So, we obtain the statement

$$f_\xi(t_1; c_1) = f_\xi(t_2; c_2), \tag{20}$$

where (more general, cf. (14)) f_ξ is given by the *integral*

$$f_\xi(t_a; c_a) := \oint_{c_a} (i_\xi \hat{\sigma} - \chi_\xi) \qquad a = 1, 2. \tag{21}$$

In words: Given a symmetry ξ take, at time t_1, an arbitrary one-cycle (loop) c_1. Compute the line integral

$$\int_{c_1} (i_\xi \hat{\sigma} - \chi_\xi). \tag{22}$$

Then, let each point of c_1 evolve by Nambu flow up to time t_2. You get another one-cycle (loop), c_2. Compute, again, the line integral

$$\int_{c_2} (i_\xi \hat{\sigma} - \chi_\xi). \tag{23}$$

The conservation law says: You get the same number.

4. Conserved quantities as relative integral invariants

In Nambu mechanics, conserved quantity associated with symmetry ξ turns out to be a *relative integral invariant.* This is, by definition, a differential p-form α such that, when integrated over a p*-cycle*, it gives an invariant w.r.t. the dynamical flow. Put in another way, if a dynamical vector field V generates the flow Φ_t (time evolution) and if c_2 is the Φ_t-image of an *arbitrary* p*-cycle* c_1, then,

$$\oint_{c_1} \alpha - \oint_{c_2} \alpha \tag{24}$$

(see, e.g., [4, 11] and [12]).

In our case, the result (19) may be regarded as the statement that on Nambu extended phase space endowed with the dynamical vector field V defined by

$$i_V d\hat{\sigma} = 0 \tag{25}$$

(see (2)) we get, as a consequence of existence of a symmetry ξ, a relative integral invariant. Namely, (24) holds for the *one-form*

$$\alpha = i_\xi \hat{\sigma} - \chi_\xi. \tag{26}$$

Of course, as is always the case, our relative integral invariant then automatically yields an *absolute* integral invariant, integral of the *exterior derivative* $d\alpha$ of α over *any* two-chain (two-dimensional surface) s. So, taking into account (17),

$$\int_{s_1} i_\xi d\hat{\sigma} = \int_{s_2} i_\xi d\hat{\sigma}. \tag{27}$$

5. More Nambu Hamiltonians

Already in the original paper [6] Nambu pointed out that the idea of *three*-dimensional phase space and *two* Nambu "Hamiltonians", H_1 and H_2, may be straightforwardly generalized to more dimensions, n-dimensional (Nambu) phase space and $n-1$ Nambu "Hamiltonians", $H_1, \dots, H_{n-1}$. (There are also other generalizations, see Refs. [6, 8].)

And it is easily seen that all constructions discussed in this paper work equally well in the n-dimensional version. In particular, $\hat{\sigma}$ becomes $(n-1)$-form, c_1 becomes $(n-2)$-cycle, Σ is $(n-1)$-dimensional surface and so on (see [8, 9]). Conserved quantities are still integral invariants (formally equally looking formulas (19) and (27) hold, where c_a are $(n-2)$-cycles and s_a are $(n-1)$-chains).

6. Conclusions

Both Hamiltonian and Nambu mechanics study motion of *points* in phase space along their *trajectories*. Therefore it is natural to expect conserved quantities to be *functions* on phase space. Once we study particular motion, we evaluate the function at the time t_1 at the point where the motion begins, and then we profit from the fact that, at the future points of the trajectory, the same value of the function is guaranteed by the conservation law.

In Hamiltonian mechanics it is really so. In Nambu mechanics, there are conserved *functions* as well. Namely, the two "Hamiltonians" H_1 and H_2 are conserved.

However, as we have seen, *these* conserved functions *do not* directly follow from symmetries, as we might expect from the Hamiltonian case. Instead, in the case of symmetries, application of more or less standard machinery results, because of a peculiar situation with the action integral, in conserved quantities which have the character of *integral invariants* rather then usual conserved functions. (The machinery leads to *higher-degree* forms rather than usual zero-forms, that is, functions.) As a reward for finding a symmetry, the conserved *number* is only obtained as *integral* of the form over a one-cycle.

We stress again that the reason lies in the peculiar structure of the *action* integral: Since we only can associate the action with a *family* of trajectories, conserved quantities also reflect properties *of the family* and they are, therefore, constructed using *integration* "over the family".

Let us note that there is the whole series of well-known *Poincaré–Cartan* integral invariants in Hamiltonian mechanics, where numbers only come out from integration "over (an appropriate) family" of trajectories. These integral invariants, however, have nothing to do with symmetries of *particular* Hamiltonian system (they hold in general, irrespective of the concrete form of the Hamiltonian).

Acknowledgment

This research was supported by VEGA project 1/0985/16.

References

[1] E. Nöther, *Invariante Variationsprobleme.* Göttinger Nachrichten, (1918), 235–257.

[2] Y. Kosmann-Schwarzbach, *The Nöther Theorems.* Springer, 2011.

[3] L.D. Landau, E.M. Lifshitz, *Mechanics.* 3rd edition, Butterworth-Heinemann Ltd, 1995.

[4] V. Arnold, *Mathematical Methods of Classical Mechanics.* Springer-Verlag, 1989.

[5] M. Fecko, *Differential Geometry and Lie Groups for Physicists.* Cambridge University Press, 2006.

[6] Y. Nambu, *Generalized Hamiltonian dynamics.* Phys. Rev. **D 7** (1973), 2405–2412.

[7] M. Fecko, *On a variational principle for the Nambu dynamics.* J. Math. Phys. **33** (1992), 930–933.

[8] L. Takhtajan, *On Foundation of the Generalized Nambu Mechanics.* Comm. Math. Phys., **160** (1994), 295–315. (arXiv: hep-th/9301111)

[9] M. Fecko, *On p-form vortex lines equations on extended phase space.* arXiv: math-ph/1305.3167v1 (2013).

[10] M. Fecko, *On symmetries and conserved quantities in Nambu mechanics.* J. Math. Phys. **54** (2013), 102901. (arXiv:1306.5969 [math-ph])

[11] E. Cartan, *Leçons sur les invariants intégraux.* Hermann, 1922.

[12] M. Fecko, *Modern geometry in not-so-high echelons of physics: Case studies.* Acta Physica Slovaca **63**, No.5 (2013), 261–359. (arXiv:1406.0078 [physics.flu-dyn])

Marián Fecko
Department of Theoretical Physics and Didactics of Physics
Comenius University in Bratislava
Mlynská dolina F2
84248 Bratislava, Slovakia
e-mail: `fecko@fmph.uniba.sk`

Geometric Methods in Physics. XXXV Workshop 2016
Trends in Mathematics, 35–49

The Triple Reduced Product and Hamiltonian Flows

L. Jeffrey, S. Rayan, G. Seal, P. Selick and J. Weitsman

Abstract. In this paper we study the triple reduced product of three coadjoint orbits of $SU(3)$ and show that, under suitable hypotheses on the parameters, it is homeomorphic to S^2. Hence by Moser's method it is symplectomorphic to a copy of S^2 whose symplectic volume equals that of the triple reduced product.

We outline a method to find a Hamiltonian function on this S^2 (with its non-standard symplectic form) which is the moment map for a circle action. In other words the period of the Hamiltonian flow is constant except at fixed points.

Mathematics Subject Classification (2010). 58F07.

Keywords. Symplectic quotient, moment map, coadjoint orbit, symplectic geometry.

1. Introduction

Throughout, G will refer to the Lie group $SU(3)$; T, to its maximal torus; and $\mathfrak{g}$, to its Lie algebra. Let λ, μ, ν be diagonal 3×3 traceless matrices with real eigenvalues, and let $i\lambda, i\mu, i\nu \in \mathfrak{g}$ so that $\mathcal{O}_{i\lambda}$, $\mathcal{O}_{i\mu}$, $\mathcal{O}_{i\nu}$ are the corresponding orbits (under the adjoint action on $\mathfrak{g}$). We define the *triple reduced product* to be the quotient

$$\mathbb{P}(\lambda, \mu, \nu) := (\mathcal{O}_{i\lambda} \times \mathcal{O}_{i\mu} \times \mathcal{O}_{i\nu})//G.$$

Here, G acts diagonally on the product of orbits (via the adjoint action). The notation // indicates that we are taking the symplectic quotient $\Phi^{-1}(0)/G$ of the product of the three orbits, where

$$\Phi(X, Y, Z) = X + Y + Z$$

is the moment map for the diagonal adjoint action of G.

LJ and PS were partially supported by a grant from NSERC
SR was partially supported by University of Toronto
JW was partially supported by NSF grant DMS-12/11819.

It is straightforward to conclude on general grounds that, at regular values of the moment map, the triple reduced product is diffeomorphic to S^2 (see Theorem 1 below). Nonetheless we want to find explicit coordinates on this 2-sphere, and so we give an explicit construction. We shall also describe the symplectic form on the triple reduced product (see §3.1, particularly (25) and the paragraph following this equation).

The main objective of our program is to find a Hamiltonian function on the triple reduced product whose Hamiltonian flow generates an S^1 action on it. Our main result is in §3.3. In §3.3 we take an arbitrary function f whose level sets are circles and construct an S^1 action using it. Later we identify the Hamiltonian whose Hamiltonian vector field is the fundamental vector field associated to this circle action.

Guillemin and Sternberg [6] showed that on a coadjoint orbit O_λ of $SU(n)$, there exists a collection $\mu_1, \ldots, \mu_{n(n-1)/2}$ of continuous functions, smooth on an open subset $U \subset O_{i\lambda}$, the complement of a collection of high codimensional submanifolds of $O_{i\lambda}$, which are moment maps for a torus action on U. (The formula for one of these functions is used below, see (60).) The image of the function $\mu_1, \ldots, \mu_{n(n-1)/2} : O_{i\lambda} \to R^{n(n-1)/2}$ is a convex polytope studied by Gelfand and Cetlin, which is our motivation for considering the Gelfand–Cetlin function later.

Among these functions are the $n-1$ moment maps for the maximal torus $T \subset SU(n)$. Therefore, the remaining functions are $\dfrac{(n-1)(n-2)}{2}$ T-invariant functions on $O_{i\lambda}$. These functions therefore give rise to $\dfrac{(n-1)(n-2)}{2}$ functions on the reduced spaces $\Big(O_{i\lambda} \times O_{i\mu} \times O_{i\nu}\Big)//G$ when λ, μ, ν are generic: we have

$$\Big(O_{i\lambda} \times O_{i\mu} \times O_{i\nu}\Big)//G = \{x \in O_{i\lambda}, z \in O_{i\nu} : x + i\mu + z = 0\}/T.$$

The reason we need only quotient by the maximal torus is that we can assume Y is in the Lie algebra of the maximal torus, and we can also assume the stabilizer of $i\mu$ is the maximal torus. This space is called the triple reduced product, since it is the reduced space at 0 of the product of three orbits $\mathcal{O}_{i\lambda}$, $\mathcal{O}_{i\mu}$ and $\mathcal{O}_{i\nu}$.

The restriction of any of the $\mu_i : O_\lambda \to \mathbb{R}$ then gives a function on $\Big(O_{i\lambda} \times O_{i\mu} \times O_{i\nu}\Big)//G$. Since $\dfrac{(n-1)(n-2)}{2} = (1/2)\dim \Big(O_{i\lambda} \times O_{i\mu} \times O_{i\nu}\Big)//G$, it is plausible to suspect that these functions are moment maps for a densely defined torus action on $\Big(O_{i\lambda} \times O_{i\mu} \times O_{i\nu}\Big)//G$.

When $G = SU(3)$, we see that $(O_{i\lambda} \times O_{i\mu} \times O_{i\nu})//G = S^2$ (see Theorem 1 below). This is a toric variety, so a global torus action does in fact exist. It seems worthwhile to check whether in this case, the Guillemin–Sternberg functions are moment maps for torus actions on $(O_{i\lambda} \times O_{i\mu} \times O_{i\nu})//G$.

One purpose of this paper is to carry out this check for $SU(3)$. Unfortunately it appears that this conjecture is false, as we show below.

This paper is organized as follows. §2 explains why (for suitable choices of λ, μ, ν) the triple reduced product is a 2-sphere. §3.1 computes the symplectic structure on the triple reduced product. Our main result is obtained in §3.3, where we give an integral formula for the moment map for a Hamiltonian S^1 action on the triple reduced product.

§4 explains a method to compute the period of the Hamiltonian flow of a Hamiltonian function.

All numerical work was done using Mathematica. The code was written by one of us (P.S.), with assistance from Jesse Bettencourt, who improved and documented it. It it is availabe at
`http://github.com/reducedproduct/triple.`

We also thank Jacques Hurtubise for useful discussions.

2. The triple reduced product is a 2-sphere

2.1. Notation conventions

The variable X is a member of the Lie algebra of $SU(3)$, viewed as a 3×3 complex matrix.

$$X = \begin{pmatrix} X_{11} & X_{12} & X_{13} \\ X_{21} & X_{22} & X_{23} \\ X_{31} & X_{32} & X_{33} \end{pmatrix} = \begin{pmatrix} ia & r = p + iq & s = u + iv \\ -p + iq & ib & t = x + iy \\ -u + iv & -x + iy & ic \end{pmatrix}. \tag{1}$$

Our variables are $r = p + iq$, $s = u + iv$, $t = x + iy$, ia and ib, where a, b, p, q, u, v, x, y are real and $c = -a - b$.

Recall λ, μ, ν are constants in $\mathbb{R}^3$, where $i\mathbb{R}^3$ is the Lie algebra of the maximal torus T of $SU(3)$. We write $\lambda = (\lambda_1, \lambda_2, \lambda_3)$, and similarly for μ and ν. We impose the condition that $\lambda_1 + \lambda_2 + \lambda_3 = 0$ (similarly for μ and ν), so that $i\lambda, i\mu, i\nu$ will lie in $\mathfrak{t}$, the Lie algebra of T.

2.2. Equations characterizing the triple reduced product

We have

$$\det(X) = X_{11}(X_{22}X_{33} - X_{23}X_{32}) - X_{12}(X_{21}X_{33} - X_{23}X_{31}) + X_{13}(X_{21}X_{32} - X_{22}X_{31}).$$

Also the second elementary symmetric polynomial $\tau_2(X)$ is

$$\tau_2(X) = X_{22}X_{11} + X_{33}X_{11} + X_{22}X_{33} + \mid X_{12} \mid^2 + \mid X_{13} \mid^2 + \mid X_{23} \mid^2 .$$

We are interested in

$$\mathcal{E} := \mathcal{M}/G, \tag{2}$$

where

$$\mathcal{M} := \{(X, Y, Z) \in \mathcal{O}_{i\lambda} \times \mathcal{O}_{i\mu} \times \mathcal{O}_{i\nu} : X + Y + Z = 0\}. \tag{3}$$

Note that if 0 is a regular value for the moment map of the diagonal, then $\mathcal{E}$ is a compact smooth manifold.

Because we are studying the quotient by the diagonal conjugation action of G, we have assume Y has been conjugated into the element $i\mu$ in the Lie algebra of the maximal torus. After this conjugation choice we are left with

$$X \in \mathcal{O}_{i\lambda} \quad \text{and} \quad Z = -X - i\mu \in \mathcal{O}_{i\nu}.$$

Henceforth we will write simply X for the triple $(X, i\mu, -X - i\mu) \in \mathcal{M}$. The statement

$$X \in \mathcal{O}_{i\lambda}$$

is equivalent to

$$i \det(X) = \lambda_1\lambda_2\lambda_3, \qquad \tau_2(X) = \lambda_1\lambda_2 + \lambda_1\lambda_3 + \lambda_2\lambda_3. \tag{4}$$

2.3. The triple reduced product is S^2

Theorem 1. *Assume 0 is a regular value of the moment map. Then the triple reduced product is either empty or homeomorphic to* S^2.

Proof. By a dimension count, the triple reduced product for $SU(3)$ has real dimension 2, and we shall demonstrate that it is homeomorphic to a 2-sphere, for suitable restrictions on λ, μ, ν. For generic values of the moment map, the zero level set of the moment map is a manifold, and a direct computation shows that the G action is free, so that the quotient is a manifold. Since the Kirwan map is surjective, dim $H^0 \leq 1$, dim $H^1 = 0$, and dim $H^2 \leq 1$. Therefore the reduced space, for regular values of the moment map, is either empty or S^2. □

The 4-dimensional subvariety $\mathcal{M} \subset \mathfrak{su}(3)$ is defined by four equations obtained by $X \approx i\lambda$ and $-X - i\mu \approx i\nu$, where the symbol "$\approx$" denotes "is conjugate to". The variety $\mathcal{M}$ is then determined by the equations:

$$-\tau_2(-X - i\mu) + \tau_2(X) = \tau_2(\nu) - \tau_2(\lambda), \tag{5}$$
$$i \det(X + i\mu) - i \det(X) = \det(\nu) - \det(\lambda), \tag{6}$$
$$i \det(X) = \det(\lambda), \tag{7}$$
$$-\tau_2(X) = \tau_2(\lambda). \tag{8}$$

Expanding the left-hand side of (5) gives

$$a\mu_2 + a\mu_3 + b\mu_1 + b\mu_3 + c\mu_1 + +c\mu_2 + \mu_1\mu_2 + \mu_1\mu_3 + \mu_2\mu_3,$$

which could be simplified by using $\mu_1 + \mu_2 + \mu_3 = 0$. Combined with $a + b + c = 0$, this gives a and b as linear functions of c.

We introduce the notation

$$\mu_{i,j} = \mu_j - \mu_i \text{ for } i, j \in \{1, 2, 3\}. \tag{9}$$

The maximal torus T acts on $\mathfrak{g}$ by

$$\text{diag}(e^{i\theta_1}, e^{i\theta_2}, e^{i\theta_3}) : X_{ij} \mapsto e^{i(\theta_i - \theta_j)} X_{ij}.$$

Hence it is always possible to choose an element of T that conjugates X into the set where the matrix elements $X_{13} \in \mathbb{R}_{\geq 0}$ and $X_{23} \in (1+i)\mathbb{R}_{\geq 0}$. In other words, we obtain a global transversal $\mathcal{T}R$ with $\mathcal{T}R \cong \mathcal{E}$ by setting

$$v = 0,\ y = x, \quad u \geq 0, \quad x \geq 0. \tag{10}$$

Lemma 2. *Given a norm-preserving action on a Riemannian manifold, there is a canonical Riemannian metric on the quotient.*

Proof. Because the action is norm-preserving, the metric on the quotient is independent of the choice of representatives. □

We apply this to the action of the maximal torus T on $\mathcal{M}$ by diagonal conjugation. Note that this action is norm-preserving.

Since elements in the image of the tangent space of $\mathcal{T}R$ can be used as the chosen representatives in computing the metric, the Riemannian metric on the subspace $\mathcal{T}R$ is the same as that on the quotient.

Our space $\mathcal{T}R$ has generic dimension two and is given by real variables p, q, x, u, a satisfying the three equations (11), (12) and (13) below.

After implementing the transversal, the equations become

$$-\mu_3|r|^2 - \mu_2 u^2 - 2\mu_1 x^2 + P = 0, \tag{11}$$

$$2ux(p+q) + abc - c|r|^2 - bu^2 - 2ax^2 - \lambda_1\lambda_2\lambda_3 = 0, \tag{12}$$

$$R - |r|^2 - u^2 - 2x^2 = 0, \tag{13}$$

where P and R are polynomials in c whose coefficients depend on the parameters λ, μ and ν. Specifically, P and R are quadratic. Noting that a and b are linear functions of c, observe that (12) has the form

$$2ux(p+q) + Q_c + Q_r|r|^2 + Q_u u^2 + Q_x x^2 = 0$$

where Q_c is cubic and Q_r, Q_u and Q_x are linear. Our transversal makes the equations (5), (6), (7) and (8) (equivalently (11)–(13)) invariant under $p \leftrightarrow q$.

We define a flow on $\mathcal{E}$. We fix a value of c. Let $\Psi : \mathcal{E} \to \mathbb{R}$ be defined by $\Psi(X) = c$. Define

$$S(c) := \Psi^{-1}(c). \tag{14}$$

If c is a regular value of Ψ, $S(c)$ is a 1-dimensional manifold, which is preserved by the involution $i : p \leftrightarrow q$. Let $m(c)$ and $M(c)$ be respectively the minimum and maximum values of $|r|^2$ on $S(c)$.

Via equations (11) and (13), we can eliminate the variables x and u. Then $\mathcal{E}$ is parametrized by $|r|$ and c subject to (12). That is, if (12′) is the equation obtained from (12) after replacing u and x from (11) and (13), then p and q are the solutions to the simultaneous equations (12′) and

$$p^2 + q^2 = |r|^2$$

after which u and x are determined by (11) and (13).

Define

$$V(c,k) := \{(p,q,c) \in \mathcal{E} \mid p^2 + q^2 = k\}. \tag{15}$$

Given c, solving equation (12') gives an equation $p + q = f(|r|^2)$. The circle $p^2 + q^2 = |r|^2$ intersects $p + q = f(|r|^2)$ in at most two points, called $A(c, |r|^2)$ and $B(c, |r|^2)$, which are interchanged by the involution $p \leftrightarrow q$. Take $A(c, |r|^2)$ to have the smaller value of $|r|^2$. At the extreme values, $|r|^2 = m(c)$ and $|r|^2 = M(c)$, the pair $A(c, m(c)), B(c, m(c))$ (resp. $A(c, M(c)), B(c, M(c))$) reduces to a single point lying on the axis of symmetry $p = q$. That is, $A(c, m(c)) = B(c, m(c))$ and $A(c, M(c)) = B(c, M(c))$. Thus $S(c)$ is a closed curve which joins $A(c, m(c))$ to $A(c, M(c))$ through the points $\{A(c, |r|^2) \mid m(c) \leq |r|^2 \leq M(c)\}$ and comes back through the points $B(c, |r|^2)$.

We define a flow on $S(c)$ by sending

$$A(c, m) \mapsto A(c, m + t) \tag{16}$$

at time t, where by convention we set $A(M + t) = B\big(M + t(M - m)\big)$.

Lemma 3. *$S(c)$ is a topological circle.*

Proof. Write $S(c) = S_A(c) \cup S_B(c)$ where $S_A(c) := \{A(c, |r|^2) \mid m(c) \leq |r|^2 \leq M(c)\}$ and $S_B(c)$ is its image under the axis of symmetry $p = q$. Since the circle $p^2 + q^2 = |r|^2$ intersects the line $p + q = f(|r|^2)$ in at most two points, each of S_A, S_B is a simple curve. Thus $S(c)$ is a union of n topological circles which intersect the axis of symmetry $p = q$ at the critical points. To show that $n = 1$, we show that there are only two critical points.

Setting $p = q$ gives $p^2 = |r|^2/2$. Solving (11) and (13) for x^2 and u^2 gives polynomials in c and $|r|^2$ which are linear in $|r|^2$. Thus (12) becomes $\pm\sqrt{2}|r| = f(|r|^2)$ where $f(z)$ has the form

$$\frac{L_1(z)}{\sqrt{L_2(z)L_3(z)}},$$

where L_1, L_2, L_3 are linear in $|r|^2$. The solutions of $\pm\sqrt{2z} = f(z)$ are symmetric under $z \mapsto -z$ and are contained in the solutions of the quintic $2L_2^2 L_3^2 z = L_1^2$. Therefore there cannot be more than two positive roots. That is, there cannot be more than two solutions for $|r|$.

□

3. Symplectic form and Hamiltonian vector field

3.1. Symplectic form related to inner product

Let X be a point in the Lie algebra of $SU(3)$. Recall that the tangent space T_X is given by

$$T_X = \{[X, V] \mid V \in \mathfrak{su}(3)\}.$$

For $A, B \in \mathfrak{su}(3)$ let $\langle A, B\rangle$ denote the Euclidean inner product, which equals $-\mathrm{Trace}(AB)$. Also let e be a point in $\mathfrak{su}(3)$. Suppose that

$$\langle [X, e], t\rangle = 0 \tag{17}$$

for all diagonal t, in other words $[X, \cdot] \in it$. We wish to find a $\beta(e)$ which satisfies

$$[X, e] = [i\mu, \beta(e)]. \tag{18}$$

Recall that e_{ij} denotes the ij-entry of the matrix e whereas $\mu_{i,j}$ means $\mu_j - \mu_i$. If $e \in su(3)$, set

$$\tilde{\beta}(e) := -\begin{pmatrix} 0 & -\dfrac{e_{12}}{i\mu_{1,2}} & -\dfrac{e_{13}}{i\mu_{1,3}} \\ * & 0 & -\dfrac{e_{23}}{i\mu_{2,3}} \\ * & * & 0 \end{pmatrix} \tag{19}$$

where the asterisks are chosen to make $\tilde{\beta}$ an element of $su(3)$. The element $\tilde{\beta}(e)$ is defined for all e, but only has the properties we need for e satisfying (17). This defines a vector field $\beta(e)$ which also depends on the variable X, assumed to be in $\mathcal{E}$.

We can define

$$\beta(e) = \tilde{\beta}([X, e]), \tag{20}$$

for all e. If also $\langle [X, e], t\rangle = 0$ for all t, then the function $\beta : \mathfrak{su}(3) \to \mathfrak{su}(3)$ satisfies (18). Calculation is required to see that (19) is the appropriate value for the function $\tilde{\beta}$.

Also let the vector field $\alpha(e)$ be defined as

$$\alpha(e) := -e - \beta(e).$$

Now define

$$V(e) := [X, \alpha(e)] \in T_X. \tag{21}$$

If $\langle [X, e], t\rangle = 0$ for all t then $V(e)$ also satisfies

$$V(e) = [X, \alpha] = -[X, e] - [X, \beta] = -[i\mu, \beta] - [X, \beta] = [-X - i\mu, \beta]. \tag{22}$$

Notice that if $\langle [X, e], t\rangle = 0$ for all $t \in \mathfrak{t}$, the tangent vector $V(e)$ is a commutator with both X and $-X - i\mu$ and thus lies in the tangent space at X to the triple reduced product, regarded as a submanifold of $\mathfrak{su}(3)$.

Given $Y = [X, y]$ for some y – in other words Y is in the tangent space at X to the orbit of the adjoint action – we have

$$\begin{aligned} \omega_X^{KKS}(V(e), Y) &= -\omega_X^{KKS}(Y, V(e)) = -\langle X, [y, \alpha(e)]\rangle \\ &= -\langle [X, y], \alpha(e)\rangle = -\langle Y, \alpha\rangle, \end{aligned} \tag{23}$$

where we denote the Kirillov–Kostant–Souriau symplectic form on coadjoint orbits by ω^{KKS}. This is based on the above expression (22) for $[X, \alpha]$.

If also Y is of the form $Y = [-X - i\mu, \tilde{y}]$ for some $\tilde{y}$, then

$$\omega_{-X-i\mu}^{KKS}(V(e), Y) = -\langle -X - i\mu, [\tilde{y}, \beta]\rangle = -\langle [-X - i\mu, \tilde{y}], \beta\rangle = -\langle Y, \beta(e)\rangle. \tag{24}$$

Let $\Omega(\cdot, \cdot)$ denote the symplectic form on the triple reduced product. For $Y \in T_X\mathcal{M}$

$$\begin{aligned} \Omega(V(e), Y) &= \omega_X^{KKS}(V(e), Y) + \omega_{-X-i\mu}^{KKS}(V(e), Y) \\ &= -\langle Y, \alpha(e)\rangle - \langle Y, \beta(e)\rangle = -\langle Y, \alpha(e) + \beta(e)\rangle = \langle Y, e\rangle. \end{aligned} \tag{25}$$

3.2. Vector fields on $\mathcal{M}$

To show that every vector field on the 4-dimensional variety $\mathcal{M}$ is $V(e)$ for some e, we find a suitable basis of this form for the tangent space $T_X(\mathcal{M})$.

With r and s as in §2, set

$$e_1 := \begin{pmatrix} 0 & r & 0 \\ -\bar{r} & 0 & 0 \\ 0 & 0 & 0 \end{pmatrix} \quad e_2 := \begin{pmatrix} 0 & 0 & s \\ 0 & 0 & 0 \\ -\bar{s} & 0 & 0 \end{pmatrix}.$$

Then it is easy to see that $V(e_1)$, $V(e_2)$, $V(i\mu)$, $V(iD)$ are linearly independent, where D is a vector in $\mathfrak{t}$ linearly independent from μ.

3.3. Equation satisfied by diffeomorphisms preserving the symplectic form

Let $f : \mathcal{E} \to \mathbb{R}$ be a function whose image is an interval $[z_{\min}, z_{\max}]$ such that $f^{-1}(z)$ is a topological circle for all but two points $z_{\min}$ and $z_{\max}$ in its image, and $f^{-1}(z_{\min})$ and $f^{-1}(z_{\max})$ are single points, with $z_{\min}$ and $z_{\max}$ as minimum and maximum respectively. One example is $f(X) = GC(X)$ (as shown in §3.6). Parametrize $f^{-1}(z)$ by $\gamma_z : [\alpha_z, \beta_z] \to f^{-1}(z)$ for some α_z, β_z with $\gamma_z(\alpha_z) = \gamma_z(\beta_z)$. Alternatively we may think of γ_z as a periodic function $\gamma_z : \mathbb{R} \to f^{-1}(z)$ with period $\beta_z - \alpha_z$. By abuse of notation we sometimes refer to the circle γ_z.

Our goal is to define an action $\bar{\phi}$ of S^1 on $\mathcal{E}$ preserving ω and keeping the value of $f(z)$ constant, and rotating the topological circle γ_z. We want to define an action

$$\bar{\phi}(e^{2\pi is}, \gamma_z(k)) = e^{2\pi is} \cdot \gamma_z(k)$$

such that

$$(\bar{\phi_s})^*\omega = \omega.$$

Write $\bar{\phi}_s(\gamma_z(k)) = \gamma_z(\phi_s(k))$ for some $\phi_s : \mathbb{R} \to \mathbb{R}$ for $k \in [\alpha_z, \beta_z]$.

Set

$$F(z,k) := \omega_{\gamma_z(k)}\left(\frac{\partial}{\partial k}, \frac{\partial}{\partial z}\right).$$

In other words, in the local coordinates k and z,

$$\omega_{\gamma_z(k)} = F(z,k)dk \wedge dz.$$

Our requirement

$$(\bar{\phi_s})^*\omega = \omega,$$

becomes

$$F_z(k)dk = F_z(\phi_s(k))\frac{\partial \phi_s}{\partial k}dk. \tag{26}$$

Note that the subscript z means the value of a function at a specific value of the parameter z – it does not refer to differentiation with respect to z. Let

$$\frac{\partial G_z}{\partial k} = F_z(k). \tag{27}$$

Solving the differential equation (26) gives

$$C_s + G_z(k) = G_z\big(\phi_s(k)\big) \tag{28}$$

for some constant C_s depending on the real parameter s but independent of k. Equation (28) is an implicit formula for ϕ_s. It yields the explicit formula

$$\phi_{s,z}(k) = g_z(G_z(k) + C_s) \tag{29}$$

for ϕ_s, where

$$g_z := {G_z}^{-1}. \tag{30}$$

(This means g_z is the inverse function of G_z, not the reciprocal.) G_z is invertible since its derivative is the symplectic form in the coordinates k and z, which is nondegenerate. Fix z and write $\phi_s(k)$ for $\phi_{s,z}(k)$.

We have

$$\phi_{s+t} = \phi_s \circ \phi_t, \tag{31}$$

$$\phi_0 = 1, \tag{32}$$

$$\phi_0 = \phi_1. \tag{33}$$

Equations (31) and (32) imply that C_s is a linear function of s.

We define an action $\bar{\phi} : S^1 \times \mathcal{E} \to \mathcal{E}$ by

$$\bar{\phi}_s(\gamma_z(k)) = e^{2\pi i s} \cdot \gamma_z(k) = \gamma_z\big(g_z(G_z(k) + \lambda_z s)\big). \tag{34}$$

The parameter λ_z is determined by the condition $\phi_0(z) = \phi_1(z)$ as follows.

Lemma 4.

$$\lambda_z = G_z(\beta_z) - G_z(\alpha_z). \tag{35}$$

Proof. When $s = 0$ we have

$$\phi_0(k) = g_z(G_z(k)) = k,$$

while for $s = 1$ we have

$$\phi_1(k) = g_z(G_z(k) + \lambda_z).$$

The condition $\phi_0 = \phi_1$ implies that

$$k = g_z(G_z(k) + \lambda_z), \tag{36}$$

by taking $s = 1$ in (34). This in turn is equivalent to

$$g_z(G_z(k)) + \beta_z - \alpha_z = k + \beta_z - \alpha_z. \tag{37}$$

We see this because from (34) (with $s = 1$) we have that

$$\gamma_z(k) = \gamma_z\big(g_z(G_z(k) + \lambda_z)\big).$$

Recall that γ_z is periodic, and so $\phi_0(k) = k$ but $\phi_1(k) = k + \beta_z - \alpha_z$.

Because g_z is injective it follows using (34) and (37) that

$$G_z(k) + \lambda_z = G_z(k + \beta_z - \alpha_z) \tag{38}$$

for all k. Rearranging, we get

$$\lambda_z = G_z(k + \beta_z - \alpha_z) - G_z(k)$$

for all k. Letting $k = \alpha_z$, we obtain

$$\lambda_z = G_z(\beta_z) - G_z(\alpha_z). \tag{39}$$

□

Let the moment map for the S^1 action be denoted by

$$\Phi : \mathcal{E} \to \mathbb{R} \cong \mathfrak{t}^*. \tag{40}$$

Let $s \in \mathfrak{t} \cong \mathbb{R}$. Let $s_\#$ denote the fundamental vector field generated by the action of $\mathbb{R}$.

In what follows, when f is a function of k and z, we write $f_z(k)$ for $f(k,z)$, and let primes denote differentiation by k.

Lemma 5. *The vector field $s_\#$ is given by*

$$s_\# = g_z' \lambda_z s \frac{\partial}{\partial k}, \tag{41}$$

with λ_z as in (35) and $X = \gamma_z(k)$.

Proof. As was proved in Lemma 5, the vector field generated by the action of $\mathbb{R}$ is

$$s_\# = g_z' \lambda_z s \frac{\partial}{\partial k} \tag{42}$$

in local coordinates. On $\mathcal{E}$, it is $s_\# = \gamma_z'(g_z' \lambda_z s \frac{\partial}{\partial k})$; in other words $d\gamma_z(g_z' \lambda_z s \frac{\partial}{\partial k})$. We have

$$i_{s_\#}\omega(\frac{\partial}{\partial z}) = F g' \lambda_z s = \lambda_z s \tag{43}$$

since the second equation comes from the fact that

$$g'(z) = \frac{dg_z}{dk} = \frac{1}{dG_z/dk} = \frac{1}{F_z}. \tag{44}$$

The vector field $s_\#$ is

$$(s_\#)_X = \lim_{\epsilon \to 0} \frac{e^{2\pi i s \epsilon} \cdot X - X}{\epsilon} = \lim_{\epsilon \to 0} \frac{\gamma_z\Big(g_z\big(G_z(k) + \lambda_z s \epsilon\big)\Big) - \gamma_z(k)}{\epsilon}, \tag{45}$$

$$\frac{\partial}{\partial s}(\gamma_z \circ \phi_s)|_{s=0} = \lim_{\epsilon \to 0} \frac{(\gamma_z \circ \phi_s)(\epsilon) - (\gamma_z \circ \phi)(0)}{\epsilon}, \tag{46}$$

which equals the above expression (45) for $(s_\#)_X$. In other words

$$\begin{aligned} \gamma_z'(k) g_z'(k) &= \gamma_z'(k)\Big(\lim_{\epsilon \to 0} \frac{g_z\Big(G_z(k) + \lambda_z s \epsilon\Big) - k}{\epsilon}\Big) \\ &= s\gamma_z'(k) \lim_{s\epsilon \to 0} \frac{g_z(G_z(k) + \lambda_z s\epsilon - g_z(G_z(k))}{s\epsilon} \end{aligned} \tag{47}$$

since $k = g_z(G_z(k))$. (The second equality is obtained by multiplying by s in the numerator and denominator.) The above quantity is equal to

$$s\gamma_z'(k) g'{}_z(k) \lim_{u \to 0} \frac{G_z(k) + \lambda_z u - G_z(k)}{u} = s\lambda_z \gamma_z'(k) g'(G_z(k)), \tag{48}$$

with the quantity on the left-hand side obtained by substituting $s\epsilon = u$, while the equality is obtained by canceling two factors $G_z(k)$) and using the chain rule. The above quantity is then equal to $\lambda_z s\frac{\gamma_z'(k)}{G_z'(k)}$. This last follows from (44). This completes the proof. □

We next want to compute the Hamiltonian and moment map. Let $\mathcal{H}_s$ be the Hamiltonian generating the vector field $s_\#$.

$$d\mathcal{H}_s = d(s\Phi) = s\frac{\partial \Phi}{\partial z}dz, \tag{49}$$

$$sd\Phi = d\mathcal{H}_s = i_{s_\#}\omega = \lambda_z s dz. \tag{50}$$

Hence the moment map is

$$\Phi(z) = \int_{z_{\min}}^{z} \lambda_h dh, \tag{51}$$

where we integrate along the trajectory γ_z. Note that by analogy with elliptic integrals, the moment map is defined as the integral of a function, not in closed form.

We have

$$\phi_s{}'(k) = g'(G(k)+s)G'(k), \tag{52}$$

$$\phi_s{}'(k) = \frac{G'(k)}{G'(G(k)+s)} = \frac{F(k)}{F(G(k)+s)}, \tag{53}$$

$$\omega = F_z(k)dz \wedge dk, \tag{54}$$

$$G_z'(k) = F_z(k)g_z. \tag{55}$$

The vector field is given by

$$(s_\#)_{\gamma_z(k)} = \lambda_z \frac{s}{G_z'(k)}\frac{\partial}{\partial k} = \frac{s}{F_z(k)}\frac{\partial}{\partial k} \tag{56}$$

from (35) and (54).

The Hamiltonian is

$$\mathcal{H}_s = s\Phi = s\int_{z_{\min}}^{z} \lambda_h dh. \tag{57}$$

We check via the previous calculation (57) that

$$d\mathcal{H}_s = i_{s_\#}\omega = s\lambda_z dz. \tag{58}$$

3.4. General formula for the period

We wish to test whether a given function f satisfying the hypothesis of §3.3 is the Hamiltonian for the circle action defined in that section. Let H be the actual Hamiltonian, and let X_H be the Hamiltonian vector field of H. The level sets of H are the same as the level sets of f.

Let $I = [z_{\min}, z_{\max}]$.

The period, which is independent of z, is given by

$$\tau = \int_{\gamma_z} \frac{1}{|X_H|} d\gamma_z \tag{59}$$

for $z \in I$.

One way to see that this is the equation for the period is to observe that if we have an equation for the Hamiltonian flow

$$\frac{dz}{dt} = H(z),$$

then this ODE is separable and we can integrate

$$\frac{dz}{H(z)} = dt.$$

Then the period is the value τ for which t completes a circuit, or

$$\int \frac{dc}{H(c)} = \tau.$$

3.5. Method to get Hamiltonian vector field

Let f be as in §3.3. Let H be the Hamiltonian associated to the circle action coming from the foliation $\{f^{-1}(z)\}$ described in the previous section. Let X_H be its associated Hamiltonian vector field.

By taking directional derivatives of f in various two linearly independent directions $V(e_1)$, $V(e_2)$ we find a matrix $e \in su(3)$ such that $df\big(V(e)\big) = 0$. By construction, H is constant along the circles $f =$ constant. Since we are in two dimensions, this implies that $V(e)$ is a multiple of X_H.

Since we are in two dimensions, $\Omega(\ ,\ )$ is determined by knowledge of $\Omega(Y, Z)$ for any Y and Z. Set $\hat{\Omega} := \big(\Omega(V(e), T_c)\big) = -\langle e, T_c\rangle$. If we knew H, comparing $\big(\Omega(V(e), T_c)\big)$ with $\Omega(X_H, T_c) = dH(T_c)$ would tell us that $X_H = \frac{\hat{\Omega}}{dH(T_c)} V(e)$. This gives us the following method of determining whether or not f is the Hamiltonian associated to the circle action its foliation determines. Set $\chi_f := \frac{\hat{\Omega}}{df(T_c)} V(e)$, and for $z \in (z_{\min}, z_{\max})$ set $\tau_z := 2\int_{\gamma_z} \frac{1}{\|\chi_f\|}\, d\gamma_z$. If $f = H$, then (as shown in the previous section) τ_z would be the period of the action, and in particular the value of τ_z would be independent of z. This method is used later to show that some candidates for H are not in fact the Hamiltonians for their associated circle action.

Our knowledge of $\hat{\Omega}$ allows us to write Ω in any local coordinate system. Set $\mathcal{E}_{z_{\min}}^{z_{\max}} := f^{-1}[z_{\min}, z_{\max}])$. Using our formula for $\Omega(\ ,\ )$ in local coordinates we can compute the symplectic volume

$$SVOL_{z_{\min}}^{z_{\max}} := \int_{\mathcal{E}_{z_{\min}}^{z_{\max}}} \Omega.$$

This allows us to construct a table of values for $H(z)$, normalized so that $H(z_{\min}) = 0$, as the solution to the equation $\frac{H(z) - z_{\min}}{z_{\max} - z_{\min}} = \frac{SVOL_{z_{\min}}^{z}}{SVOL}$. Of course, this gives another way to test if values of the starting function f agree with those of H. This algorithm is implemented in our Mathematica code.

3.6. Gelfand–Cetlin satisfies conditions of §3.3

We now check that this class of functions f from §3.3 includes the Gelfand–Cetlin function. For M a skew-Hermitian square matrix of dimension 3 in a specific adjoint orbit, the Gelfand–Cetlin function is defined as

$$H_{GC}(M) = -M_{11} + \sqrt{-(M_{22} - M_{33})^2 + 4M_{23}M_{32}}. \tag{60}$$

With our chosen parameters the Gelfand–Cetlin function becomes

$$GC(X) = -a + \sqrt{-(b-c)^2 + 4x^2}.$$

As noted earlier, a and b are linear functions of c, using equation (5) and the condition $a+b+c=0$. Thus the curve $\mathrm{GC}(X) = C$ is given by a quadratic function of x and a. Its level sets are conics. For appropriate values of the constants λ, μ, ν and C, its level sets are therefore circles.

The Gelfand–Cetlin function is obviously not suitable, since it is not symmetric under $X \mapsto X + i\mu$. Another related function which is symmetric under this operation is the average Gelfand–Cetlin function, obtained by averaging with respect to this operation.

4. The period of the average Gelfand–Cetlin function

Let GC_{ave} be the average Gelfand–Cetlin function

$$GC_{\text{ave}}(X) := (GC(X) + GC(-X - i\mu))\,/2.$$

In this section we will simply denote GC_{ave} by f. If the level sets of f are not circles, then f clearly is not the moment map for a circle action on $\mathcal{E}$. We show that there are values of the parameters for which f is not the Hamiltonian of a circle action.

Appropriate values of parameters have been chosen so that the level sets are circles. We compute the period of the Hamiltonian flow. We find numerically that the period is not constant, hence f is not the moment map for a circle action.

4.1. Generic constants

The integral for the period is too complicated to compute in closed form. To simplify the figures, we studied particular values of μ, λ, ν. To investigate whether or not the period is independent of the value z of f, we choose generic values of our parameters. We used the following values

$$\lambda = (-7/2, 3, 1/2),$$
$$\mu = (-3, 0, 3),$$
$$\text{and } \nu = (-11/2, 4, 3/2).$$

These values are chosen somewhat randomly, with a view to simplifying the algebra. For no particular reason we chose $\mu_1 = 3$ and $\mu_2 = 0$. The values of λ and ν above were chosen so that the coefficients in the equations defining $\mathcal{M}$ would come out to be integers.

With these values of our parameters, we found that $z_{\text{min}} \approx 3.899$ and $z_{\text{max}} \approx 5.179$, where as before z_{min} and z_{max} are the minimum and maximum values of f.

These can be regarded respectively as the north and south poles of our S^2.

We note that our algorithm and our Mathematica code enable us to produce a table of values for the moment map for a circle action. Some representative values of the period are given by Figure 1.

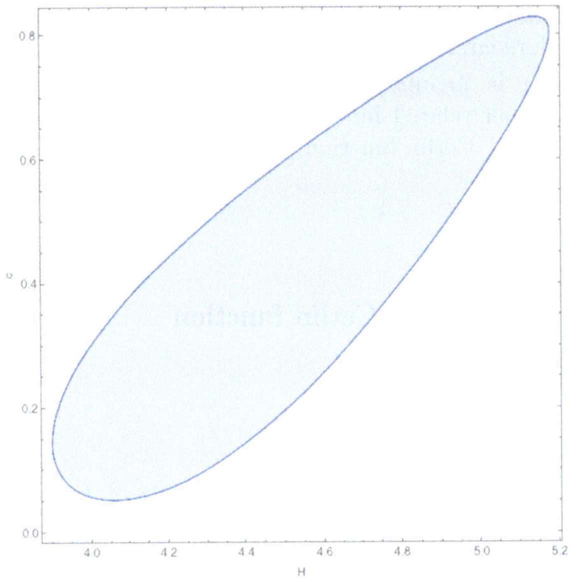

FIGURE 1. Image of the parameter space in H and c for $\mathcal{E}$.

TABLE 1. GC_{ave}

f	$\tau/2$
5.079	6.332
4.979	5.072
4.879	4.648
4.779	4.576
4.679	4.806
4.579	4.975
4.479	3.578
4.379	2.786
4.279	2.285
4.179	1.941
4.079	1.695
3.979	1.527

As can be seen from Table 1, the period depends on the value of f and increases monotonically with decreasing f. The period is not constant, which shows that f is not the Hamiltonian for a circle action. For the Gelfand–Cetlin function, the period is also not constant – for example we computed the period for two level sets and found they are different.

References

[1] M. Atiyah, Convexity and commuting Hamiltonians. *Bull. London Math. Soc.* **14** (1982) 1–15.

[2] L. Bégin, A.N. Kirillov, P. Mathieu, and M. Walton, Berenstein–Zelevinsky triangles, elementary couplings, and fusion rules. *Lett. Math. Phys.* **28** (1993), no. 4, 257–268.

[3] A. Berenstein, A. Zelevinsky, Triple multiplicities for $sl(r+1)$ and the spectrum of the exterior algebra of the adjoint representation. *J. Algebraic Comb.* **1** (1): 7–22 (1992).

[4] W. Goldman, Invariant functions on Lie groups and Hamiltonian flows of surface group representations. *Invent. Math.* **85** (1986), no. 2, 263–302.

[5] V. Guillemin, S. Sternberg, Convexity properties of the moment mapping I. *Invent. Math.* **67** (1982) 491–513.

[6] V. Guillemin, S. Sternberg, The Gelfand–Cetlin system and quantization of the complex flag manifolds. *J. Func. Analysis*, **52** (1983), 106–28.

[7] F. Kirwan, Convexity properties of the moment mapping. III. *Invent. Math.* **77** (1984) 547–552.

[8] D. McDuff, D. Salamon, *Introduction to Symplectic Topology.* Oxford University Press, 1999.

[9] T. Suzuki, T. Takakura, Symplectic Volumes of Certain Symplectic Quotients Associated with the Special Unitary Group of Degree Three. *Tokyo J. Math.* **31** (2008) 1–26.

L. Jeffrey and P. Selick
Department of Mathematics
University of Toronto
Toronto, Ontario, Canada
e-mail: `jeffrey@math.toronto.edu`
`selick@math.toronto.edu`

S. Rayan
Department of Mathematics and Statistics
University of Saskatchewan
Saskatoon, SK, Canada
e-mail: `rayan@math.usask.ca`

G. Seal and J. Weitsman
Department of Mathematics
Northeastern University
Boston, MA 02115, USA
e-mail: `seal.g@husky.neu.edu`
`j.weitsman@neu.edu`

Geometric Methods in Physics. XXXV Workshop 2016
Trends in Mathematics, 51–59

The Puzzle of Empty Bottle in Quantum Theories

Bogdan Mielnik

Abstract. We discuss an extremely simple case of 'shadowing' when the very existence of quantum detector deforms the behavior of quantum particles even if the detection is never performed. In spite of known statistical interpretations, it may support some recent doubts about the completeness of quantum theory.

Mathematics Subject Classification (2010). 81P05, 81P15, 81P16.

Keywords. Quantum philosophy, measurements, state collapse.

1. Introduction

While the basic idea of quantum theory, about the linear 'navigation' of pure states in Hilbert spaces owes its origin to Schrödinger's thought [1], the resulting measurement axioms with the unavoidable *state collapses* were encrusted as an additional element, against Schrödinger himself. "*Diese verdammte Quantenspringerei*" [2] were never completely understood (for was it a sudden jump or some microobject instability [3], or decoherence [4], some subtle unknown effects [5] or sudden non-linear catastrophe?). The interpretational doubts inspired the famous anecdote of Schrödinger's cat [6], surviving until today. The subsequent discussions [7–12], illustrate the peculiarity of the problem, without offering a convincing solution. Almost all hide the Wheeler paradox of "delayed choice experiment" at the bottom [13], treated however as a dark anecdote, which should not frustrate an intense effort to follow the promising trends. Hence, the multiple decades focused their interest on the non-locality problems, starting from the historical Einstein–Rosen–Podolski (EPR) work [14] accepted with difficulty by Einstein himself. Later on, the teleportation effects [15, 16] exhibited new empirical perspectives. Yet, almost from the beginning, the doubt existed, whether the unique probabilistic interpretation can be formulated for quantum states defined on the relativistic spacelike surfaces, independently on the measurements performed in the future. The simplest doubt is reported in Figure 1.

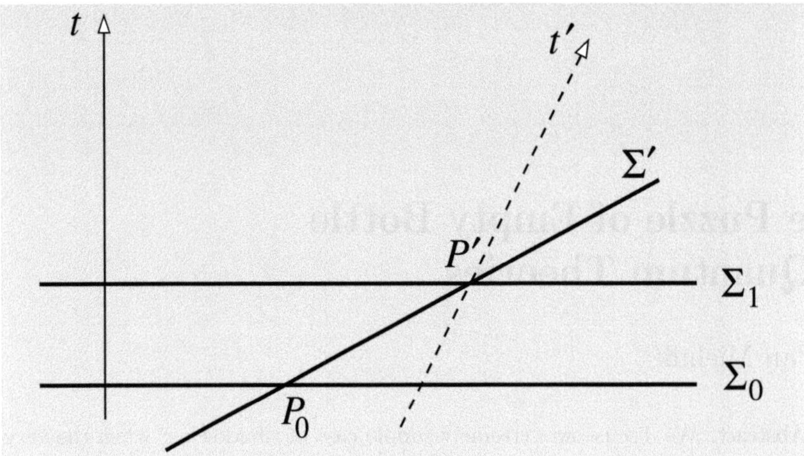

FIGURE 1. The measurement performed at point P' of the plane Σ_1 by a moving observer with a simultaneity hyperplane Σ' can affect the probability distribution around the point P_0 of the hyperplane Σ_0.

The story turned even more challenging in an inspired article of Elitzur and Vaidman (EV) on the 'interaction free measurement' [17, 18]. While the (EV) idea resists too realistic interpretations, it opens unsuspected perspectives to 'see in the dark' [19], with hopes for new age in quantum information permitting the use of powerful quantum computing [20]. In the parallel development, some ambitious trends in Quantum Field Theories (QFT) started already to tell about the "Theory of Everything" as if the end of fundamental research was not too far away.

Yet, some symptoms indicate that the doctrine of quantum theory might be not so universal as generally believed. To illustrate this, it is enough to consider again some details of the 'interaction free measurement'.

2. Interaction free detection?

In their idealized experiment Elitzur and Vaidman consider a photon in a system of optical fibers, with beam splitters and mirrors of Mach–Zehnder interferometer (Fig. 2). The photon wave function is divided into two coherent parts by the first beam splitter, then reflected by two mirrors toward the second splitter, where they unify again, recovering their original state of motion.

So, if there is no obstacle, the photon recovers its original propagation momentum, falling into the detector **D**. However, if one of the branches is blocked, e.g., by a 'perfectly absorbing obstacle' (the terminology used in [17]), then the system performs the first state reduction. Either the obstacle detects (by absorbing) the photon, which therefore arrives neither to **D** nor to **E**. Or it reduces the whole propagation branch, canceling the blocked trajectory. The second splitter

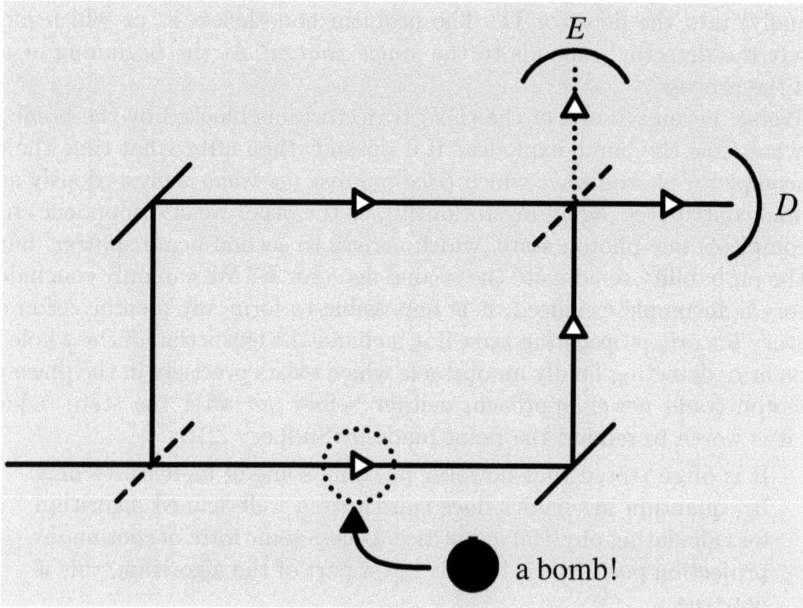

FIGURE 2. Elitzur and Vaidman 'interaction free experiment'. An absorbing obstacle (e.g., a bomb) eliminates one of possible photon trajectories. The photon, moving only along the obstacle-free way, has the probability 1/4 to fall to the detector **E**, thus revealing the existence of an obstacle with which it never collided

will then receive the photon which propagated only along the free path (as if the other one from the beginning did not exist). Its photonic state is then decomposed by the second splitter into the superposition of two parts, tentatively reaching **D** or **E**. The choice of one of them is the second state reduction. The peculiar effect is that the first reduction in 1/2 cases eliminates completely one of the trajectories, while the second one makes equitative choice between **D** and **E**. So in 1/4 of cases the photon appears in **E**, thus revealing the existence of an obstacle, to which it has never approached.

Elitzur and Vaidman choose still a more challenging version of the experiment, assuming that the obstacle is a supersensitive bomb, which would explode immediately under any contact with the photon. Hence, if the detector **E** clicks, it would mean that the bomb was detected (without exploding) by a single photon which could pass hundred kilometers away!

All this seems quite suggestive, if the photon was just an instantaneous pulse. However, what is precisely the *single photon*? Must it propagate always as an infinitely short pulse? Or perhaps, it can also form a very long, narrow wave divided by the first splitter into a pair of still weaker but as long components which laboriously reconstruct their initial form at the second splitter, falling then (gradually) into the detector **D**? The problem nonetheless is, at which moment precisely the detector responds to the *single photon*? At the beginning or at the end of the process?

Worse, because if one of the (EV) trajectories is blocked by the bomb, then after what time the bomb explodes? If it doesn't, then after what time the (large but incomplete) photon wave which *tried to cross the bomb* is mysteriously annihilated and contributes (again mysteriously), to the other weak component creating the (complete) one-photon state, which arrives to second beam splitter, but now with the probability to activate the second detector **E**? We can only conclude that the story is incomplete: indeed, it is impossible to form any mental vision of the obligatory *linearly propagating wave* if it includes the extinction of the whole propagation arm, detecting finally an obstacle which exists precisely in the place which the photon could newer approach, neither before nor after the state reduction! Here, it is worth to remind the point made by Sudbery [21]:

> It is often stated that however puzzling some of its features may be, quantum mechanics does constitute a well-defined algorithm for calculating physical quantities. Unless some form of continuous projection postulate is included as a part of the algorithm, this is not true.

While this seems true, it does not yet offer any concrete image of the 'slow reduction' which could explain the Elitzur–Vaidman effect (if it indeed occurs!). However, if the story is incomplete at this fundamental point, then, except for some simple cases, it can be as questionable to solve the photon behavior in topologically complicated nets, associated with some macroscopic detectors. Incidentally, the description of photon waves propagating in fibers is already known from the paper of I. Białynicki-Birula [22], described not by plane, but by Bessel waves (indeed, a significant progress comparing to the visions of quarks as the plane waves inside of the nucleon surfaces – mind you, without any credible model to explain the quark confinement!). Yet, even this might be insufficient to solve the problem of the linear propagation corrected by the sudden collapses. The intense combinatorial studies to attend the challenge are developed with hopes to program the efficient quantum computing in the topologically nontrivial net of the optical fibers [23–25].

Meanwhile, a sequence of studies of the *imperfect* cases of (EV) bombs was also undertaken [26–29]. Yet, at least one [26] indicates that the experiments with linearly propagating entangled states can affect the past. The similar paradoxical conclusion on *quantum steering into the past* seems to emerge from the 2012 study of Vienna-Innsbruck group [30]. In spite of the 'benign' teleportation without the causality dangers, one might wonder whether the insistence about the linear navigation of the entangled states in the tensor product spaces does not cross some consistency limits. In what follows, our aim is to postpone the locality problems, returning to the traditional quantum paradoxes still waiting for credible solutions.

3. Half full, half empty

Our story concerns a quantum system in a superposed energy state, which will be reduced – though not when the experimentalist decides, but when the system itself decides by emitting a photon (compare with the 'time of arrival' [31–34]). As a simplified model, let us consider a bottle containing an atom in a state of equitative *superposition* of two lowest energy levels, ground state ϕ_0 and an excited state ϕ_1.

In some distant past, the experimentalists examining the spectral lines imagined an atom always in one of the energy eigenstates. Today the picture changed. The existence of the superposed (but pure) energy states is (or seems?) unavoidable if one takes seriously the quantum mechanical formalism. Now, if the atom is in the excited state, we shall say that 'the bottle is full', but if in the ground state, 'the bottle is empty'. The bottle is just to assure that the atom is left in peace, isolated from the external perturbations, as well as from the other bottles. It should be ample enough to neglect the influence of its surface onto the atom behavior, but be able to detect the events of radiation. So, at the top of each bottle, at some safe distance, there is a sensitive screen, prepared to detect the photon, should the atom radiate. If it does, the top of its cell turns black (it is burned!). For purely illustrative reasons, the bottles in Figure 3, are painted hexagonal, resembling the bee hive. By observing the detectors which turned black, we can see, how many photons already 'incubated'.

Now, in almost all studies of the atom radiation one can find the description of the process starting from the excited state ϕ_1 but not from the superposed one. This includes the suggestive representation of the excited states as some narrow superpositions of slightly different energy eigenstates, forming an unstable composition, with the average lifetime τ inverse to the (little) energy width δE, in agreement with the time-energy uncertainty (even though, the last point awoke a lot of unfinished discussions [31–33]). Anyhow, by reading the literature you can always find the considerations in which the beginning of the decay process is an *excited state*, with a slightly diffused spectral line, which seems to confirm the validity of the idea. However, what about the decay starting from the superposition of two very distant levels? Perhaps, the difference is superfluous, but it may be worth to examine.

To fix attention, let us thus assume that our initial state ϕ is an equitative superposition $\phi = a_0\phi_0 + a_1\phi_1$, with $|a_0|^2 = |a_1|^2 = 1/2$ (bottle half full, half empty). From a credible phenomenology we know the behavior of an atom in its ground state ϕ_0. If unperturbed, it just remains in ϕ_0 forever, $\phi_0(t) = exp(-itE_0)\phi_0$. We also know something about the behavior of the excited state ϕ_1. On the level of purely quantum mechanical approximation, this state is as stationary

$$\phi_1(t) = \exp(-itE_1)\phi_1 \tag{1}$$

In reality, though, the stationary evolution is corrected by an unpredictable photon emission with a transition to the ground state ϕ_0. The probabilities of these events include some sequence of calculations which I skip. What can be noticed,

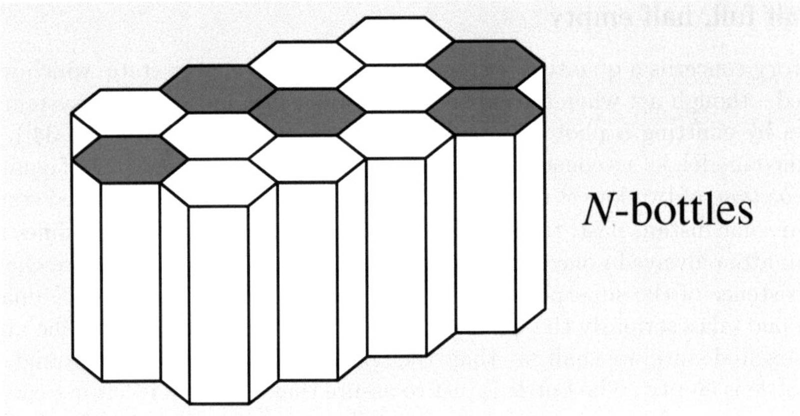

FIGURE 3. Idealized bottles, half full – half empty

however, is that the energy and momentum balance for each *single* radiation act obeys the conservation laws, which makes reasonable to describe the evolution of each *single atom* (with or without the sudden radiation).

However, what happens for atoms with an initial superposed state $\phi = a_0\phi_0 + a_1\phi_1$ (the bottle *neither full, nor empty*)? At the first sight, it may seem that there is hardly any problem here: just apply another 'standard calculations', and the problem is over. However, what if we try to visualize the story, trying to test once again the idea, that the quantum system in presence of a detector performs first a unitary evolution (an extremely linear picture?), until bam!, it is interrupted by the sudden act of detection (an extreme non-linear picture?). So, the evolution until some moment would obey the simple minded law

$$\phi(t) = a_0 e^{-iE_0 t}\phi_0 + a_1 e^{-iE_1 t}\phi_1, \tag{2}$$

granted by the superposition principle, until emitting a photon of energy $E_1 - E_0$, falling into the ground state ϕ_0. This picture seems extremely naive, but remember that the whole quantum theory was conceived by naive pictures of linear navigation interrupted by sudden collapses. Naive or not, our picture contains this time certain additional information. Indeed, in the superposed initial state ϕ the average energy is $\frac{1}{2}(E_1 - E_0)$ but nobody observed the photons emitted with (incomplete) energies smaller than $\Delta E = E_1 - E_0$. Does it mean that before radiating, the atoms must perform first of all a spontaneous (*introspective?*) state reduction, making up (or making down) their minds whether they are or are not in the excited state ϕ_1? The question then is, whether they must ask for some energy credit from their detector? If so, is the detector's favor due to its very existence, even if the photon was not yet emitted [16], or is it a kind of *shadowing* [5], or some friendly help of the 'polarized vacuum'? Yet, let us remind an ambiguous sense of the *polarized vacuum* used, perhaps, too abundantly to heal all QFT emergencies, see R. Penrose [35].

Inversely, if the *spontaneous reduction* failed to bring some extra energy – locating the semi-excited atom on the ground state ϕ_0 then it will stay there forever without emitting anything. Even if the total energy balance is not violated, the single atom behavior hides still some mystery.

4. The principle of vanishing hope? ...

It may be interesting to imagine a population of N atoms in the initial state $\phi = a_0\phi_0 + a_1\phi_1$, each closed in its own bottle, in form of a little, mesoscopic cell. We still assume, that the top surface of each cell is simultaneously a detector, sensitive to the photons of the particular energy $\hbar\omega = E_1 - E_0$. By calculating the (increasing) number of the black cells, we know how many atoms have already radiated (Fig. 3). If all atoms are initially in an identical superposed state $\phi = a_0\phi_0 + a_1\phi_1$, then if somebody performed a check at the very beginning, he would find 50% of them in the ground state ϕ_0, and henceforth, unable to radiate. However, if no initial test was performed at $t = 0$, then anyhow 50% of the atoms will never radiate. Thus, for $t \to +\infty$ all atoms must end up in the ground state ϕ_0, though for different reasons: 1/2 of them, since they have radiated and settled down in ϕ_0; the remaining 1/2, even though no photon was emitted.

Even if the global energy balance is not affected, the situation seems extremely strange. While the atoms which have radiated cause already some trouble, the ones which didn't contain a puzzle! Their superposed energy state vanished, giving place to ϕ_0. The bottle was *half full, half empty*, nothing escaped, and the bottle is *empty!* What has caused the state collapse in this last case, was not any active external intervence. No detector *clicked*, neither the top of any bottle turned black. The only external factor was our *vanishing hope* (take it as a rhetoric figure if you dislike!). Indeed, supposing that the average lifetime of the atom in the excited state ϕ_1 is, e.g., 10^{-10} sec., but the atom in the initial state ϕ did not radiate over 10 years, then, we can be certain that it will never radiate. According to quite orthodox statistical interpretation, this certainty means that the atom state can no longer correspond to ϕ, but it must be practically identical to ϕ_0.

The way to avoid the trouble would be to answer that we have *postselected* our ensemble. The principle of 'don't postselect' (equivalently, 'don't retrospect') is reasonable if some microobjects were submitted to a measurement. Their initial states (in general) were reduced and it makes no sense to look for their past. The principle is not so clear if the micro-objects *escaped* the detection. In their fascinating paper the Swiss group considers the non-orthogonal transformations of pure $\to$ pure states for an ensemble whose particles escaped absorption [16]. The (EV) story of interaction free experiment also contains an obvious element of retrospection. The same, the famous 'delayed choice measurements' of J.A. Wheeler [13]. Similarly, all paradoxes implying the reduction of the past states. So have we to restrict our theory only to strictly pragmatic rules, like *ensemble and only ensem-*

ble, correlation and only correlations, or other 'don't think principles?' Moreover, the naive ideas are still the true source of our sophisticated theories!

Note also that all difficulties would vanish if we simply assumed that no coherent superposition of two distant bound states can exist (remember Einstein boxes [36]?) Yet, all this might be premature conclusions. They show only that our theories are still not close to the proud image of the *Theory of Everything*. They seem closer to the *'Shadows of the mind'* [37], perhaps a hidden allusion of R. Penrose to Platonic cave, containing our mental images (like the linear navigation, etc.). Why they sometimes help and sometimes not, we still ignore!

Acknowledgment

The author is indebted to David Fernández C. and Piotr Kielanowski for skeptical discussions; to Mikhail Zelikin for a valuable information about the cyclones and magnetic vortices, and to Anatol Odzijewicz for his interest in the subject. The assistance of ing. Erasmo Gómez and M.J. Rivero-Monroy. is highly appreciated. The support by the Conacyt project 49253-F was helpful.

References

[1] E. Schrödinger, *Collected papers on wave mechanics*, New York, Chelsea (1978), see also [13]

[2] A. Sudbery, "Diese verdammte Quantenspringerei" in *Studies in Hist. and Philos. Mod. Phys.* **33**, 387–411 (2002)

[3] S. Haroche, Phys. Today, **51**, 36 (1998)

[4] C. Deutsch, F. Ramirez-Martinez, C. Lacroute, F. Reinhard, T. Schneider, J.N. Fuchs, F. Piechon, F. Laloe, J. Reichel and P. Rosenbusch, Phys. Rev. Lett., **105**, 020401 (2010)

[5] Ph. Blanchard and A. Jadczyk, Helv. Phys. Acta, **69**, 613 (1996)

[6] E. Schrödinger, *Naturwiss.* **23**, 807 (1935), see also in [13]

[7] Y. Aharonov and D.Z. Albert, Phys. Rev. D, **24**, 359 (1981)

[8] Y. Aharonov and D.Z. Albert, Phys. Rev. D, **29**, 228 (1984)

[9] B. Mielnik, Found.Phys., **20**, 745 (1990)

[10] B. Mielnik, Found. Phys., **21**, 869 (1991)

[11] J. Finkelstein, Found. Phys. Lett., **5**, 383 (1992)

[12] S.N. Mosley, Phys. Lett. A, **182**, 1 (1993)

[13] J.A. Wheeler and W.H. Żurek, *Quantum theory of measurement*, Princeton Univ. Press (1983)

[14] A. Einstein, B. Podolsky, N. Rosen, Phys. Rev., **47**, 777 (1935)

[15] C.H. Bennett, G. Brassard and C. Crepeau, Phys. Rev. Lett., **70**, 1895 (1993)

[16] B. Huttner, A. Muller, J.D. Gautier, H. Zbinden, and N. Gisin, Phys. Rev. A, **54**, 3783 (1996)

[17] A.C. Elitzur and L. Vaidman, Found. Phys., **23**, 987 (1993), (Tel Aviv preprint TAUP 1865-91)

[18] L. Vaidman, arXiv:quant-ph/0006077v1, 16 June 2000
[19] P. Kwiat, H. Weinfurter, and A. Zeilinger, Sci. Am., **278**, 52 (1996)
[20] S. Lloyd, M. Mosheni, P. Rebentrost, Nature Physics, **10**, 631 (2014)
[21] A. Sudbery, *Continuous state reduction*, in Quantum Concepts in Space and Time, eds. R. Penrose, C.J. Isham, OUP (1986)
[22] I. Białynicki-Birula, *Coherence and Quantum Optics VII*, eds. Eberly et al., Plenum, New York (1996), p. 313
[23] N. Gisin, arXiv:quant-ph/0512168v1 (Dec. 2005)
[24] J.W. Pan, D. Bouwmeester, M. Damell, H. Weinfurter, A. Zeilinger, Nature, **403** 515 (2000)
[25] T. Jennewein, C. Simon, G. Weights, H. Weinfurter, A. Zeilinger, Phys. Rev. Lett., **84** 4729 (2000)
[26] Seth Lloyd, Science, **338** 621 (2012)
[27] A. Peruzzo, P. Shadbold, and N. Brunner, Science, **338** 634 (2012)
[28] F. Kaiser, T. Coudreau, and P. Milman, Science, **338** 637 (2012)
[29] Z. Blanco-Garcia and O. Rosas-Ortiz, J. Phys. Conf., **698** 012013 (2016)
[30] X.S. Ma, S. Zotter, J. Kofler, R. Ursin, T. Jennewein, C. Brukner, A. Zeilinger, Nature Physics, **8** 479 (2012)
[31] B. Mielnik, Found. Phys., **24**, 1113 (1994)
[32] C.R. Leavens, Phys. Lett. A, **303**, 154 (2005)
[33] C.R. Leavens, Phys. Lett. A, **338**, 19 (2005)
[34] J.B. Mackrory, K. Jacobs, D.A. Stek, New J. Phys, **12** 113023 (2919)
[35] R. Penrose, *The Road to Reality*, Publ. Jonathan Cape G.B. (2004), Ch. 26.5 'Alternative vacua'.
[36] T. Norsen, *Einstein Boxes*, Am. J. Phys., **70** (7), 664 (2002), see also arXiv:quant-ph/0404016v2 (Feb. 2005)
[37] R. Penrose, *Shadows of the Mind*, Oxford Univ. Press (1994)

Bogdan Mielnik
Departamento de Física
Centro de Investigación y de Estudios Avanzados del I.P.N.
Av. Instituto Politécnico Nacional 2508
07360 Ciudad de México, Mexico
e-mail: `bogdan@fis.cinvestav.mx`

Geometric Methods in Physics. XXXV Workshop 2016
Trends in Mathematics, 61–67

Poisson Transforms for Tensor Products in Compact Picture

Vladimir F. Molchanov

Abstract. We write explicitly differential operators (Poisson transforms) which intertwine irreducible representations of the group $\mathrm{SL}(2,\mathbb{R})$ with the tensor product of two irreducible representations, one of them is infinite-dimensional and the other is finite-dimensional

Mathematics Subject Classification (2010). Primary 43A85; Secondary 22E46.

Keywords. Lie groups and Lie algebras; representations of Lie groups; tensor products; intertwining operators.

We continue the study of tensor products for the group $G = \mathrm{SL}(2,\mathbb{R})$. Earlier [1] we considered the product of two infinite-dimensional representations, then in [2, 3] we studied the product of two finite-dimensional representations. Now we write explicitly operators (Poisson transforms) which intertwine the tensor product $T_{\sigma,\varepsilon} \otimes \pi_m$ and its irreducible constituents $T_{\tau,\nu}$. These transforms turn out to be differential operators. The representation $T_{\sigma,\varepsilon}$ is infinite-dimensional, the representation π_m is finite-dimensional (of dimension $2m+1$). We use eigenvectors of the product of raising and lowering operators.

Let us introduce some notation and conventions.

We use the following notation for a character of the group $\mathbb{R}^* = \mathbb{R} \setminus \{0\}$:

$$t^{\lambda,\nu} = |t|^\lambda (\mathrm{sgn}\, t)^\nu, \quad t \in \mathbb{R}^*, \ \lambda \in \mathbb{C}, \ \nu \in \mathbb{Z}.$$

This character depends on ν modulo 2 rather than ν itself. For a manifold M, $\mathcal{D}(M)$ denotes the space of compactly supported infinitely differentiable complex-valued functions on M, with the usual topology. For a representation of a Lie group, we retain the same symbol for the corresponding representations of its Lie algebra. We use the following notation for "generalized powers":

$$a^{[m]} = a(a+1)\dots(a+m-1), \quad a^{(m)} = a(a-1)\dots(a-m+1),$$

where a is a number or an operator. The congruence $\equiv$ means congruence modulo 2.

Supported by the grant of Minobrnauki 2014/285, No. 2476.

Let us recall some material on representations of the group G. This group consists of real 2×2 matrices with determinant 1. The Lie algebra $\mathfrak{g}$ of G consists of real 2×2 trace zero matrices. Let us take in its complexification $\mathfrak{g}^{\mathbb{C}}$ the following basis:

$$L^0 = \begin{pmatrix} 0 & -1 \\ 1 & 0 \end{pmatrix}, \quad E^+ = -\frac{1}{2}\begin{pmatrix} 1 & -i \\ -i & -1 \end{pmatrix}, \quad E^- = -\frac{1}{2}\begin{pmatrix} 1 & i \\ i & -1 \end{pmatrix}.$$

The center of the universal enveloping algebra $\mathrm{Env}(\mathfrak{g})$ is generated by the Casimir element

$$\Delta_{\mathfrak{g}} = -\frac{1}{4}(L^0)^2 + \frac{1}{2}\left(E^+E^- + E^-E^+\right).$$

Let $\sigma \in \mathbb{C}$, $\varepsilon = 0, 1$. We consider the principal series of representations $T_{\sigma,\varepsilon}$ of the group G in compact picture. This group acts on the plane $\mathbb{R}^2$ from the right, so we write vectors in $\mathbb{R}^2$ in row form. Let $|x| = \sqrt{x_1^2 + x_2^2}$ be the Euclidean length of a vector $x = (x_1, x_2)$. Let us denote by S the circle $|x| = 1$. Let $\mathcal{D}_\varepsilon(S)$ be the subspace in $\mathcal{D}(S)$ of functions φ on S of parity ε:

$$\varphi(-s) = (-1)^\varepsilon \varphi(s).$$

The representation $T_{\sigma,\varepsilon}$ acts on $\mathcal{D}_\varepsilon(S)$ as follows:

$$(T_{\sigma,\varepsilon}(g)\varphi)(s) = \varphi\left(\frac{sg}{|sg|}\right)|sg|^{2\sigma}.$$

Let us take on S the coordinate α: a point $s \in S$ is $s = (\sin\alpha, \cos\alpha)$. Sometimes we write $\varphi(\alpha)$ instead of $\varphi(s)$.

Introduce in $\mathcal{D}_\varepsilon(S)$ an operator $A_{\sigma,\varepsilon}$:

$$(A_{\sigma,\varepsilon}\varphi)(\alpha) = \frac{1}{2}\int_0^{2\pi} [\sin(\alpha-\beta)]^{-2\sigma-2,\,\varepsilon}\, \varphi(\beta)\, d\beta.$$

It intertwines $T_{\sigma,\varepsilon}$ and $T_{-\sigma-1,\,\varepsilon}$. The integral converges absolutely when $\mathrm{Re}\,\sigma < -1/2$ and can be extended to the whole σ-plane as a meromorphic function.

Let us take in $\mathcal{D}_\varepsilon(S)$ a basis consisting of exponents:

$$\varphi_r(\alpha) = e^{ir\alpha}, \quad r \in \mathbb{Z}, \quad r \equiv \varepsilon.$$

The operator $A_{\sigma,\varepsilon}$ moves φ_r in φ_r with a factor:

$$A_{\sigma,\varepsilon}\varphi_r = a(\sigma,\varepsilon;\, r)\,\varphi_r,$$

where

$$\begin{aligned} a(\sigma,\varepsilon;\, r) = {} & i^{-r}\,\pi^{-1}\,2^{2\sigma+1}\left[(-1)^\varepsilon - \cos 2\sigma\pi\right]\Gamma(-2\sigma-1) \\ & \times \Gamma\left(\sigma - r/2 + 1\right)\Gamma\left(\sigma + r/2 + 1\right). \end{aligned}$$

Differential operators corresponding to elements of the Lie algebra $\mathfrak{g}$ in the representation $T_{\sigma,\varepsilon}$ are independent on ε, so we omit index ε here. For elements

L^0, $E^{\pm}$ and $\Delta_{\mathfrak{g}}$ we have:

$$T_\sigma(L^0) = \frac{d}{d\alpha},$$
$$T_\sigma(E^+) = e^{2i\alpha}\cdot\left(\sigma + \frac{i}{2}\frac{d}{d\alpha}\right),$$
$$T_\sigma(E^-) = e^{-2i\alpha}\cdot\left(\sigma - \frac{i}{2}\frac{d}{d\alpha}\right),$$
$$T_\sigma(\Delta_{\mathfrak{g}}) = \sigma(\sigma+1)\cdot \mathrm{id}.$$

Exponents φ_r are eigenfunctions for the operator $T_\sigma(L^0)$:

$$T_\sigma(L^0)\varphi_r = ir\,\varphi_r\,,$$

and operators $T_\sigma(E^{\pm})$ act on them as follows:

$$T_\sigma(E^{\pm})\varphi_r = (\sigma \mp r/2)\,\varphi_{r\pm 2}\,. \tag{1}$$

Representations $T_{\sigma,\varepsilon}$ are *irreducible* except when $2\sigma \in \mathbb{Z}$ and $2\sigma \equiv \varepsilon$. Let $2m \in \mathbb{N} = \{0,1,2,\ldots\}$. Then the representation $T_{m,\varepsilon}$ has an invariant irreducible finite-dimensional subspace

$$V_m = \{\varphi_r : -2m \leqslant r \leqslant 2m,\ r \equiv \varepsilon\},$$

so that $\dim V_\sigma = 2m+1$. Denote by π_m the restriction to V_m of the representation $T_{m,\varepsilon}$. Representations π_m, $2m \in \mathbb{N}$, exhaust all irreducible finite-dimensional representations of G.

Consider the tensor product $T_{\sigma,\varepsilon}\otimes\pi_m$. For definiteness, we take *generic* σ, i.e., $2\sigma \notin \mathbb{N}$. We suppose that k ranges over the set $\{0,1,2,\ldots,2m\}$. Let us denote

$$\tau = \sigma - m + k. \tag{2}$$

The tensor product $T_{\sigma,\varepsilon}\otimes\pi_m$ acts on the space $\mathcal{D}_\varepsilon(S)\otimes V_m$ consisting of functions $f(s,t)$ of parity

$$\nu \equiv \varepsilon + 2m$$

on the direct product $S\times S$ of two circles: $f(-s,-t) = (-1)^\nu f(s,t)$. Denote

$$v = \alpha - \beta.$$

Theorem 1. *The tensor product $T_{\sigma,\varepsilon}\otimes\pi_m$ decomposes into the direct multiplicity free sum of irreducible representations:*

$$T_{\sigma,\varepsilon}\otimes\pi_m = \sum_{k=0}^{2m} T_{\tau,\nu}. \tag{3}$$

Accordingly, $\mathcal{D}_\varepsilon(S)\otimes V_m$ decomposes into the direct sum of irreducible subspaces $W_k^{(\sigma)}$. The subspace $W_k^{(\sigma)}$ is the image of a Poisson transform $M_k^{(\sigma)}$ mapping the *space $\mathcal{D}_\nu(S)$ to the space $\mathcal{D}_\varepsilon(S)\otimes V_m$ and intertwining the representation $T_{\tau,\nu}$ with the tensor product $T_{\sigma,\varepsilon}\otimes\pi_m$. The Poisson transform is a differential operator:*

$$\begin{aligned} M_k^{(\sigma)} &= (\sin v)^{2m-k}\,(-1)^k\sum_{r=0}^{k}\binom{k}{r}(2\tau-r)^{(k-r)} \\ &\quad\times e^{i(k-r)v}\,(-2i\sin v)^r\cdot\left(\tau-\frac{i}{2}\frac{d}{d\alpha}\right)^{(r)}. \end{aligned} \tag{4}$$

It can be present as a product of k linear differential operators (they do not commute):

$$\begin{aligned} M_k^{(\sigma)} = (\sin v)^{2m-k} &\left[(\sin v)\,\frac{\partial}{\partial\alpha}-(2\tau-k+1)\,\cos v\right] \\ &\times\left[(\sin v)\,\frac{\partial}{\partial\alpha}-(2\tau-k+2)\,\cos v\right] \\ &\quad\dots \\ &\times\left[(\sin v)\,\frac{\partial}{\partial\alpha}-2\tau\,\cos v\right]. \end{aligned} \tag{5}$$

Proof. For simplicity, we denote $\mathcal{D}_\varepsilon(S)\otimes V_m=\widetilde{\mathcal{D}}$ and $T_{\sigma,\varepsilon}\otimes\pi_m=\widetilde{T}$, we do not show indices σ, ε, m. For points $(s,t)\in S\times S$ we take parameters α and β respectively, so that $s=(\sin\alpha,\cos\alpha)$ and $t=(\sin\beta,\cos\beta)$.

To elements $X\in\mathfrak{g}$, the representation $\widetilde{T}$ assigns operators $\widetilde{T}(X)$, for brevity we denote them $\widetilde{X}$, we have $\widetilde{X}=T_{\sigma,\varepsilon}(X)\otimes 1+1\otimes\pi_m(X)$. Let us take in $\widetilde{\mathcal{D}}$ a basis consisting of exponents:

$$\varphi_{r,h}(\alpha,\beta)=e^{ir\alpha}e^{ih\beta},$$

so that $r+h\equiv\nu$. These exponents are eigenvectors for $\widetilde{L}^0$:

$$\widetilde{L}^0\varphi_{r,h}=i(r+h)\varphi_{r,h}.$$

Operators $\widetilde{E}^\pm$ act as follows:

$$\widetilde{E}^\pm\varphi_{r,h}=(\sigma\mp r/2)\cdot\varphi_{r\pm2,h}+(m\mp h/2)\cdot\varphi_{r,h\pm2}.$$

Denote by H_p the subspace in $\widetilde{\mathcal{D}}$ spanned by $\varphi_{r,h}$, $r+h=2p$, so that $2p\equiv\nu$ and $\dim H_p=2m+1$. This subspace is an eigenspace for $\widetilde{L}^0$ with eigenvalue 2π. The whole space $\widetilde{\mathcal{D}}$ decomposes into the direct multiplicity free sum of subspaces H_p, $2p\in\mathbb{Z}$, $2p\equiv\nu$. Denote by $\widetilde{E}_p^\pm$ the restriction to H_p of operators $\widetilde{E}^\pm$.

The $\widetilde{E}_p^\pm$ maps H_p to $H_{p\pm1}$. The composition $R_p=\widetilde{E}_{p+1}^-\widetilde{E}_p^+$ (first we do $\widetilde{E}_p^+$ and then $\widetilde{E}_{p+1}^-$) maps H_p to H_p. We want to find eigenvectors and eigenvalues of R_p.

Let us take in H_p the following basis $\{\xi_k^{(p)}\}$:

$$\xi_k^{(p)}=e^{2pi\alpha}\,e^{ikv}\,\left[e^{iv}-e^{-iv}\right]^{2m-k},\quad k=0,1,\dots,2m. \tag{6}$$

In it, operators $\widetilde{E}_p^{\pm}$ act as follows (we use notation (2)):

$$\widetilde{E}_p^+ \, \xi_k^{(p)} = (\tau - p)\, \xi_k^{(p+1)} - k\, \xi_{k-1}^{(p+1)}\,, \tag{7}$$

$$\widetilde{E}_p^- \, \xi_k^{(p)} = (\tau + p)\, \xi_k^{(p-1)}. \tag{8}$$

It follows from (7) and (8) that

$$R_p\, \xi_k^{(p)} = (\tau + p + 1)(\tau - p)\xi_k^{(p)} - k(\tau + p)\xi_{k-1}^{(p)}\,.$$

It means that in the basis $\{\xi_k^{(p)}\}$ in H_p the operator R_p has upper triangular two-diagonal matrix. Therefore, the eigenvector of R_p with number k is

$$w_k^{(p)} = b_k \sum_{r=0}^{k} (-1)^{k-r} \binom{k}{r} (\tau + p)^{(k-r)} (2\tau - k + 1)^{[r]}\, \xi_r^{(p)}, \tag{9}$$

where b_k is some coefficient (notice that the superscript (p) *is not* a generalized power), and the corresponding eigenvalue is

$$\lambda_k^{(p)} = (\tau - p)(\tau + p + 1),$$

so that

$$R_p w_k^{(p)} = \lambda_k^{(p)}\, w_k^{(p)}.$$

Let us substitute in (9) expressions (6) for $\xi_r^{(p)}$ and take the following coefficient $b_k = (-1)^k (2i)^{-2m+k}$, we obtain

$$\begin{aligned} w_k^{(p)} = e^{2pi\alpha}\, (\sin v)^{2m-k}\, (-1)^k \sum_{r=0}^{k} \binom{k}{r} (2\tau - r)^{(k-r)} \\ \times\, e^{i(k-r)v}\, (-2i \sin v)^r \cdot (\tau + p)^{(r)}\,. \end{aligned} \tag{10}$$

Let us find how operators $\widetilde{E}_p^{\pm}$ act on $w_k^{(p)} \in H_p$.

Let $w \in H_p$ be an eigenvector of R_p, i.e., $R_p w = \lambda w$. From the commutation relation $[E^+, E^-] = -iL^0$ we obtain

$$E^+E^-E^+ = E^-E^+E^+ - iL^0E^+.$$

Hence $\widetilde{E}_p^+ R_p w = R_{p+1} \widetilde{E}_p^+ w - i\widetilde{L}^0 \widetilde{E}_p^+ w$. The vector $\widetilde{E}_p^+ w$ belongs to the space H_{p+1}, it is an eigenspace for $i\widetilde{L}^0$ with the eigenvalue $2p+2$. Therefore, $\lambda \widetilde{E}_p^+ w = R_{p+1} \widetilde{E}_p^+ w + (2p+2) \widetilde{E}_p^+ w$. Hence

$$R_{p+1} \widetilde{E}_p^+ w = (\lambda - 2p - 2) \widetilde{E}_p^+ w.$$

Thus, the operator $\widetilde{E}_p^+$ maps an eigenvector of R_p to an eigenvector of R_{p+1}. For eigenvalues λ we have

$$\lambda_k^{(p)} - 2p - 2 = \lambda_k^{(p+1)},$$

so $\widetilde{E}_p^+$ maps $w_k^{(p)}$ just to $w_k^{(p+1)}$ – with some factor. In order to find this factor, one has to trace to the summand of $w_k^{(p)}$ containing $\xi_k^{(p)}$; here a coefficient does not depend on p. Similarly we deal with $\widetilde{E}_p^-$. Finally we have

$$\widetilde{E}_p^{\pm} w_k^{(p)} = (\tau \mp p) w_k^{(p\pm 1)}. \tag{11}$$

Let W_k be the subspace in $\widetilde{\mathcal{D}}$ spanned by $w_k^{(p)}$, $2p \in \mathbb{Z}$, $2p \equiv \nu$. It is isomorphic to $\mathcal{D}_\nu(S)$. The whole space $\widetilde{\mathcal{D}}$ decomposes into the direct multiplicity free sum of subspaces W_k.

Formulae (11) show that operators $\widetilde{E}^{\pm}$ act on vectors $w_k^{(p)}$ exactly as operators $T_{\tau,\nu}(E^{\pm})$ act on exponents $\varphi_{2p}(s)$ in $\mathcal{D}_\nu(S)$, see (1). The operator corresponding to the Casimir element multiplies $w_k^{(p)}$ by $\tau(\tau+1)$. Therefore, the restriction of $\widetilde{T}$ to W_k is equivalent to $T_{\tau,\nu}$, and $\widetilde{T}$ itself is the direct multiplicity free sum given by (3).

Now let us construct an operator $M_k^{(\sigma)} : \mathcal{D}_\nu(S) \to W_k$ intertwining $T_{\tau,\nu}$ and $\widetilde{T}$. This operator has to move the exponent $\varphi_{2p}(\alpha)$ to the function $w_k^{(p)}$. Since

$$p\, e^{2pi\alpha} = -\frac{i}{2}\,\frac{d}{d\alpha}\, e^{2pi\alpha},$$

then (10) gives (4).

Finally let us prove (5). Introduce a differential operator

$$Z = 2\tau - i\frac{d}{d\alpha}.$$

Besides it, introduce the 2-step generalized power:

$$x^{\langle\langle n\rangle\rangle} = x(x-2)(x-4)\dots(x-2n+2)$$

(n factors). Then the operator (4) can be written as follows:

$$M_k^{(\sigma)} = (\sin v)^{2m-k}\, Q_k^{(\sigma)},$$

where

$$Q_k^{(\sigma)} = \sum_{r=0}^{k} (2\tau - k + 1)^{[k-r]} \cdot \binom{k}{r} (-1)^{k-r} \cdot (i \sin v)^r e^{i(k-r)v} \cdot Z^{\langle\langle r\rangle\rangle}$$

The operator $Q_k^{(\sigma)}$ can be decomposed into the following product of k linear operators (not commuting):

$$\begin{aligned} Q_k^{(\sigma)} &= \big[i(\sin v)\,(Z-k+1) - (2\tau-k+1)\,e^{iv}\big] \\ &\quad\times \big[i(\sin v)\,(Z-k+2) - (2\tau-k+2)\,e^{iv}\big] \\ &\quad\quad \cdots \\ &\quad\times \big[i(\sin v)\,(Z-1) - (2\tau-1)\,e^{iv}\big] \\ &\quad\times \big[i(\sin v)\,Z - 2\tau\, e^{iv}\big]. \end{aligned} \tag{12}$$

This statement is proved by induction on k. The inductive step means:

$$Q_{k+1}^{(\sigma)} = [i(\sin v)(Z-k) - (2\tau - k)\, e^{iv}]\, Q_k^{(\sigma)}\,.$$

To compute the right-hand side we use the following relation of *operators*:

$$\{i(\sin v)(Z-k)\} \circ \left\{(i\sin v)^r e^{i(k-r)v}\right\}$$
$$= r\,(i\sin v)^r e^{i(k+1-r)v} + (i\sin v)^{r+1} e^{i(k-r)v}(Z-2r).$$

Therefore, the factor in (12) with number $q = 0, 1, \dots, k-1$ is

$$i(\sin v)(Z-q) - (2\tau - q)\, e^{iv} = \sin v \frac{\partial}{\partial \alpha} - (2\tau - q)\cos v.$$

It proves (5). □

In conclusion we notice the interaction of Poisson transforms with intertwining operators:

$$(A_{\sigma,\varepsilon} \otimes 1)\, M_k^{(\sigma)} = b(\sigma, k) \cdot M_{2m-k}^{(-\sigma-1)}\, A_{\tau,\nu}\,,$$

where

$$b(\sigma, k) = 2^{\,4m-4k}\, \Gamma(-2\sigma - 1)/\Gamma(-2\tau - 1)\,.$$

References

[1] V.F. Molchanov, *Tensor products of unitary representations of the three-dimensional Lorentz group*, Izvestiya Akad. nauk USSR. Ser. matem., 1979, **43**, No. 4, 860–891.

[2] V.F. Molchanov, *Poisson and Fourier transforms for tensor products and an overalgebra*, Geometric Methods in Physics. XXXIII Workshop 2014, Trends in Mathematics, Springer Basel AG, 2015, 195–203.

[3] V.F. Molchanov, *Poisson and Fourier transforms for tensor products*, Funkts. Analiz Prilozh., 2015, **49**, issue 4, 50–60.

Vladimir F. Molchanov
Derzhavin Tambov State University
Internatsionalnaya, 33
392000 Tambov, Russia
e-mail: v.molchanov@bk.ru

Geometric Methods in Physics. XXXV Workshop 2016
Trends in Mathematics, 69–85

Conformal Symmetry Breaking Operators for Anti-de Sitter Spaces

Toshiyuki Kobayashi, Toshihisa Kubo and Michael Pevzner

Abstract. For a pseudo-Riemannian manifold X and a totally geodesic hypersurface Y, we consider the problem of constructing and classifying all linear differential operators $\mathcal{E}^i(X) \to \mathcal{E}^j(Y)$ between the spaces of differential forms that intertwine multiplier representations of the Lie algebra of conformal vector fields. Extending the recent results in the Riemannian setting by Kobayashi–Kubo–Pevzner [Lecture Notes in Math. 2170, (2016)], we construct such differential operators and give a classification of them in the pseudo-Riemannian setting where both X and Y are of constant sectional curvature, illustrated by the examples of anti-de Sitter spaces and hyperbolic spaces.

Mathematics Subject Classification (2010). Primary 53A30, 53Z05; Secondary 53C10, 22E70.

Keywords. Conformal geometry, conformal group, symmetry breaking operator, branching law, holography, space form, pseudo-Riemannian geometry, hyperbolic manifold.

1. Introduction

Let X be a manifold endowed with a pseudo-Riemannian metric g. A vector field Z on X is called *conformal* if there exists $\rho(Z,\cdot) \in C^\infty(X)$ (*conformal factor*) such that

$$L_Z g = \rho(Z,\cdot) g,$$

where L_Z stands for the Lie derivative with respect to the vector field Z. We denote by $\mathfrak{conf}(X)$ the Lie algebra of conformal vector fields on X.

Let $\mathcal{E}^i(X)$ be the space of (complex-valued) smooth i-forms on X. We define a family of multiplier representations of the Lie algebra $\mathfrak{conf}(X)$ on $\mathcal{E}^i(X)$ ($0 \leq i \leq \dim X$) with parameter $u \in \mathbb{C}$ by

$$\Pi_u^{(i)}(Z)\alpha := L_Z\alpha + \frac{1}{2}u\rho(Z,\cdot)\alpha \quad \text{for } \alpha \in \mathcal{E}^i(X). \tag{1}$$

For simplicity, we write $\mathcal{E}^i(X)_u$ for the representation $\Pi_u^{(i)}$ of $\mathfrak{conf}(X)$ on $\mathcal{E}^i(X)$.

For a submanifold Y of X, conformal vector fields along Y form a subalgebra

$$\mathfrak{conf}(X;Y) := \{Z \in \mathfrak{conf}(X) : Z_y \in T_yY \text{ for all } y \in Y\}.$$

If the metric tensor g is nondegenerate when restricted to the submanifold Y, then Y carries a pseudo-Riemannian metric $g|_Y$ and there is a natural Lie algebra homomorphism $\mathfrak{conf}(X;Y) \to \mathfrak{conf}(Y)$, $Z \mapsto Z|_Y$. In this case we compare the representation $\Pi_u^{(i)}$ of the Lie algebra $\mathfrak{conf}(X)$ on $\mathcal{E}^i(X)$ with an analogous representation denoted by the lowercase letter $\pi_v^{(j)}$ of the Lie algebra $\mathfrak{conf}(Y)$ on $\mathcal{E}^j(Y)$ for $u, v \in \mathbb{C}$. For this, we analyze *conformal symmetry breaking operators*, that is, linear maps $T\colon \mathcal{E}^i(X) \to \mathcal{E}^j(Y)$ satisfying

$$\pi_v^{(j)}(Z|_Y) \circ T = T \circ \Pi_u^{(i)}(Z) \quad \text{for all } Z \in \mathfrak{conf}(X;Y). \tag{2}$$

Some of such operators are given as differential operators (*e.g.*, [3, 6, 12, 14, 15]), and others are integral operators and their analytic continuation (*e.g.*, [16]). We denote by $\mathrm{Diff}_{\mathfrak{conf}(X;Y)}(\mathcal{E}^i(X)_u, \mathcal{E}^j(Y)_v)$ the space of differential operators satisfying (2).

In the case $X = Y$ and $i = j = 0$, the Yamabe operator, the Paneitz operator [18], which appears in four-dimensional supergravity [4], or more generally, the so-called GJMS operators [5] are such differential operators. Branson and Gover [1, 2] extended such operators to differential forms when $i = j$. The exterior derivative d and the codifferential d^* also give examples of such operators for $j = i + 1$ and $i - 1$, respectively. Maxwell's equations in a vacuum can be expressed in terms of conformally covariant operators on 2-forms in the Minkowski space $\mathbb{R}^{1,3}$ (see [17] for a bibliography). All these classical examples concern the case where $X = Y$. On the other hand, the more general setting where $X \supsetneq Y$ is closely related to branching laws of infinite-dimensional representations (cf. "Stage C" of branching problems in [11]). In recent years, for $(X, Y) = (S^n, S^{n-1})$, such operators in the scalar-valued case ($i = j = 0$) were classified by Juhl [6], see also [3, 10, 14] for different approaches. More generally, such operators have been constructed and classified also in the matrix-valued case (i, j arbitrary) by the authors [12]. In this paper, we give a variant of [12] by extending the framework as follows:

the group of conformal diffeomorphisms $\Longrightarrow$ the Lie algebra of conformal vector fields;
homogeneous spaces $\Longrightarrow$ locally homogeneous spaces;
Riemannian setting $\Longrightarrow$ pseudo-Riemannian setting.

Let $\mathbb{R}^{p,q}$ denote the space $\mathbb{R}^{p+q}$ endowed with the flat pseudo-Riemannian metric:

$$g_{\mathbb{R}^{p,q}} = dx_1^2 + \cdots + dx_p^2 - dy_{p+1}^2 - \cdots - dy_{p+q}^2. \tag{3}$$

For $p, q \in \mathbb{N}$, we define a hypersurface $S^{p,q}$ of $\mathbb{R}^{1+p+q}$ by

$$S^{p,q} := \begin{cases} \{(\omega_0, \omega, \eta) \in \mathbb{R}^{1+p+q} : \omega_0^2 + |\omega|^2 - |\eta|^2 = 1\} & (p > 0), \\ \{(\omega_0, \eta) \in \mathbb{R}^{1+q} : \omega_0 > 0,\ \omega_0^2 - |\eta|^2 = 1\} & (p = 0). \end{cases} \tag{4}$$

Then, the metric $g_{\mathbb{R}^{1+p,q}}$ on the ambient space $\mathbb{R}^{1+p+q}$ induces a pseudo-Riemannian structure on the hypersurface $S^{p,q}$ of signature (p,q) with constant sectional curvature $+1$, which is sometimes referred to as the (positively curved) *space form* of a pseudo-Riemannian manifold. We may regard $S^{p,q}$ also as a pseudo-Riemannian manifold of signature (q,p) with constant curvature -1 by using $-g_{\mathbb{R}^{1+p,q}}$ instead, giving rise to the negatively curved space form.

Example 1 (Riemannian and Lorentzian cases).

$$\begin{array}{llll} S^{n,0} = S^n & \text{(sphere)}, & S^{0,n} = H^n & \text{(hyperbolic space)}, \\ S^{n-1,1} = dS^n & \text{(de Sitter space)}, & S^{1,n-1} = AdS^n & \text{(anti-de Sitter space)}. \end{array}$$

In Theorems A–C below, we assume $n = p+q \geq 3$ and consider

$$(X,Y) = (S^{p,q}, S^{p-1,q}),\ (S^{p,q}, S^{p,q-1}),\ (\mathbb{R}^{p,q}, \mathbb{R}^{p-1,q}),\ \text{or } (\mathbb{R}^{p,q}, \mathbb{R}^{p,q-1}). \quad (5)$$

Example 2. $\mathfrak{conf}(X;Y) \simeq \mathfrak{o}(p,q+1)$ *if* $(X,Y) = (S^{p,q}, S^{p-1,q})$ *or* $(\mathbb{R}^{p,q}, \mathbb{R}^{p-1,q})$.

Theorem A below addresses the question if any conformal symmetry breaking operator defined locally can be extended globally.

Theorem A (automatic continuity). *Let V be any open set of X such that $V \cap Y$ is connected and nonempty. Suppose $u, v \in \mathbb{C}$. Then the map taking the restriction to V induces a bijection:*

$$\mathrm{Diff}_{\mathfrak{conf}(X;Y)}(\mathcal{E}^i(X)_u, \mathcal{E}^j(Y)_v) \xrightarrow{\sim} \mathrm{Diff}_{\mathfrak{conf}(V,V\cap Y)}(\mathcal{E}^i(V)_u, \mathcal{E}^j(V\cap Y)_v).$$

We recall from [19, Chap. II] that the pseudo-Riemannian manifolds $\mathbb{R}^{p,q}$ and $S^{p,q}$ have a common conformal compactification:

$$\begin{array}{ccc} \mathbb{R}^{p,q} & & S^{p,q} \\ & \searrow \quad \swarrow & \\ & (S^p \times S^q)/\mathbb{Z}_2 & \end{array}$$

where $(S^p \times S^q)/\mathbb{Z}_2$ denotes the direct product of p- and q-spheres equipped with the pseudo-Riemannian metric $g_{S^p} \oplus (-g_{S^q})$, modulo the direct product of antipodal maps, see also [13, II, Lem. 6.2 and III, Sect. 2.8]. For $X = \mathbb{R}^{p,q}$ or $S^{p,q}$, we denote by $\overline{X}$ this conformal compactification of X.

Theorem B.

(1) (Automatic continuity to the conformal compactification.) *Suppose $u, v \in \mathbb{C}$ and $0 \leq i \leq n$, $0 \leq j \leq n-1$. Then the map taking the restriction to X is a bijection*

$$\mathrm{Diff}_{\mathfrak{conf}(\overline{X};\overline{Y})}(\mathcal{E}^i(\overline{X})_u, \mathcal{E}^j(\overline{Y})_v) \xrightarrow{\sim} \mathrm{Diff}_{\mathfrak{conf}(X;Y)}(\mathcal{E}^i(X)_u, \mathcal{E}^j(Y)_v).$$

(2) *If $n \geq 3$, all these spaces are isomorphic to each other for (X,Y) in (5) as far as (p,q) satisfies $p+q = n$.*

By Theorems A and B, we see that all conformal symmetry breaking operators given locally in some open sets in the pseudo-Riemannian case (5) are derived from the Riemannian case (*i.e.*, $p = 0$ or $q = 0$). We note that our representation (1) is normalized in a way that $\Pi_u^{(i)}$ coincides with the differential of the representation $\varpi_{u,\delta}^{(i)}$ ($\delta \in \mathbb{Z}/2\mathbb{Z}$) of the conformal group $\mathrm{Conf}(X)$ introduced in [12, (1.1)]. In particular, we can read from [12, Thms. 1.1 and 2.10] the dimension of $\mathrm{Diff}_{\mathrm{conf}(X;Y)}(\mathcal{E}^i(X)_u, \mathcal{E}^j(Y)_v)$ for any i, j, u, v. For simplicity of exposition, we present a coarse feature as follows.

Theorem C. *Suppose (X, Y) is as in (5), and V any open set of X such that $V \cap Y$ is connected and nonempty. Let $u, v \in \mathbb{C}$, $0 \leq i \leq n$, and $0 \leq j \leq n-1$.*

(1) *For any $u, v \in \mathbb{C}$ and $0 \leq i \leq n$, $0 \leq j \leq n-1$,*

$$\dim_{\mathbb{C}} \mathrm{Diff}_{\mathrm{conf}(V;V\cap Y)}(\mathcal{E}^i(V)_u, \mathcal{E}^j(V \cap Y)_v) \leq 2.$$

(2) $\mathrm{Diff}_{\mathrm{conf}(V;V\cap Y)}(\mathcal{E}^i(V)_u, \mathcal{E}^j(V \cap Y)_v) \neq \{0\}$ *only if u, v, i, j satisfy*

$$(v+j)-(u+i) \in \mathbb{N} \quad \text{and} \quad \big(-1 \leq i-j \leq 2 \ \text{ or } \ n-2 \leq i+j \leq n+1\big). \tag{6}$$

A precise condition when the equality holds in Theorem C (1) will be explained in Section 7 in the case $n = 4$. We shall give explicit formulæ of generators of $\mathrm{Diff}_{\mathrm{conf}(X;Y)}(\mathcal{E}^i(X)_u, \mathcal{E}^j(Y)_v)$ in Theorem D in Section 2 for the flat pseudo-Riemannian manifolds, and in Theorem E in Section 3 for positively (or negatively) curved space forms. These operators (with "renormalization") and their compositions by the Hodge star operators with respect to the pseudo-Riemannian metric exhaust all differential symmetry breaking operators (Remark 4). The proof of Theorems A–C will be given in Section 5.

Notation. $\mathbb{N} = \{0, 1, 2, \dots\}$, $\mathbb{N}_+ = \{1, 2, \dots\}$.

2. Conformally covariant symmetry breaking operators – flat case

In this section, we give explicit formulæ of conformal symmetry breaking operators in the flat pseudo-Riemannian case $(X, Y) = (\mathbb{R}^{p,q}, \mathbb{R}^{p-1,q})$ or $(\mathbb{R}^{p,q}, \mathbb{R}^{p,q-1})$. This extends the results in [12] that dealt with the Riemannian case $(X, Y) = (\mathbb{R}^n, \mathbb{R}^{n-1})$.

We note that the signature of the metric restricted to nondegenerate hyperplanes of $\mathbb{R}^{p,q}$ is either $(p-1, q)$ or $(p, q-1)$. Thus it is convenient to introduce two types of coordinates in $\mathbb{R}^{p+q}$ accordingly. We set

$$\mathbb{R}_+^{p,q} = \{(y, x) \in \mathbb{R}^{q+p}\} \quad \text{with} \quad -dy_1^2 - \cdots - dy_q^2 + dx_{q+1}^2 + \cdots + dx_{p+q}^2,$$
$$\mathbb{R}_-^{p,q} = \{(x, y) \in \mathbb{R}^{p+q}\} \quad \text{with} \quad dx_1^2 + \cdots + dx_p^2 - dy_{p+1}^2 - \cdots - dy_{p+q}^2.$$

Then by letting the last coordinate to be zero, we get hypersurfaces of $\mathbb{R}^{p,q}$ of two types:

$$\mathbb{R}_+^{p-1,q} \subset \mathbb{R}_+^{p,q} \ (p \geq 1), \quad \mathbb{R}_-^{p,q-1} \subset \mathbb{R}_-^{p,q} \ (q \geq 1).$$

For $\ell \in \mathbb{N}$ and $\mu \in \mathbb{C}$, we define a family of differential operators on $\mathbb{R}^{p+q}$ by using the above coordinates:

$$(\mathcal{D}_\ell^\mu)_+ \equiv (\mathcal{D}_\ell^\mu)_{\mathbb{R}_+^{p,q}} := \sum_{k=0}^{[\frac{\ell}{2}]} a_k(\mu,\ell) \left(\sum_{j=1}^{q} \frac{\partial^2}{\partial y_j^2} - \sum_{j=q+1}^{n-1} \frac{\partial^2}{\partial x_j^2} \right)^k \left(\frac{\partial}{\partial x_n} \right)^{\ell-2k} \quad \text{on } \mathbb{R}_+^{p,q},$$

$$(\mathcal{D}_\ell^\mu)_- \equiv (\mathcal{D}_\ell^\mu)_{\mathbb{R}_-^{p,q}} := \sum_{k=0}^{[\frac{\ell}{2}]} a_k(\mu,\ell) \left(\sum_{j=1}^{p} \frac{\partial^2}{\partial x_j^2} - \sum_{j=p+1}^{n-1} \frac{\partial^2}{\partial y_j^2} \right)^k \left(\frac{\partial}{\partial y_n} \right)^{\ell-2k} \quad \text{on } \mathbb{R}_-^{p,q},$$

where we set for $k \in \mathbb{N}$ with $0 \le 2k \le \ell$

$$a_k(\mu,\ell) := \frac{(-1)^k 2^{\ell-2k}\Gamma(\ell-k+\mu)}{\Gamma(\mu+[\frac{\ell+1}{2}])k!(\ell-2k)!}. \tag{7}$$

In the case $(p,q,\varepsilon) = (n,0,+)$, $(\mathcal{D}_\ell^\mu)_{\mathbb{R}_\varepsilon^{p,q}}$ coincides with the differential operator $\mathcal{D}_\ell^\mu$ in [12, (1.2)], which was originally introduced in [6] (up to scalar).

The coefficients $a_k(\mu,\ell)$ arise from a hypergeometric polynomial

$$\widetilde{C}_\ell^\mu(t) := \sum_{k=0}^{[\frac{\ell}{2}]} a_k(\mu,\ell) t^{\ell-2k}.$$

This is a "renormalized" Gegenbauer polynomial [15, II, (11.16)] in the sense that $\widetilde{C}_\ell^\mu(t)$ is nonzero for all $\mu \in \mathbb{C}$ and $\ell \in \mathbb{N}$ and satisfies the Gegenbauer differential equation:

$$\left((1-t^2)\frac{d^2}{dt^2} - (2\mu+1)t\frac{d}{dt} + \ell(\ell+2\mu) \right) f(t) = 0.$$

We set $\mu =: u + i - \frac{1}{2}(n-1)$ and $\gamma(\mu,a) := 1$ (a: odd), $\mu + \frac{a}{2}$ (a: even).

For parameters $u \in \mathbb{C}$ and $\ell \in \mathbb{N}$, we define a family of linear operators

$$(\mathcal{D}_{u,\ell}^{i\to j})_- : \mathcal{E}^i(\mathbb{R}^{p,q}) \to \mathcal{E}^j(\mathbb{R}^{p,q-1})$$

in the coordinates $(x_1,\ldots,x_p,y_{p+1},\ldots,y_{p+q})$ of $\mathbb{R}_-^{p,q}$ as follows: For $j = i-1$ or i,

$$(\mathcal{D}_{u,\ell}^{i\to i-1})_{\mathbb{R}_-^{p,q}} := \operatorname{Rest}_{y_n=0} \circ \Big((\mathcal{D}_{\ell-2}^{\mu+1})_-\, dd^* \iota_{\frac{\partial}{\partial y_n}} + \gamma(\mu,a)(\mathcal{D}_{\ell-1}^{\mu+1})_-\, d^* + \frac{u+2i-n}{2}(\mathcal{D}_\ell^\mu)_-\, \iota_{\frac{\partial}{\partial y_n}} \Big),$$

$$(\mathcal{D}_{u,\ell}^{i\to i})_{\mathbb{R}_-^{p,q}} := \operatorname{Rest}_{y_n=0} \circ \Big(-(\mathcal{D}_{\ell-2}^{\mu+1})_-\, dd^* - \gamma(\mu-\frac{1}{2},\ell)(\mathcal{D}_{\ell-1}^{\mu})_-\, d\iota_{\frac{\partial}{\partial y_n}} + \frac{u+\ell}{2}(\mathcal{D}_\ell^\mu)_- \Big).$$

Here $d^* : \mathcal{E}^i(\mathbb{R}_-^{p,q}) \to \mathcal{E}^{i-1}(\mathbb{R}_-^{p,q})$ is the codifferential $d^*_{\mathbb{R}_-^{p,q}} = (-1)^i *^{-1} d*$, where $* \equiv *_{\mathbb{R}_-^{p,q}}$ is the Hodge operator with respect to the pseudo-Riemannian structure on $\mathbb{R}_-^{p,q}$, $\iota_{\frac{\partial}{\partial y_n}}$ is the interior multiplication by the vector field $\frac{\partial}{\partial y_n}$, and $(\mathcal{D}_\ell^\mu)_-$ acts on $\mathcal{E}^i(\mathbb{R}_-^{p,q})$ as a scalar differential operator.

In contrast to the case $j = i - 1$ or i where the family of operators $\mathcal{D}_{u,\ell}^{i\to j}$ contains a continuous parameter $u \in \mathbb{C}$ and discrete one $\ell \in \mathbb{N}$, it turns out that the remaining case where $j \notin \{i-1, i\}$ or its Hodge dual $j \notin \{n-i+1, n-i\}$ is not abundant in conformal symmetry breaking operators. Actually, for $j \in \{i-2, i+1\}$, we define $(\mathcal{D}_{u,\ell}^{i\to j})_{\mathbb{R}_-^{p,q}}$ only for special values of (i, u, ℓ) as follows:

$$
\begin{aligned}
(\mathcal{D}_{n-2i,1}^{i\to i-2})_{\mathbb{R}_-^{p,q}} &:= -\mathrm{Rest}_{y_n=0} \circ \iota_{\frac{\partial}{\partial y_n}} d^* && (2 \le i \le n-1),\\
(\mathcal{D}_{1-n-\ell,\ell}^{n\to n-2})_{\mathbb{R}_-^{p,q}} &:= -\mathrm{Rest}_{y_n=0} \circ \left(\mathcal{D}_{\ell-1}^{\frac{3-n}{2}-\ell}\right)_- \iota_{\frac{\partial}{\partial y_n}} d^* && (\ell \in \mathbb{N}_+),\\
(\mathcal{D}_{0,1}^{i\to i+1})_{\mathbb{R}_-^{p,q}} &:= \mathrm{Rest}_{y_n=0} \circ d && (1 \le i \le n-2),\\
(\mathcal{D}_{1-\ell,\ell}^{0\to 1})_{\mathbb{R}_-^{p,q}} &:= \mathrm{Rest}_{y_n=0} \circ \left(\mathcal{D}_{\ell-1}^{\frac{3-n}{2}-\ell}\right)_- d && (\ell \in \mathbb{N}_+).
\end{aligned}
$$

Likewise, for $\mathbb{R}_+^{p,q}$, we define a family of linear operators

$$(\mathcal{D}_{u,\ell}^{i\to j})_+ \colon \mathcal{E}^i(\mathbb{R}^{p,q}) \to \mathcal{E}^j(\mathbb{R}^{p-1,q})$$

in the coordinates $(y_1, \ldots, y_q, x_{q+1}, \ldots, x_{p+q})$ of $\mathbb{R}_+^{p,q}$ with parameters $u \in \mathbb{C}$ and $\ell \in \mathbb{N}$. In this case, the formulæ are essentially the same as those in the Riemannian case ($q = 0$) which were introduced in [12, (1.4)–(1.12)]. (The changes from $(\mathcal{D}_{u,\ell}^{i\to j})_{\mathbb{R}_-^{p,q}}$ to $(\mathcal{D}_{u,\ell}^{i\to j})_{\mathbb{R}_+^{p,q}}$ are made by replacing $y_n = 0$ with $x_n = 0$, $\frac{\partial}{\partial y_n}$ with $\frac{\partial}{\partial x_n}$, and $d^*_{\mathbb{R}_-^{p,q}}$ with $-d^*_{\mathbb{R}_+^{p,q}}$.) For the convenience of the reader, we give formulæ for $j = i - 1$ or i and omit the case $j = i - 2$ and $i + 1$:

$$
\begin{aligned}
(\mathcal{D}_{u,\ell}^{i\to i-1})_+ &:= \mathrm{Rest}_{x_n=0} \circ \Big(-(\mathcal{D}_{\ell-2}^{\mu+1})_+ dd^* \iota_{\frac{\partial}{\partial x_n}} - \gamma(\mu, \ell)(\mathcal{D}_{\ell-1}^{\mu+1})_+ d^* \\
&\qquad + \frac{u + 2i - n}{2} (\mathcal{D}_\ell^{\mu})_+ \iota_{\frac{\partial}{\partial x_n}} \Big),\\
(\mathcal{D}_{u,\ell}^{i\to i})_+ &:= \mathrm{Rest}_{x_n=0} \circ \Big((\mathcal{D}_{\ell-2}^{\mu+1})_+ dd^* - \gamma(\mu - \frac{1}{2}, \ell)(\mathcal{D}_{\ell-1}^{\mu})_+ d\iota_{\frac{\partial}{\partial x_n}} \\
&\qquad + \frac{u + \ell}{2} (\mathcal{D}_\ell^{\mu})_+ \Big).
\end{aligned}
$$

If $i = j = 0$, the operators $(\mathcal{D}_{u,\ell}^{i\to j})_{\mathbb{R}_+^{p,q}}$ reduce to scalar-valued differential operators that are proportional to $(\mathcal{D}_\ell^{\mu})_+$ because d^* and $\iota_{\frac{\partial}{\partial x_n}}$ are identically zero on $\mathcal{E}^0(X) = C^\infty(X)$.

Theorem D below gives conformal symmetry breaking operators on the flat pseudo-Riemannian manifolds:

Theorem D. *Let $p+q \ge 3$, $0 \le i \le p+q$, $0 \le j \le p+q-1$, and $u, v \in \mathbb{C}$. Assume $j \in \{i-2, i-1, i, i+1\}$ and $\ell := (v + j) - (u + i) \in \mathbb{N}$. (For $j \in \{i-2, i+1\}$, we need an additional condition on the quadruple (i, j, u, v), or equivalently, on*

(i,j,u,ℓ) as indicated in the $\mathbb{R}^{p,q}_-$ case.) Then

$$(\mathcal{D}^{i\to j}_{u,\ell})_+ \in \mathrm{Diff}_{\mathrm{conf}(\mathbb{R}^{p,q};\mathbb{R}^{p-1,q})}(\mathcal{E}^i(\mathbb{R}^{p,q})_u, \mathcal{E}^j(\mathbb{R}^{p-1,q})_v) \quad \text{for } p \geq 1,$$
$$(\mathcal{D}^{i\to j}_{u,\ell})_- \in \mathrm{Diff}_{\mathrm{conf}(\mathbb{R}^{p,q};\mathbb{R}^{p,q-1})}(\mathcal{E}^i(\mathbb{R}^{p,q})_u, \mathcal{E}^j(\mathbb{R}^{p,q-1})_v) \quad \text{for } q \geq 1.$$

Remark 3. In recent years, special cases of Theorem D have been obtained as below.

1. $i = j = 0$, $\varepsilon = +$, $q = 0$: [6], see also [3, 10, 14] for different approaches.
2. $i = j = 0$, $\varepsilon = +$, p and q are arbitrary: [14, Thm. 4.3].
3. i and j are arbitrary, $\varepsilon = +$, $q = 0$: [12, Thms. 1.5, 1.6, 1.7 and 1.8].

The main machinery of finding symmetry breaking operators in various geometric settings in [12], [14], and [15, II] is the "algebraic Fourier transform of generalized Verma modules" (*F-method* [9]), see [15, I] for a detailed exposition of the F-method.

The proof of Theorem D will be given in Section 6.

Remark 4. There are a few values of parameters (u, ℓ, i, j) for which $(\mathcal{D}^{i\to j}_{u,\ell})_+$ or $(\mathcal{D}^{i\to j}_{u,\ell})_-$ vanishes, but we can define nonzero conformal symmetry breaking operators for such values by "renormalization" as in [12, (1.9), (1.10)]. The "renormalized" operators $(\widetilde{\mathcal{D}}^{i\to j}_{u,\ell})_\pm$ and the compositions $* \circ (\widetilde{\mathcal{D}}^{i\to j}_{u,\ell})_\pm$ by the Hodge operator $*$ for $\mathbb{R}^{p-1,q}$ or $\mathbb{R}^{p,q-1}$ exhaust all conformal differential symmetry breaking operators in our framework, as is followed from Theorem B (2) and from the classification theorem [12, Thms. 1.1 and 2.10] in the Riemannian setting.

3. Symmetry breaking operators in the space forms

In this section we explain how to transfer the formulæ for symmetry breaking operators in the flat case (Theorem D) to the ones in the space form $\mathrm{S}^{p,q}$ (see Theorem E). In particular, Theorem E gives conformal symmetry breaking operators in the anti-de Sitter space (Example 6).

We consider the following open dense subsets of the flat space $\mathbb{R}^{p,q}$ and the space form $\mathrm{S}^{p,q}$ (see (4)), respectively:

$$(\mathbb{R}^{p,q}_-)' := \{(x,y) \in \mathbb{R}^{p+q} : |x|^2 - |y|^2 \neq -4\},$$
$$(\mathrm{S}^{p,q})' := \{(\omega_0, \omega, \eta) \in \mathrm{S}^{p,q} : \omega_0 \neq -1\} \qquad \subset \mathbb{R}^{1+p+q}.$$

We define a variant of the stereographic projection and its inverse by

$$\Psi\colon (\mathrm{S}^{p,q})' \longrightarrow (\mathbb{R}^{p,q}_-)', \quad (\omega_0, \omega, \eta) \mapsto \frac{2}{1+\omega_0}(\omega, \eta),$$

$$\Phi\colon (\mathbb{R}^{p,q}_-)' \longrightarrow (\mathrm{S}^{p,q})', \quad (x,y) \mapsto \frac{1}{|x|^2 - |y|^2 + 4}(4 - |x|^2 + |y|^2, 4x, 4y).$$

Lemma 5. *The map* Φ *is a conformal diffeomorphism from* $(\mathbb{R}^{p,q}_-)'$ *onto* $(\mathrm{S}^{p,q})'$ *with its inverse* Ψ, *and the conformal factor is given by*

$$\Phi^* g_{\mathrm{S}^{p,q}} = \frac{16}{(|x|^2 - |y|^2 + 4)^2} g_{\mathbb{R}^{p,q}_-}, \qquad \Psi^* g_{\mathbb{R}^{p,q}_-} = \frac{4}{(1+\omega_0)^2} g_{\mathrm{S}^{p,q}}.$$

Proof. See [13, I, Lem. 3.3], for instance. □

The pseudo-Riemannian spaces $\mathbb{R}^{p,q}_+$ and $\mathbb{R}^{p,q}_-$ are obviously isomorphic to each other by switch of the coordinates

$$s\colon \mathbb{R}^{p,q}_+ \xrightarrow{\sim} \mathbb{R}^{p,q}_- \quad (y,x) \mapsto (x,y).$$

We set

$$\Phi_- := \Phi, \quad \Phi_+ := \Phi \circ s, \quad \Psi_- := \Psi, \quad \Psi_+ := s \circ \Psi.$$

For $v \in \mathbb{C}$, we define the "twisted pull-back" of differential forms according to [13, I, (2.3.2)]:

$$(\Phi_\pm)^*_v \colon \mathcal{E}^j\left((\mathbb{R}^{p,q}_\pm)'\right) \longrightarrow \mathcal{E}^j((\mathrm{S}^{p,q})'), \quad \alpha \mapsto \left(\frac{1+\omega_0}{2}\right)^{-v} \Phi^*\alpha, \tag{8}$$

$$(\Psi_\pm)^*_v \colon \mathcal{E}^j((\mathrm{S}^{p,q})') \longrightarrow \mathcal{E}^j((\mathbb{R}^{p,q}_\pm)'), \quad \beta \mapsto \left(\frac{|x|^2 - |y|^2 + 4}{4}\right)^{-v} \Psi^*\beta. \tag{9}$$

Then $(\Psi_\pm)^*_v$ is the inverse of $(\Phi_\pm)^*_v$ in accordance with $\Psi_\pm = (\Phi_\pm)^{-1}$.

We realize the space forms $\mathrm{S}^{p-1,q}$ $(p \geq 1)$ and $\mathrm{S}^{p,q-1}$ $(q \geq 1)$ as totally geodesic hypersurfaces of $\mathrm{S}^{p,q}$ by letting $\omega_p = 0$ and $\eta_q = 0$, respectively. Then $\Phi_\pm$ induce the following diffeomorphisms between hypersurfaces.

$$\begin{array}{ccccccc}
(\mathbb{R}^{p,q}_-)' & \xrightarrow[\Phi_-]{\sim} & (\mathrm{S}^{p,q})' & \qquad & (\mathbb{R}^{p,q}_+)' & \xrightarrow[\Phi_+]{\sim} & (\mathrm{S}^{p,q})' \\
\cup & & \cup & & \cup & & \cup \\
(\mathbb{R}^{p,q-1}_-)' & \xrightarrow{\sim} & (\mathrm{S}^{p,q-1})', & & (\mathbb{R}^{p-1,q}_+)' & \xrightarrow{\sim} & (\mathrm{S}^{p-1,q})'
\end{array}$$

We are ready to transfer the formulæ of conformal symmetry breaking operators for the flat case (Theorem D) to those for negatively (or positively) curved spaces:

Theorem E. *For* $\varepsilon = \pm$, *let* $(\mathcal{D}^{i\to j}_{u,\ell})_\varepsilon$ *be as in Theorem* D. *Then the operator* $(\Phi_\varepsilon)^*_v \circ (\mathcal{D}^{i\to j}_{u,\ell})_\varepsilon \circ (\Psi_\varepsilon)^*_u$, *originally defined in the open dense set* $(\mathrm{S}^{p,q})'$ *of the space form* $\mathrm{S}^{p,q}$, *extends uniquely to the whole* $\mathrm{S}^{p,q}$ *and gives an element in*

$$\mathrm{Diff}_{\mathrm{conf}(X;Y)}(\mathcal{E}^i(X)_u, \mathcal{E}^j(Y)_v)$$

where $(X,Y) = (\mathrm{S}^{p,q}, \mathrm{S}^{p,q-1})$ *for* $\varepsilon = -$ *and* $(X,Y) = (\mathrm{S}^{p,q}, \mathrm{S}^{p-1,q})$ *for* $\varepsilon = +$.

Here, by a little abuse of notation, we have used the symbol $(\Phi_\varepsilon)^*_v$ to denote the operator in the $(n-1)$-dimensional case.

Admitting Theorem A, we give a proof of Theorem E.

Proof of Theorem E. Similarly to [12, Prop. 11.3] in the Riemannian case ($q = 0$ and $\varepsilon = +$), the composition $(\Phi_\varepsilon)_v^* \circ (\mathcal{D}_{u,\ell}^{i\to j})_\varepsilon \circ (\Phi_\varepsilon)_u^*$ gives an element in $\mathrm{Diff}_{\mathfrak{conf}(V;V\cap Y)}(\mathcal{E}^i(V)_u, \mathcal{E}^j(V\cap Y)_v)$ for $V = (\mathrm{S}^{p,q})'$. Then this operator extends to the whole $X = \mathrm{S}^{p,q}$ by Theorem A. □

The n-dimensional anti-de Sitter space $\mathrm{AdS}^n (= \mathrm{S}^{1,n-1})$ contains the hyperbolic space $\mathrm{H}^{n-1} (= \mathrm{S}^{0,n-1})$ and the anti-de Sitter space $\mathrm{AdS}^{n-1} (= \mathrm{S}^{1,n-2})$ as totally geodesic hypersurfaces.

Example 6 (hypersurfaces in the anti-de Sitter space). *For* $(p,q) = (1, n-1)$, *the formulæ in Theorem* E *give conformal symmetry breaking operators as follows.*

$$\mathcal{E}^i(\mathrm{AdS}^n)_u \longrightarrow \mathcal{E}^j(\mathrm{H}^{n-1})_v \qquad \text{for } \varepsilon = +,$$
$$\mathcal{E}^i(\mathrm{AdS}^n)_u \longrightarrow \mathcal{E}^j(\mathrm{AdS}^{n-1})_v \quad \text{for } \varepsilon = -.$$

4. Idea of holomorphic continuation

In this section we explain an idea of holomorphic continuation that will bridge between differential symmetry breaking operators in the Riemannian setting and those in the non-Riemannian setting.

We begin with an observation from Example 2 that, for any p, q with $p \geq 1$, the Lie algebras

$$\mathfrak{conf}(\mathrm{S}^{p,q}; \mathrm{S}^{p-1,q}) \simeq \mathfrak{conf}(\mathbb{R}^{p,q}; \mathbb{R}^{p-1,q}) \simeq \mathfrak{o}(p, q+1)$$

have the same complexification $\mathfrak{o}(n+1, \mathbb{C})$ as far as $p+q = n$. In turn to geometry, we shall compare (real) conformal vector fields on pseudo-Riemannian manifolds $\mathrm{S}^{p,q}$ or $\mathbb{R}^{p,q}$ of various signatures (p,q) via holomorphic vector fields on a complex manifold which contains $\mathrm{S}^{p,q}$ or $\mathbb{R}^{p,q}$ as totally real submanifolds.

Let $X_\mathbb{C}$ be a connected complex manifold, and $\Omega^i(X_\mathbb{C})$ the space of holomorphic i-forms on $X_\mathbb{C}$. If X is a totally real submanifold, then the restriction map

$$\mathrm{Rest}_X \colon \Omega^i(X_\mathbb{C}) \longrightarrow \mathcal{E}^i(X)$$

is obviously injective.

Definition-Lemma 7. Suppose $D_\mathbb{C} \colon \Omega^i(X_\mathbb{C}) \to \Omega^j(X_\mathbb{C})$ is a holomorphic differential operator. Then there is a unique differential operator $E \colon \mathcal{E}^i(X) \to \mathcal{E}^j(X)$, such that

$$E|_{V\cap X} \circ \mathrm{Rest}_{V\cap X}\alpha = \mathrm{Rest}_{V\cap X} \circ D_\mathbb{C}|_V \alpha$$

for any open set V of $X_\mathbb{C}$ with $V \cap X \neq \emptyset$ and for any $\alpha \in \Omega^i(V)$. We say that $D_\mathbb{C}$ is the *holomorphic extension* of E. We write $(\mathrm{Rest}_X)_* D_\mathbb{C}$ for E.

If X is a real analytic, pseudo-Riemannian manifold with complexification $X_\mathbb{C}$, then a holomorphic analogue of the action (1) makes sense by analytic continuation for $Z \in \mathfrak{conf}(X) \otimes_\mathbb{R} \mathbb{C}$: L_Z being understood as the holomorphic Lie derivative with respect to a holomorphic extension of the vector field Z in a complex neighbourhood U of X, which acts on $\alpha \in \Omega^i(U)$; and the conformal factor

$\rho(\cdot,\cdot)$ being understood as its holomorphic extension (complex linear in the first argument). Likewise for the pair $X \supset Y$ of pseudo-Riemannian manifolds with complexification $X_{\mathbb{C}} \supset Y_{\mathbb{C}}$, we may consider a holomorphic analogue of the covariance condition (2). Then we have:

Lemma 8. *Suppose* $D_{\mathbb{C}} \colon \Omega^i(X_{\mathbb{C}}) \to \Omega^j(Y_{\mathbb{C}})$ *is a holomorphic differential operator, and* $D = (\mathrm{Rest}_X)_*(D_{\mathbb{C}})$*. Then* $D \colon \mathcal{E}^i(X) \to \mathcal{E}^j(Y)$ *satisfies the conformal covariance* (2) *if and only if*

$$\pi_v^{(j)}(Z|_{Y_{\mathbb{C}}}) \circ D_{\mathbb{C}}\, \alpha = D_{\mathbb{C}} \circ \Pi_u^{(i)}(Z)\alpha$$

for any $Z \in \mathfrak{conf}(X;Y) \otimes_{\mathbb{R}} \mathbb{C}$*, any open subset* U *of* $X_{\mathbb{C}}$ *with* $U \cap Y_{\mathbb{C}} \neq \emptyset$ *and any* $\alpha \in \Omega^i(U)$.

We define a family of totally real vector spaces of $\mathbb{C}^n$ by embedding the space $\mathbb{R}^n = \mathbb{R}^p_x \oplus \mathbb{R}^q_y$ as

$$\iota_+ \colon \mathbb{R}^q_y \oplus \mathbb{R}^p_x \overset{\sim}{\to} \sqrt{-1}\mathbb{R}^q \oplus \mathbb{R}^p = \{(\sqrt{-1}y_1,\dots,\sqrt{-1}y_q, x_{q+1},\dots,x_{p+q}) : x_j, y_j \in \mathbb{R}\},$$
$$\iota_- \colon \mathbb{R}^p_x \oplus \mathbb{R}^q_y \overset{\sim}{\to} \mathbb{R}^p \oplus \sqrt{-1}\mathbb{R}^q = \{(x_1,\dots,x_p, \sqrt{-1}y_{p+1},\dots,\sqrt{-1}y_{p+q}) : x_j, y_j \in \mathbb{R}\}.$$

Let us apply Lemma 8 to the following setting where $n = p + q$.

$$\begin{array}{ccccccc} \mathbb{R}^n & \simeq \mathbb{R}^{p,q}_+ & \underset{\iota_+}{\hookrightarrow} & X_{\mathbb{C}} = \mathbb{C}^n & \underset{\iota_-}{\hookleftarrow} & \mathbb{R}^{p,q}_- & \simeq \mathbb{R}^n \\ & \cup & & \cup & & \cup & \\ \mathbb{R}^{n-1} & \simeq \mathbb{R}^{p-1,q}_+ & \hookrightarrow & Y_{\mathbb{C}} = \mathbb{C}^{n-1} & \hookleftarrow & \mathbb{R}^{p,q-1}_- & \simeq \mathbb{R}^{n-1} \end{array}$$

The holomorphic symmetric 2-tensor

$$ds^2 = dz_1^2 + \cdots + dz_n^2$$

on $\mathbb{C}^n$ induces a flat pseudo-Riemannian structure on $\mathbb{R}^n$ of signature (p,q) by restriction via $\iota_\pm$. The resulting pseudo-Riemannian structures (and coordinates) on $\mathbb{R}^n$ are nothing but those of $\mathbb{R}^{p,q}_+$ and $\mathbb{R}^{p,q}_-$ given in Section 2.

5. Proof of Theorems A, B, and C

This section gives a proof of Theorems A, B, and C. The key machinery for differential symmetry breaking operators (SBOs for short) is in threefold:

(1) holomorphic extension of differential SBOs (Section 4);
(2) duality theorem between differential SBOs and homomorphisms for generalized Verma modules that encode branching laws [15, I, Thm. 2.9];
(3) automatic continuity theorem of differential SBOs in the Hermitian symmetric setting [15, I, Thm. 5.3].

We note that both (1) and (2) indicate the independence of real forms as formulated in Theorem B (2), whereas (3) appeals to the theory of admissible restrictions of real reductive groups [7] for a specific choice of real forms of complex reductive Lie groups.

Let G be $SO_0(p+1,q+1)$, the identity component of the indefinite orthogonal group $O(p+1,q+1)$, $P=LN$ a maximal parabolic subgroup of G with Levi subalgebra $\mathrm{Lie}(L)\simeq\mathfrak{so}(p,q)+\mathbb{R}$, and H the identity component of P. Then G acts conformally on $G/H\simeq \mathrm{S}^p\times\mathrm{S}^q$ equipped with the pseudo-Riemannian structure $g_{\mathrm{S}^p}\oplus(-g_{\mathrm{S}^q})$. Similarly, H' is defined by taking $G':=SO_0(p,q+1)$ ($\varepsilon=+$) or $SO_0(p+1,q)$ ($\varepsilon=-$).

Applying the duality theorem [15, I, Theorem 2.9] to the quadruple

$$(G,H,G',H'),$$

we see that any element in

$$\mathrm{Hom}_{\mathfrak{g}'_{\mathbb{C}}}(U(\mathfrak{g}'_{\mathbb{C}})\otimes_{U(\mathfrak{p}'_{\mathbb{C}})}(\textstyle\bigwedge^{n-1-j}(\mathbb{C}^{n-1})\otimes\mathbb{C}_{-v-j}),U(\mathfrak{g}_{\mathbb{C}})\otimes_{U(\mathfrak{p}_{\mathbb{C}})}(\bigwedge^{n-i}(\mathbb{C}^n)\otimes\mathbb{C}_{-u-i})) \tag{10}$$

with notation as in [12, Sect. 2.6] induces a differential symmetry breaking operator $\overline{D}\in\mathrm{Diff}_{\mathfrak{conf}(\overline{X};\overline{Y})}(\mathcal{E}^i(\overline{X})_u,\mathcal{E}^j(\overline{Y})_v)$ on the conformal compactification $\overline{X}$, and hence the one on any open subset V of X with $V\cap Y\neq\emptyset$ by restriction. In order to prove Theorem A and Theorem B (1), it is then sufficient to show the following converse statement.

Claim 9. *Any $D\in\mathrm{Diff}_{\mathfrak{conf}(V;V\cap Y)}(\mathcal{E}^i(V)_u,\mathcal{E}^j(V\cap Y)_v)$ is derived from an element in* (10).

Let us prove Claim 9.

- Step 1. Reduction to the flat case

By using the twisted pull-backs $(\Phi_\pm)^*_v$ and $(\Psi_\pm)^*_v$ (see (8)), we may and do assume that $X=\mathbb{R}^{p,q}$ ($\simeq\mathbb{R}^n$) and that Y is the hypersurface $\mathbb{R}^{n-1}$ given by the condition that the last coordinate is zero. By replacing V with an open subset V' of $\mathbb{R}^n$ with $V\cap\mathbb{R}^{n-1}=V'\cap\mathbb{R}^{n-1}$ if necessary, we may further assume that V is a convex neighbourhood of $V\cap\mathbb{R}^{n-1}$ in $\mathbb{R}^n$.

- Step 2. Holomorphic extension

With the coordinates $x=(x',x_n)\equiv(x_1,\dots,x_{n-1},x_n)$ of $X=\mathbb{R}^n$, any differential operator $D\colon\mathcal{E}^i(\mathbb{R}^n)\to\mathcal{E}^j(\mathbb{R}^{n-1})$ takes the form

$$D=\mathrm{Rest}_{x_n=0}\circ\sum_{\alpha\in\mathbb{N}^n}a_\alpha(x')\frac{\partial^{|\alpha|}}{\partial x_1^{\alpha_1}\cdots\partial x_n^{\alpha_n}}$$

where $a_\alpha\in C^\infty(\mathbb{R}^{n-1})\otimes\mathrm{Hom}_{\mathbb{C}}\left(\bigwedge^i(\mathbb{C}^n),\bigwedge^j(\mathbb{C}^{n-1})\right)$ (see [15, I, Ex. 2.4]). Since $Z_k:=\frac{\partial}{\partial x_k}$ $(1\leq k\leq n-1)$ is a Killing vector filed, namely, $Z_k\in\mathfrak{conf}(X;Y)$ with $\rho(Z_k,\cdot)\equiv 0$, the conformal covariance (2) reduces to $L_{\frac{\partial}{\partial x_k}}\circ D=D\circ L_{\frac{\partial}{\partial x_k}}$, which implies that the matrix-valued function $a_\alpha(x')$ is independent of x' for every α. We shall denote $a_\alpha(x')$ simply by a_α. Then D extends to a holomorphic differential operator $D_{\mathbb{C}}\colon\Omega^i(\mathbb{C}^n)\to\Omega^j(\mathbb{C}^{n-1})$, by setting

$$D_{\mathbb{C}}:=\mathrm{Rest}_{z_n=0}\circ\sum_{\alpha\in\mathbb{N}^n}a_\alpha\frac{\partial^{|\alpha|}}{\partial z_1^{\alpha_1}\cdots\partial z_n^{\alpha_n}}.$$

If D satisfies the conformal covariance condition (2) on $\mathcal{E}^i(V)$ for all $Z \in \mathfrak{conf}(V; V \cap Y) \simeq \mathfrak{o}(p, q+1)$ or $\mathfrak{o}(p+1, q)$, then by Lemma 8, $D_{\mathbb{C}}$ satisfies the holomorphic extension of the condition (2) on $\Omega^i(\mathbb{C}^n)$ for all $Z \in \mathfrak{conf}(V; V \cap Y) \otimes_{\mathbb{R}} \mathbb{C} \simeq \mathfrak{o}(n+1, \mathbb{C})$.

• Step 3. Automatic continuity in the Hermitian symmetric spaces $G_{\mathbb{R}}/K_{\mathbb{R}} \supset G'_{\mathbb{R}}/K'_{\mathbb{R}}$

The automatic continuity theorem is known for holomorphic differential SBOs in the Hermitian symmetric setting [15, I, Thm. 5.3]. Then our strategy to prove Claim 9 is to utilize the automatic continuity theorem in the Hermitian symmetric setting by embedding a pair $(G_{\mathbb{R}}/K_{\mathbb{R}}, G'_{\mathbb{R}}/K'_{\mathbb{R}})$ of Hermitian symmetric spaces into the pair $(\mathbb{C}^n, \mathbb{C}^{n-1})$ of the affine spaces as in Step 2. For this, we shall choose a specific real form $G_{\mathbb{R}}$ of $G_{\mathbb{C}} := SO(n+2, \mathbb{C})$ such that $G_{\mathbb{R}}$ is the group of biholomorphic transformations of a bounded symmetric domain in $\mathbb{C}^n$ as below.

Let $Q(\tilde{x}) := -x_0^2 + x_1^2 + \cdots + x_n^2 - x_{n+1}^2$ be the quadratic form on $\mathbb{R}^{n+2}$, and $G_{\mathbb{R}}$ the identity component of the isotropy group

$$\{h \in GL(n+2, \mathbb{R}) : Q(h \cdot \tilde{x}) = Q(\tilde{x}) \text{ for all } \tilde{x} \in \mathbb{R}^{n+2}\}.$$

Then $K_{\mathbb{R}} := G_{\mathbb{R}} \cap SO(n+2)$ is a maximal compact subgroup of $G_{\mathbb{R}} \simeq SO_0(n, 2)$ such that $G_{\mathbb{R}}/K_{\mathbb{R}}$ is the Hermitian symmetric space of type IV in the É. Cartan classification. We take $G'_{\mathbb{R}}$ to be the stabilizer of x_n. Then $G'_{\mathbb{R}} \simeq SO_0(n-1, 2)$.

We use the notation as in [15, II, Sect. 6], and identify $\mathbb{C}^n$ with the open Bruhat cell of the complex quadric

$$\mathbb{Q}^n\mathbb{C} = \{\tilde{z} \in \mathbb{C}^{n+2} \setminus \{0\} : Q(\tilde{z}) = 0\}/\mathbb{C}^\times \simeq G_{\mathbb{C}}/P_{\mathbb{C}}.$$

Then $G_{\mathbb{R}}/K_{\mathbb{R}}$ is realized as the Lie ball

$$U = \{z \in \mathbb{C}^n : |z\, {}^t z|^2 + 1 - 2\bar{z}\, {}^t z > 0, \ |z\, {}^t z| < 1\}.$$

We compare the real form G of $G_{\mathbb{C}}$ with Lie algebra $\mathfrak{conf}(X) \simeq \mathfrak{o}(p+1, q+1)$ in Step 1 and another real form $G_{\mathbb{R}} \simeq SO_0(n, 2)$ in Step 3 ($n = p+q$). The point here is that the G-orbit $G \cdot o \simeq G/P$ through the origin $o = eP_{\mathbb{C}} \in G_{\mathbb{C}}/P_{\mathbb{C}}$ is closed in $G_{\mathbb{C}}/P_{\mathbb{C}}$, while the $G_{\mathbb{R}}$-orbit $G_{\mathbb{R}} \cdot o \simeq G_{\mathbb{R}}/K_{\mathbb{R}}$ is open in $G_{\mathbb{C}}/P_{\mathbb{C}}$, as is summarized in the figure below.

$$\begin{array}{ccccc}
G_{\mathbb{R}}/K_{\mathbb{R}} \simeq U \underset{\text{open}}{\subset} \mathbb{C}^n & \underset{\text{Bruhat cell}}{\subset} & \mathbb{Q}^n\mathbb{C} & \simeq & G_{\mathbb{C}}/P_{\mathbb{C}} \\
\cup & & \cup & & \cup \ \text{totally real} \\
\mathbb{R}^{p,q} & \underset{\text{conformal compactfication}}{\subset} & (S^p \times S^q)/\mathbb{Z}_2 & \simeq & G/P
\end{array}$$

We note that the $G'_{\mathbb{R}}$-orbit $G'_{\mathbb{R}} \cdot o \simeq G'_{\mathbb{R}}/K'_{\mathbb{R}}$ is realized as the subsymmetric domain $U \cap \{z_n = 0\} \simeq \mathbb{C}^{n-1}$. Since the holomorphic differential operator $D_{\mathbb{C}}$ is defined on $\Omega^i(\mathbb{C}^n)$, $D_{\mathbb{C}}$ induces a holomorphic differential operator

$$D_{\mathbb{C}}|_{G_{\mathbb{R}}/K_{\mathbb{R}}} : \Omega^i(G_{\mathbb{R}}/K_{\mathbb{R}}) \longrightarrow \Omega^j(G'_{\mathbb{R}}/K'_{\mathbb{R}}) \tag{11}$$

via the inclusion $G_{\mathbb{R}}/K_{\mathbb{R}} \simeq U \subset \mathbb{C}^n$.

Then the automatic continuity theorem [15, I, Thm. 5.3 (2)] (and its proof), applied to (11) implies that $D_{\mathbb{C}}|_{G_{\mathbb{R}}/K_{\mathbb{R}}}$ is derived from an element of (10) via the duality theorem in the holomorphic setting (see [15, I, Thm. 2.12]). Thus the proof of Claim 9 is completed. Therefore Theorem A and Theorem B (1) follow from [15, I, Thm. 2.9].

Since (10) is independent of the choice of real forms, Theorem B (2) is now clear.

Proof of Theorem C. Owing to Theorems A and B, Theorem C is reduced to the Riemannian case $p=0$ and $\varepsilon=-$ or $q=0$ and $\varepsilon=+$. Then the assertion follows from the classification results [12, Thm. 1.1] for the (disconnected) conformal group and from a discussion on the connected group case (see [12, Thm. 2.10]). □

6. Proof of Theorem D

In this section, we give a proof of Theorem D in Section 2 by reducing it to the Riemannian case $(p,q,\varepsilon)=(n,0,+)$ or $(0,n,-)$ which was established in [12, Thms. 1.5, 1.6, 1.7 and 1.8]. For this, we apply Definition-Lemma 7 to the totally real embedding $\iota_\pm\colon \mathbb{R}^{p,q}_\pm \hookrightarrow \mathbb{C}^{p+q}$.

With the coefficients $a_k(\mu,\ell)$ given in (7), we define a family of (scalar-valued) holomorphic differential operators on $\mathbb{C}^n$ by

$$(\mathcal{D}^\mu_\ell)_{\mathbb{C}} := \sum_{k=0}^{[\frac{\ell}{2}]} a_k(\mu,\ell)\left(-\sum_{j=1}^{n-1}\frac{\partial^2}{\partial z_j^2}\right)^k\left(\frac{\partial}{\partial z_n}\right)^{\ell-2k},$$

which are the holomorphic extensions of the operators $(\mathcal{D}^\mu_\ell)_{\mathbb{R}^{n,0}_+}$ defined in the Riemannian case, that is, $(\mathrm{Rest}_{\mathbb{R}^{n,0}_+})_*((\mathcal{D}^\mu_\ell)_{\mathbb{C}}) = (\mathcal{D}^\mu_\ell)_{\mathbb{R}^{n,0}_+}$. Likewise, we extend $(\mathcal{D}^{i\to j}_{u,\ell})_{\mathbb{R}^{n,0}_+}$ to a (matrix-valued) holomorphic differential operator

$$(\mathcal{D}^{i\to j}_{u,\ell})_{\mathbb{C}}\colon \Omega^i(\mathbb{C}^n)\longrightarrow \Omega^j(\mathbb{C}^{n-1})$$

in such a way that $(\mathrm{Rest}_{\mathbb{R}^{n,0}_+})_*(\mathcal{D}^{i\to j}_{u,\ell})_{\mathbb{C}}$ coincides with $(\mathcal{D}^{i\to j}_{u,\ell})_{\mathbb{R}^{n,0}_+}$. By definition of $(\mathcal{D}^{i\to j}_{u,\ell})_{\mathbb{R}^{p,q}_+}$, it is readily seen that $(\mathrm{Rest}_{\mathbb{R}^{p,q}_+})_*(\mathcal{D}^{i\to j}_{u,\ell})_{\mathbb{C}} = (\mathcal{D}^{i\to j}_{u,\ell})_{\mathbb{R}^{p,q}_+}$ for all (p,q) with $p+q=n$. Concerning the other real form $\mathbb{R}^{p,q}_-$, we have the following.

Lemma 10. $(\mathrm{Rest}_{\mathbb{R}^{p,q}_-})_*(\mathcal{D}^{i\to j}_{u,\ell})_{\mathbb{C}} = e^{-\frac{\pi\sqrt{-1}(\ell+i-j)}{2}}(\mathcal{D}^{i\to j}_{u,\ell})_{\mathbb{R}^{p,q}_-}$.

Proof. The assertion is deduced from the formulæ of $(\mathrm{Rest}_{\mathbb{R}^{p,q}_-})_*$ for the following basic operators. (For the convenience of the reader, we also list the cases $\mathbb{R}^{p,q}_+$ as well.)

	$d_{\mathbb{C}}$	$d^*_{\mathbb{C}}$	$\frac{\partial}{\partial z_n}$	$\iota_{\frac{\partial}{\partial z_n}}$	$(\mathcal{D}^\mu_\ell)_{\mathbb{C}}$
$(\mathrm{Rest}_{\mathbb{R}^{p,q}_+})_*$	$d_{\mathbb{R}^{p,q}_+}$	$d^*_{\mathbb{R}^{p,q}_+}$	$\frac{\partial}{\partial x_n}$	$\iota_{\frac{\partial}{\partial x_n}}$	$(\mathcal{D}^\mu_\ell)_{\mathbb{R}^{p,q}_+}$
$(\mathrm{Rest}_{\mathbb{R}^{p,q}_-})_*$	$d_{\mathbb{R}^{p,q}_-}$	$d^*_{\mathbb{R}^{p,q}_-}$	$\frac{1}{\sqrt{-1}}\frac{\partial}{\partial y_n}$	$\frac{1}{\sqrt{-1}}\iota_{\frac{\partial}{\partial y_n}}$	$e^{-\frac{\pi\sqrt{-1}\ell}{2}}(\mathcal{D}^\mu_\ell)_{\mathbb{R}^{p,q}_-}$

□

We are ready to complete the proof of Theorem D.

Proof of Theorem D. Since $(\mathcal{D}_{u,\ell}^{i\to j})_{\mathbb{R}_+^{n,0}} \in \mathrm{Diff}_{\mathfrak{conf}(\mathbb{R}^n;\mathbb{R}^{n-1})}(\mathcal{E}^i(\mathbb{R}^n)_u, \mathcal{E}^j(\mathbb{R}^{n-1})_v)$ by [12, Thms. 1.5, 1.6, 1.7, 1.8], the holomorphic differential operator $(\mathcal{D}_{u,\ell}^{i\to j})_{\mathbb{C}}$ satisfies the holomorphic and covariance condition by Lemma 8. In turn, we conclude Theorem D by Lemmas 10 and 8. □

7. Four-dimensional example

In contrast to the multiplicity-free theorem ([12, Thm. 1.1]) for differential SBOs for (disconnected) conformal *groups* $(\mathrm{Conf}(X), \mathrm{Conf}(X;Y))$ when

$$(X,Y) = (\mathrm{S}^n, \mathrm{S}^{n-1}) \quad (n \geq 3),$$

it may happen that an analogous statement for the *Lie algebras*

$$(\mathfrak{conf}(X), \mathfrak{conf}(X;Y))$$

does not hold anymore. In fact, for some u, v, i, j, one has

$$\dim_{\mathbb{C}} \mathrm{Diff}_{\mathfrak{conf}(X;Y)}(\mathcal{E}^i(X)_u, \mathcal{E}^j(Y)_v) > 1 \tag{12}$$

(or equivalently, $= 2$).

In this section we first address the question when and how (12) happens and then describe the corresponding generators when $(X,Y) = (\mathbb{R}^{p,q}, \mathbb{R}^{p-1,q})$ with $p+q$ $(= n) = 4$.

As we have seen in Theorems C and D, there are two types of conditions on (i,j), namely,

$$-1 \leq i-j \leq 2 \quad \text{or} \quad n-2 \leq i+j \leq n+1,$$

for which nontrivial differential symmetry breaking operators $\mathcal{E}^i(X)_u \to \mathcal{E}^j(Y)_v$ exist for some $u, v \in \mathbb{C}$. (The latter inequality arises from the composition of the Hodge star operator with respect to the pseudo-Riemannian metric.) It turns out that (12) happens only if these two conditions are simultaneously fulfilled, that is, only if

$$-1 \leq i-j \leq n \quad \text{and} \quad n-2 \leq i+j \leq n+1.$$

The four-dimensional case is illustrative to understand (12) for the arbitrary dimension n. We give a complete list of parameters (i,j,u,v) for which (12) happens together with explicit generators of $\mathrm{Diff}_{\mathfrak{conf}(X;Y)}(\mathcal{E}^i(X)_u, \mathcal{E}^j(Y)_v)$.

Let $X = \mathbb{R}_+^{p,q}$ and $Y = \mathbb{R}_+^{p-1,q}$ with $n = p+q = 4$. We shall simply write as $(X,Y) = (\mathbb{R}^{p,q}, \mathbb{R}^{p-1,q})$. (The case $(X,Y) = (\mathbb{R}_-^{p,q}, \mathbb{R}_-^{p,q-1})$ is essentially the same and we omit it.) We set

$$A := \mathrm{Rest}_{x_4=0} \circ d, \qquad B := \mathrm{Rest}_{x_4=0} \circ d^*,$$
$$C := \mathrm{Rest}_{x_4=0} \circ \iota_{\frac{\partial}{\partial x_4}} d, \quad D := \mathrm{Rest}_{x_4=0} \circ \iota_{\frac{\partial}{\partial x_4}} d^*.$$

By using the formulæ in [12, Ch. 8. Sect. 5], we readily see that

$$D \circ *_{\mathbb{R}^{p,q}} = \pm *_{\mathbb{R}^{p-1,q}} \circ A, \quad C \circ *_{\mathbb{R}^{p,q}} = \pm *_{\mathbb{R}^{p-1,q}} \circ B. \tag{13}$$

Theorem F. *Suppose* $(X,Y)=(\mathbb{R}^{p,q},\mathbb{R}^{p-1,q})$ *with* $p+q=4$ *and* $p\geq 1$. *Then* (12) *occurs if and only if* (i,j,u,v) *appears in the nonempty boxes in the table below. Moreover, the pairs of operators in the table provide generators of*

$$\mathrm{Diff}_{\mathrm{conf}(X;Y)}(\mathcal{E}^i(X)_u,\mathcal{E}^j(Y)_v).$$

$i \backslash j$	0	1	2	3
0				
1		$u=0,\ v=1,$ $\ell=1$ - - - - - $*_{\mathbb{R}^{p-1,q}}\circ A,$ C	$u=v=0,$ $\ell=1$ - - - - - $A,$ $*_{\mathbb{R}^{p-1,q}}\circ C$	
2	$u=0,\ v=3,$ $\ell=1$ - - - - - $D,$ $*_{\mathbb{R}^{p-1,q}}\circ A$	$v-u\in\mathbb{N}_+,$ $\ell=v-u-1$ - - - - - $(\mathcal{D}_{u,\ell}^{2\to 1})_+,$ $*_{\mathbb{R}^{p-1,q}}\circ(\mathcal{D}_{u,\ell}^{2\to 2})_+$	$v-u\in\mathbb{N},$ $\ell=v-u$ - - - - - $(\mathcal{D}_{u,\ell}^{2\to 2})_+,$ $*_{\mathbb{R}^{p-1,q}}\circ(\mathcal{D}_{u,\ell}^{2\to 1})_+$	$u=v=0,$ $\ell=1$ - - - - - $A,$ $*_{\mathbb{R}^{p-1,q}}\circ D$
3		$u=-2,\ v=1,$ $\ell=1$ - - - - - $D,$ $*_{\mathbb{R}^{p-1,q}}\circ B$	$u=-2,\ v=0,$ $\ell=1$ - - - - - $B,$ $*_{\mathbb{R}^{p-1,q}}\circ D$	
4				

Remark 11.

(1) By the duality theorem [15, I, Thm. 9], the multiplicity in the branching laws of the generalized Verma modules, given as the dimension of (10) is also equal to 2 for the parameters in Theorem F (cf. [8]).

(2) For $(p,q)=(1,3)$, Maxwell's equations are expressed as $d\alpha=0$ and $d^*\alpha=0$ for $\alpha\in\mathcal{E}^2(\mathbb{R}^{1,3})$, see [17] for instance.

(3) $\mathcal{D}_{u,\ell}^{2\to 1}$ reduces to $-\mathrm{Rest}_{x_4=0}\circ d^*$ if $(u,\ell)=(0,1)$.

Proof of Theorem F. The assertions follow from [12, Thms. 1.1 and 2.10] owing to Theorems B and D. □

Acknowledgement

The first named author was partially supported by Grant-in-Aid for Scientific Research (A) (25247006), Japan Society for the Promotion of Science. The third named author expresses his gratitude to the organizers of the XXXVth workshop on Geometric Methods in Physics for their kind invitation to Białowieża. All three authors were partially supported by CNRS Grant PICS n° 7270.

References

[1] T.P. Branson, *Conformally covariant equations on differential forms.* Comm. Partial Differential Equations **7**, 393–431 (1982).

[2] T. Branson, A.R. Gover, *Conformally invariant operators, differential forms, cohomology and a generalisation of Q-curvature.* Comm. Partial Differential Equations **30**, 1611–1669 (2005).

[3] C. Fefferman, C.R. Graham, *Juhl's formulæ for GJMS operators and Q-curvatures.* J. Amer. Math. Soc. **26**, 1119–1207 (2013).

[4] E.S. Fradkin, A.A. Tseytlin, *Asymptotic freedom in extended conformal supergravities.* Phys. Lett. B **110**, 117–122 (1982).

[5] C.R. Graham, R. Jenne, L.J. Mason, G.A.J. Sparling, *Conformally invariant powers of the Laplacian. I. Existence.* J. London Math. Soc. (2) **46**, 557–565 (1992).

[6] A. Juhl, *Families of Conformally Covariant Differential Operators, Q-Curvature and Holography.* Progress in Mathematics, vol. 275 (Birkhäuser, Basel, 2009).

[7] T. Kobayashi, *Discrete decomposability of the restriction of $A_{\mathfrak{q}}(\lambda)$ with respect to reductive subgroups and its applications.* Invent. Math. **117**, 181–205 (1994); *II: Microlocal analysis and asymptotic K-support.* Ann. Math. **147**, 709–729 (1998); *III. Restriction of Harish-Chandra modules and associated varieties.* Invent. Math. **131**, 229–256 (1998).

[8] T. Kobayashi, *Restrictions of generalized Verma modules to symmetric pairs.* Transform. Groups **17**, 523–546 (2012).

[9] T. Kobayashi, *F-method for constructing equivariant differential operators*, in *Geometric Analysis and Integral Geometry*, edited by E.T. Quinto, F.B. Gonzalez, and J. Christensen. Contemporary Mathematics, vol. 598 (American Mathematical Society, Providence, 2013), pp. 141–148.

[10] T. Kobayashi, *F-method for symmetry breaking operators.* Differential Geom. Appl. **33**, 272–289 (2014).

[11] T. Kobayashi, *A program for branching problems in the representation theory of real reductive groups*, in *Representations of Lie Groups: In Honor of David A. Vogan, Jr. on his 60th Birthday*, edited by M. Nevins and P. Trapa. Progress in Mathematics, vol. 312 (Birkhäuser, 2015), pp. 277–322.

[12] T. Kobayashi, T. Kubo, M. Pevzner, *Conformal Symmetry Breaking Operators for Differential Forms on Spheres.* Lecture Notes in Mathematics, vol. 2170 (Springer-Nature), xiii+192 pages, 2016. ISBN: 978–981–10–2656–0.

[13] T. Kobayashi, B. Ørsted, *Analysis on the minimal representation of $O(p, q)$. I. Realization via conformal geometry.* Adv. Math. **180**, 486–512; *II. Branching laws.* Adv. Math. **180**, 513–550; *III. Ultrahyperbolic equations* on $\mathbb{R}^{p-1,q-1}$. Adv. Math. **180**, 551–595 (2003).

[14] T. Kobayashi, B. Ørsted, P. Somberg, V. Souček, *Branching laws for Verma modules and applications in parabolic geometry. I.* Adv. Math. **285**, 1796–1852 (2015).

[15] T. Kobayashi, M. Pevzner, *Differential symmetry breaking operators. I. General theory and F-method.* Selecta Math. (N.S.) **22**, 801–845; *II. Rankin–Cohen operators for symmetric pairs.* Selecta Math. (N.S.) **22**, 847–911 (2016).

[16] T. Kobayashi, B. Speh, *Symmetry Breaking for Representations of Rank One Orthogonal Groups.* Memoirs of American Mathematical Society, vol. 238 (American Mathematical Society, Providence, 2015). ISBN: 978-1-4704-1922-6.

[17] B. Kostant, N.R. Wallach, *Action of the conformal group on steady state solutions to Maxwell's equations and background radiation*, in *Symmetry: Representation Theory and Its Applications.* Progress in Mathematics, vol. 257 (Birkhäuser/Springer, New York, 2014), pp. 385–418.

[18] S. Paneitz, *A quartic conformally covariant differential operator for arbitrary pseudo-Riemannian manifolds.* SIGMA Symmetry Integrability Geom. Methods Appl. vol. 4, (2008), paper 036.

[19] I.E. Segal, *Mathematical Cosmology and Extragalactic Astronomy.* Pure and Applied Mathematics, vol. 68 (Academic Press, New York, 1976).

Toshiyuki Kobayashi
Kavli Institute for the Physics
and Mathematics of the Universe

and

Graduate School of Mathematical Sciences
The University of Tokyo
3-8-1 Komaba, Meguro
Tokyo, 153-8914, Japan
e-mail: toshi@ms.u-tokyo.ac.jp

Toshihisa Kubo
Faculty of Economics, Ryukoku University
67 Tsukamoto-cho, Fukakusa, Fushimi-ku
Kyoto 612-8577, Japan
e-mail: toskubo@econ.ryukoku.ac.jp

Michael Pevzner
Laboratoire de Mathématiques de Reims
Université de Reims-Champagne-Ardenne
FR 3399 CNRS
F-51687 Reims, France
e-mail: pevzner@univ-reims.fr

Geometric Methods in Physics. XXXV Workshop 2016
Trends in Mathematics, 87–94

Hamilton Operators and Related Integrable Differential Algebraic Novikov–Leibniz Type Structures

Anatolij K. Prykarpatski

In memory of a friend, coauthor and brilliant mathematical physicist, Professor Jerzy Zagrodziński (⋆02.10.1935, †15.03.2002)

Abstract. There is devised a general differential-algebraic approach to constructing multi-component Hamiltonian operators as classical Lie–Poisson structures on the suitably constructed adjacent loop Lie co-algebras. The related Novikov–Leibniz type algebraic structures are derived, a new non-associative right Leibniz and Riemann algebra is constructed, deeply related with infinite multi-component Riemann type integrable hydrodynamic hierarchies.

Mathematics Subject Classification (2010). Primary 35A30, 35G25, 35N10; Secondary 37K35, 58J70,58J72, 34A34.

Keywords. Poisson brackets, Hamiltonian operators, differential algebras, differentiations, loop-algebra, Novikov algebra, right Leibniz algebra, Riemann algebra, Riemann type hydrodynamic hierarchy, integrability.

1. Introduction

As it is well known [4–6], many of integrable Hamiltonian systems, discovered during the last decades, were understood owing to the Lie-algebraic properties of their internal hidden symmetry structures. A first account of the Hamiltonian operators and related differential-algebraic structures, lying in the background of integrable systems, was given by I. Gelfand and I. Dorfman [7] and later was extended by S. Novikov and A. Balinsky [1, 3]. In our work we have devised a simple algorithm allowing to construct new algebraic structures within which the corresponding Hamiltonian operators exist and generate integrable multi-component dynamical systems. We show, as examples, that the well-known Novikov algebraic structure,

---

This work was completed with the support of AGH University local grant.

obtained before in [3, 7] as a condition for a matrix differential expression to be Hamiltonian, appears within the devised approach as a classical Lie–Poisson structure on a suitably adjacent Lie co-algebra, naturally associated with the constructed non-associative and non-commutative differential loop algebra.

2. The Hamiltonian operators and related algebraic structures within the differential-algebraic approach

Assume $(\mathbb{A};\circ)$ to be a finite-dimensional algebra of the dimension $N=\dim\mathbb{A}\in\mathbb{Z}_+$ (in general non-commutative and non-associative) over an algebraically closed field $\mathbb{K}$. Based on the algebra $\mathbb{A}$ one can construct the related loop algebra $\widetilde{\mathbb{A}}$ of smooth mappings $u:\mathbb{S}^1\to\mathbb{A}$ and endow it with the suitably generalized natural convolution $\prec\cdot,\cdot\succ$ on $\widetilde{\mathbb{A}}^*\times\widetilde{\mathbb{A}}\to\mathbb{K}$, where $\widetilde{\mathbb{A}}^*$ is the corresponding adjoint to $\widetilde{\mathbb{A}}$ space.

First, we will consider a general scheme of constructing nontrivial ultra-local and local [6] Poisson structures on the adjoint space $\widetilde{\mathbb{A}}^*$, compatible with the internal multiplication in the loop algebra $\widetilde{\mathbb{A}}$. Consider a basis $\{e_s\in\mathbb{A}:s=\overline{1,N}\}$ of the algebra $\mathbb{A}$ and its dual $\{u^s\in\mathbb{A}^*:s=\overline{1,N}\}$ with respect to the natural convolution $\prec\cdot,\cdot\succ$ on $\mathbb{A}^*\times\mathbb{A}$, that is $\prec u^j,e_i\succ:==\delta_i^j$ for all $i,j=\overline{1,N}$, and such that for any $u(x)=\sum_{s=\overline{1,N}}u_s(x)u^s\in\widetilde{\mathbb{A}}^*$, $x\in\mathbb{S}^1$, the quantities $u_s(x):=\prec u(x),e_s\succ\in\mathbb{K}$ for all $s=\overline{1,N}$, $x\in\mathbb{S}^1$. Denote by $\widetilde{\mathbb{A}}\wedge\widetilde{\mathbb{A}}:=\mathrm{Skew}(\widetilde{\mathbb{A}}\otimes\widetilde{\mathbb{A}})$ and let $\vartheta^*:\widetilde{\mathbb{A}}\wedge\widetilde{\mathbb{A}}\to\mathrm{Symm}(\widetilde{\mathbb{A}})$ be a skew-symmetric bi-linear mapping. Then the expression

$$\{u_i(x)\ ,u_j(x)\}\ :=\prec u(x),\vartheta^*(e_i\wedge e_j)\succ \tag{1}$$

defines for any $x,y\in\mathbb{S}^1$ and all $i,j=\overline{1,N}$ an ultra-local *linear* skew-symmetric pre-Poisson bracket on $\widetilde{\mathbb{A}}^*$. Since the algebra $\widetilde{\mathbb{A}}$ possesses its internal multiplicative structure "$\circ$", the important problem arises: *under what conditions is the pre-Poisson bracket* (1) *a Poisson one, compatible with this internal structure on* $\widetilde{\mathbb{A}}$? To proceed with elucidating this question, let us define a co-multiplication $\Delta:\widetilde{\mathbb{A}}^*\to\widetilde{\mathbb{A}}^*\otimes\widetilde{\mathbb{A}}^*$ on any element $u\in\widetilde{\mathbb{A}}^*$ by means of the relationship

$$\prec\Delta(u),(a\otimes b)\succ:=\prec u,a\circ b\succ \tag{2}$$

for arbitrary $a,b\in\widetilde{\mathbb{A}}$. Remind also that the co-multiplication $\Delta:\widetilde{\mathbb{A}}^*\to\widetilde{\mathbb{A}}^*\otimes\widetilde{\mathbb{A}}^*$, defined this way, is a *homomorphism* of the tensor algebra $\mathrm{T}(\widetilde{\mathbb{A}}^*)$ and the linear pre-Poisson structure $\{\cdot,\cdot\}$ (1) on $\widetilde{\mathbb{A}}^*$ is called *compatible* with the multiplication "$\circ$" on the algebra $\widetilde{\mathbb{A}}$, if the following invariance condition

$$\Delta\{u_i(x),u_j(x)\}=\{\Delta(u_i(x)),\Delta(u_j(x))\}, \tag{3}$$

holds for any $x\in\mathbb{S}^1$ and all $i,j=\overline{1,N}$. Now, taking into account that multiplication in the algebra $\mathbb{A}$ is given for any $i,j=\overline{1,N}$ by the condition

$$e_i\circ e_j:=\sum_{s=\overline{1,N}}\sigma_{ij}^s e_s, \tag{4}$$

where the quantities $\sigma_{ij}^{s} \in \mathbb{K}$for all i,j and $k = \overline{1,N}$ are constant, the related comultiplication $\Delta : \mathbb{A}^* \to \mathbb{A}^* \otimes \mathbb{A}^*$ acts on the basic functionals $u^s \in \mathbb{A}^*$, $s = \overline{1,N}$, as

$$\Delta(u^s) = \sum_{i,j=\overline{1,N}} \sigma_{ij}^{s} u^i \otimes u^j. \tag{5}$$

Additionally, if the mapping $\vartheta^* : \widetilde{\mathbb{A}} \wedge \widetilde{\mathbb{A}} \to \mathrm{Symm}(\widetilde{\mathbb{A}})$ is given, for instance, in the simple linear form

$$\vartheta^* : (e_i \otimes e_j - e_j \otimes e_i) \to \sum_{s=\overline{1,N}} (c_{ij}^{s} - c_{ji}^{s}) e_s, \tag{6}$$

where quantities $c_{ij}^{s} \in \mathbb{K}$ are constant for all i,j and $s = \overline{1,N}$, then for the adjoint to (6) mapping $\vartheta : \mathrm{Symm}(\widetilde{\mathbb{A}}^*) \to \widetilde{\mathbb{A}}^* \wedge \widetilde{\mathbb{A}}^*$ one obtains the expression

$$\vartheta : u^s \to \sum_{i,j=\overline{1,N}} (c_{ij}^{s} \; - c_{ji}^{s}) u^i \otimes u^j. \tag{7}$$

For the pre-Poisson bracket (1) to be a Poisson bracket on $\widetilde{\mathbb{A}}^*$, it should satisfy additionally the Jacobi identity. To find the corresponding additional constraints on the internal multiplication "$\circ$" on the algebra $\widetilde{\mathbb{A}}$, define for any $u(x) \in \widetilde{\mathbb{A}}^*$ the skew-symmetric linear mapping

$$\vartheta(u) : \widetilde{\mathbb{A}} \to \widetilde{\mathbb{A}}^*, \tag{8}$$

called [7] by the *Hamiltonian operator*, via the identity

$$\prec \vartheta(u) a, b \succ := \prec \vartheta \; u(x), a \wedge b \succ \tag{9}$$

for any $a, b \in \widetilde{\mathbb{A}}$, where the mapping $\vartheta :\ \mathrm{Symm}(\widetilde{\mathbb{A}}^*) \to \widetilde{\mathbb{A}}^* \wedge \widetilde{\mathbb{A}}^*$ is determined by the expression (7), being adjoint to it. Then it is well known [7] that the pre-Poisson bracket (1) is a Poisson one iff the Hamiltonian operator (8) satisfies the Schouten–Nijenhuis condition:

$$[[\vartheta(u), \vartheta(u)]] = 0 \tag{10}$$

for any $u(x) \in \widetilde{\mathbb{A}}^*$. Having observed that the following action

$$\vartheta(u) e_i = \sum_{s,k=\overline{1,N}} (c_{ik}^{s} \; - c_{ki}^{s}) u_s(x) u^k \tag{11}$$

holds for any basis element $e_i \in \mathbb{A}, i = \overline{1,N}$, the resulting pre-Poisson bracket (1) becomes equal to

$$\begin{aligned} \{u_i(x)\;, u_j(x)\} &= \prec \vartheta(u) e_i, e_j \succ \\ &= \sum_{s=\overline{1,N}} (c_{ij}^{s} - c_{ji}^{s}) u_s(x) = \prec u(x), \sum_{s=\overline{1,N}} (c_{ij}^{s} - c_{ji}^{s}) e_s > \end{aligned} \tag{12}$$

for any $u(x) \in \widetilde{\mathbb{A}}^*$. If now to define on the algebra $\mathbb{A}$ the natural adjacent to the algebra $\mathbb{A}$ Lie algebra structure

$$[e_i, e_j] = e_i \circ e_j - e_j \circ e_i := \sum_{s=\overline{1,N}} (c_{ij}^s - c_{ji}^s) e_s \tag{13}$$

for any basis elements $e_i, e_j \in \mathbb{A}, i, j = \overline{1,N}$, the expression (12) yields the well-known [4–6] classical Lie-Poisson bracket

$$\{u_i(x)\ , u_j(x)\} = \prec u, [e_i, e_j] \succ . \tag{14}$$

Concerning the adjacent Lie algebra structure condition (13), it can be easily rewritten as the set of relationships,

$$\sigma_{ij}^s - \sigma_{ji}^s = c_{ij}^s - c_{ji}^s \tag{15}$$

whose evident solution is

$$c_{ij}^s = \sigma_{ij}^s \tag{16}$$

for any $i, j, s = \overline{1,N}$. As the bracket (14) is of the classical Lie-Poisson type, for the Hamiltonian operator (11) to satisfy the Schouten–Nijenhuis condition (10) is enough to check only the Jacobi identity for the Lie algebra $\mathcal{L}_{\widetilde{\mathbb{A}}}$, adjacent to the algebra $\widetilde{\mathbb{A}}$ via imposing the Lie structure (13), taking into account the relationships (16). Simple calculations for the special skew-symmetric case

$$e_i \circ e_j + e_j \circ e_i = 0 \tag{17}$$

for all $i, j = \overline{1,N}$ give rise to the constraints

$$e_i \circ e_j + e_j \circ e_i = 0, (e_i \circ e_j) \circ e_k + (e_j \circ e_k) \circ e_i + (e_k \circ e_i) \circ e_j = 0, \tag{18}$$

coinciding exactly with those stated before in [7]. The corresponding Hamiltonian operator (8) then acts as

$$\vartheta(u) e_i = \sum_{s,k=\overline{1,N}} (\sigma_{ik}^s - \sigma_{ki}^s) u_s(x) u^k \tag{19}$$

on any basis element $e_i \in \mathbb{A}, i = \overline{1,N}$. Since the bracket (14), owing to the constraints (17) and (18), satisfies the Jacobi identity and thereby the mapping $\vartheta(u) : \widetilde{\mathbb{A}} \to \widetilde{\mathbb{A}}^*$ does the Schouten–Nijenhuis condition (10), one can formulate the following theorem.

Theorem 1. *The general pre-Poisson bracket* (1) *on* $\widetilde{\mathbb{A}}^*$ *under the constraints* (17) *and* (18) *on the algebra* $\mathbb{A}$ *is a Poisson one, compatible with its internal algebraic structure.*

Remark 2. The same way one can consider a simple ultra-local quadratic pre-Poisson bracket on $\widetilde{\mathbb{A}}^*$ in the form

$$\{u_i(x)\ , u_j(x)\} := \prec u(x) \otimes u(x)\ , \vartheta^*(e_i \wedge e_j) \succ, \tag{20}$$

where the skew-symmetric mapping $\vartheta^* : \widetilde{\mathbb{A}} \wedge \widetilde{\mathbb{A}} \to \mathrm{Symm}(\widetilde{\mathbb{A}} \otimes \widetilde{\mathbb{A}})$ is given for any $i, j = \overline{1,N}$ in the quadratic form

$$\vartheta^*(e_i \otimes e_j - e_j \otimes e_i) := \sum_{k,s=\overline{1,N}} (c_{ij}^{ks} - c_{ji}^{ks})(e_k \otimes e_s + e_s \otimes e_k). \tag{21}$$

In particular, if to assume that the coefficients $c_{ij}^{ks} = \sigma_{ij}^k \alpha^s$ for some constant numbers σ_{ij}^k and $\alpha^s \in \mathbb{K}$ for all i, j and $k, s = \overline{1,N}$, where, by definition, $e_k \circ e_s := \sum_{k=\overline{1,N}} \sigma_{ij}^k e_k$, then the pre-Poisson bracket (20) yields a very compact form

$$\{u_i(x)\ , u_j(x)\} := \prec u(x) \otimes u(x),\ \alpha \otimes [e_i, e_j] + [e_i, e_j] \otimes \alpha \succ, \tag{22}$$

generalizing (14) and parametrically depending on the constant vector

$$\alpha := \sum\nolimits_{s=\overline{1,N}} \alpha^s e_s \in \mathbb{A}.$$

For the pre-Poisson bracket (22) one can formulate suitable constraints on the algebraic structure of $\widetilde{\mathbb{A}}$, similar to those obtained in [2], yet we will not stop more on this in detail.

Let now $\widetilde{\mathbb{A}}(u) \subset \widetilde{\mathbb{A}}$ denote the polynomial differential ideal, generated by an element $u \in \widetilde{\mathbb{A}}$ and its derivatives $D_x^n u \in \widetilde{\mathbb{A}}$, $n \in \mathbb{Z}_+$. The corresponding space of polynomial functions $\widetilde{\mathbb{A}}(u) \to \mathbb{K}$, constructed by means of some scalar form on $\widetilde{\mathbb{A}}(u)$, will be respectively denoted by $\mathcal{F}_{\widetilde{\mathbb{A}}}(u)$. Then the basic ultra-local and linear with respect to an independent element $u(x) \in \widetilde{\mathbb{A}}, x \in \mathbb{S}^1$, pre-Poisson bracket (1) is easily generalized to a local pre-Poisson bracket for arbitrary functions $f, g \in \mathcal{F}_{\widetilde{\mathbb{A}}}(u)$:

$$\{f, g\}(u) = \prec u(x), \vartheta^*(\nabla f(u(x)) \ \wedge \nabla g(u(x) \succ, \tag{23}$$

in which the mapping $\vartheta^* : \widetilde{\mathbb{A}} \wedge \widetilde{\mathbb{A}} \to \mathrm{Symm}(\widetilde{\mathbb{A}} \otimes \widetilde{\mathbb{A}})$ is invariantly reduced on the subspace $\widetilde{\mathbb{A}}(u) \wedge \widetilde{\mathbb{A}}(u)$ and depends nontrivially on the differentiation $D_x : \widetilde{\mathbb{A}} \to \widetilde{\mathbb{A}}$. In (23) we have denoted by sign "∇" the usual linear gradient mapping from $\mathcal{F}_{\widetilde{\mathbb{A}}}(u)$ to the ideal $\widetilde{\mathbb{A}}(u) \subset \widetilde{\mathbb{A}}$, that is for a given $h \in \mathcal{F}_{\widetilde{\mathbb{A}}}(u)$ there holds $\nabla h(u(x)) \ \in \widetilde{\mathbb{A}}(u)$ and $\prec v(x), \nabla h(u(x) \succ := dh(u + \varepsilon v)/d\varepsilon|_{\varepsilon=0}$ for any $v(x) \in \widetilde{\mathbb{A}}^*, x \in \mathbb{S}^1$. Keeping in mind the problem of finding constraints on the multiplicative structure of the algebra $\widetilde{\mathbb{A}}$ under which the pre-Poisson bracket (23) is a Poisson one, it is very interesting to construct nontrivial examples of *linear* local pre-Poisson brackets on $\mathcal{F}_{\widetilde{\mathbb{A}}}(u)$, compatible with the multiplication "$\circ$" on $\mathbb{A}$ and non trivially depending on the usual differential operator $D_x : \widetilde{\mathbb{A}} \to \widetilde{\mathbb{A}}$ for $x \in \mathbb{S}^1$. In particular, for arbitrary functions $f, g \in \mathcal{F}_{\widetilde{\mathbb{A}}}(u)$ one can consider the following non trivial and simplest linear local pre-Poisson bracket

$$\{f, g\}(u) := \prec u(x), \vartheta^*(\nabla f(u(x)) \wedge \nabla g(u(x)) \succ, \tag{24}$$

where, by definition,

$$\vartheta^* : (a(x) \wedge b(x)) \to \sum_{j,k,s=\overline{1,N}} [c_{jk}^s D_x a^j(x) b^k(x) - c_{jk}^s D_x b^j(x) a^k(x)] e_s \tag{25}$$

for any $a(x) := \sum_{j=\overline{1,N}} a^j(x)e_j,\ b(x) := \sum_{j=\overline{1,N}} b^j(x)e_j \in \widetilde{\mathbb{A}},\ x \in \mathbb{S}^1$, and some arbitrarily chosen constant quantities $c_{jk}^s \in \mathbb{K}$ for all j,k and $s = \overline{1,N}$. If to assume additionally that these constant quantities satisfy the condition (16), that is $c_{ij}^s = \sigma_{ij}^s$ for all i,j and $s = \overline{1,N}$, the mapping (25) can be equivalently rewritten as

$$\vartheta^* : (a(x) \wedge b(x)) \to D_x a(x) \circ b(x) - D_x b(x) \circ a(x), \tag{26}$$

providing the pre-Poisson bracket (24) for arbitrary functions $f, g \in \mathcal{F}_{\widetilde{\mathbb{A}}}(u)$ with the canonical Lie–Poisson form

$$\{f,g\}(u) := \prec u(x), D_x\nabla f(u(x)) \circ \nabla g(u(x)) - D_x\nabla g(u(x)) \circ \nabla f(u(x) \succ, \tag{27}$$

which was recently presented in [12]. Thus, if the Lie structure

$$[a(x), b(x)]_D := D_x a(x) \circ b(x) - D_x b(x) \circ a(x) \tag{28}$$

for any $a(x), b(x) \in \widetilde{\mathbb{A}},\ x \in \mathbb{S}^1$, renders the adjacent Lie algebra $\mathcal{L}_{\widetilde{\mathbb{A}}}$, the pre-Poisson bracket (27) will be automatically a Poisson one on the space $\mathcal{F}_{\widetilde{\mathbb{A}}}(u)$. Moreover, from the expression (27), rewritten in the tensor form

$$\begin{aligned}\{f,g\}(u) &= \prec \Delta u(x), D_x\nabla f(u(x)) \otimes \nabla g(u(x)) - D_x\nabla g(u(x)) \otimes \nabla f(u(x) \succ \\ &:= (\Delta_1 u(x) D_x + \Delta_2 u(x) D_x)\nabla f(u(x), \nabla g(u(x)) \\ &= (\vartheta(u)\nabla f(u(x)), \nabla g(u(x)))\end{aligned} \tag{29}$$

naturally defines some bi-linear form $(\cdot,\cdot)$ on the adjacent Lie algebra $\mathcal{L}_{\widetilde{\mathbb{A}}}$, allowing to determine the corresponding Hamiltonian operator $\vartheta(u) : \mathcal{L}_{\widetilde{\mathbb{A}}} \to \mathcal{L}_{\widetilde{\mathbb{A}}}$:

$$\vartheta(u) = \Delta_1 u(x) D_x + D_x \Delta_2 u(x), \tag{30}$$

where the $\Delta_j u(x) : \widetilde{\mathbb{A}} \to \widetilde{\mathbb{A}}^*, j = \overline{1,2}$, are the convolution operators of the co-multiplication with respect to its first and second tensor components, respectively. So, if the Hamiltonian operator (30) satisfies the Schouten–Nijenhuis condition (10), the pre-Poisson bracket (29) will be a Poisson one. Yet, simultaneously, if the adjacent Lie algebra structure (28) satisfies the Jacobi condition, then the equivalent to (29) pre-Poisson bracket (27) will be also a Poisson one. As the second case is easier to check, after some calculations one obtains the well-known [3, 7] Novikov algebra constraints

$$[R_a, R_b] = 0, [L_a, L_b] = L_{[a,b]} \tag{31}$$

on the multiplication structure of the algebra $\mathbb{A}$, where, by definition, for any $a, b \in \mathbb{A}$ the bracket $[a,b] := a \circ b - b \circ a$ and the mappings $L_a, R_a : \mathbb{A} \to \mathbb{A}$ are left and right multiplications, respectively: $L_a b := a \circ b = R_b a$. The next example of the bilinear, local and weakly skew-symmetric mapping

$$\vartheta^* : (a(x) \wedge b(x)) \to D_x^{-1} a(x) \circ b(x) - D_x^{-1} b(x) \circ a(x), \tag{32}$$

where, by definition, $D_x D_x^{-1} := 1 : \widetilde{\mathbb{A}} \to \widetilde{\mathbb{A}}$ is the identity mapping, generates the weak adjacent Lie algebra $\mathcal{L}_{\widetilde{\mathbb{A}}}$ structure

$$[a(x), b(x)]_D := D_x^{-1} a(x) \circ b(x) - D_x^{-1} b(x) \circ a(x) \tag{33}$$

for any $a(x), b(x) \in \widetilde{\mathbb{A}}$ iff the multiplicative structure of the algebra $\widetilde{\mathbb{A}}$ satisfies the so-called *right Leibniz algebra* constraints:

$$R_{b \circ a} = [R_a, R_b], \quad R_{a \circ b} + R_{b \circ a} = 0 \tag{34}$$

for arbitrary elements $a, b \in \mathbb{A}$. The corresponding integro-differential Hamiltonian operator on the space $\mathcal{F}_{\widetilde{\mathbb{A}}}(u)$ for this case equals

$$\vartheta(u) = \Delta_1 u(x) D_x^{-1} + D_x^{-1} \Delta_2 u(x) \tag{35}$$

for any $u(x) \in \widetilde{\mathbb{A}}^*, x \in \mathbb{S}^1$. If now to take the bilinear, local and weakly skew-symmetric mapping

$$\vartheta^* : (a(x) \wedge b(x)) \to -D_x^{-1} a(x) \circ D_x b(x)) + D_x^{-1} b(x) \circ D_x a(x)) \tag{36}$$

or any $a(x), b(x) \in \widetilde{\mathbb{A}}$, the related adjacent Lie algebra $\mathcal{L}_{\widetilde{\mathbb{A}}}$ structure is respectively given by the expression

$$[a(x), b(x)]_D := -D_x^{-1} a(x) \circ D_x b(x) + D_x^{-1} b(x) \circ D_x a(x) \tag{37}$$

and satisfies the weak Jacobi identity, iff the following so-called *Riemann algebra* $\mathbb{A}$ multiplicative structure

$$[R_a, R_b] = 0, \; L_{a \circ b} = R_{a \circ b} \; = L_{b \circ a} \tag{38}$$

holds for arbitrary elements $a, b \in \mathbb{A}$. For the related Hamiltonian operator on the space $\mathcal{F}_{\widetilde{\mathbb{A}}}(u)$ one easily obtains from (36) the integro-differential expression

$$\vartheta(u) = D_x \Delta_1 u(x) D_x^{-1} - D_x^{-1} \Delta_2 u(x) D_x \tag{39}$$

for any $u(x) \in \widetilde{\mathbb{A}}^*, x \in \mathbb{S}^1$.

3. Conclusion

In this work we succeeded in formal tensor and differential-algebraic reformulating the criteria [7, 8, 11] for a given linear in the field variables matrix differential expression to be Hamiltonian and developed an effective approach to classification of the algebraic Poisson structures lying in the background of the integrable multi-component Hamiltonian systems.

Acknowledgment

The author cordially thanks Prof. Vladimir Zakharov (Berkley, USA) and Prof. Jacek Szmigelski (Saskatoon University, Canada) for fruitful discussions during the XXXV Workshop (2016) in Białowieża. He is also very indebted to Prof. Maciej Błaszak (Poznan University, Poland), Prof. Jan Cieśliński (Białystok University, Poland), Prof. O.D. Artemovych (Kraków Polytechnical University, Poland) and Prof. Maxim Pavlov (Novosibirsk State University) for the cooperation and instrumental suggestions.

References

[1] A.A. Balinsky, A.I. Balinsky, *On the algebraic structures connected with the linear Poisson brackets of hydrodynamics type*, arXiv:hep-th/9311134v1 22 Nov 1993

[2] A.A. Balinsky, Yu. Burman, *Quadratic Poisson brackets compatible with an algebra structure*, J .Phys. **A27** (1994) L693–L696

[3] A.A. Balinski, S.P. Novikov, *Poisson brackets of hydrodynamic type, Frobenius algebras and Lie algebras*, Sov. Math. Dokl. **32** (1985) pp. 228–231

[4] D. Blackmore, A.K. Prykarpatsky and V.H. Samoylenko, *Nonlinear dynamical systems of mathematical physics*, World Scientific Publisher, NJ, USA, 2011

[5] M. Blaszak, *Bi-Hamiltonian dynamical systems.* NY, Springer, 1998

[6] L.D. Faddeev, L.A. Takhtadjan, *Hamiltonian methods in the theory of solitons*, Springer, New York, Berlin, 1986

[7] I.M. Gel'fand, I.Ya. Dorfman, *Hamiltonian operators and algebraic structures related to them*, Funct. Anal. Appl. 13 (1979) 248–262

[8] O.I. Mokhov, *Symplectic and Poisson geometry on loop spaces of smooth manifolds and integrable equations*, Computer Science Institute Publisher, Moscow, 2004

[9] Y.A. Prykarpatsky, O.D. Artemovych, M. Pavlov and A.K. Prykarpatsky, *The differential-algebraic and bi-Hamiltonian integrability analysis of the Riemann type hierarchy revisited*, J. Math. Phys. **53**, 103521 (2012); arXiv:submit/1108.0878 [nlin.SI] 20 Sep. 2011

[10] A. Prykarpatsky and I. Mykytyuk, *Algebraic integrability of nonlinear dynamical systems on manifolds: classical and quantum aspects*, Kluwer Academic Publishers, the Netherlands, 1998

[11] A. Sergyeyev, *A simple way of making a Hamiltonian systems into a bi-Hamiltonian one*, Acta Applicandae Mathematica, Volume **83**, Issue 1, (2004) pp. 183–197

[12] I.A.B. Strachan, B.M. Szablikowski, *Novikov Algebras and a Classification of Multicomponent Camassa–Holm Equations*, Studies in Appl. Mathematics, 2014, **133**, p. 84–117

[13] O.O. Vaneeva, R.O. Popovych, C. Sophocleous, *Equivalence transformation in the study of integrability*, Phys. Scr. **89**, (2014), 038003 (9 pp.)

Anatolij K. Prykarpatski
Institute of Mathematics
Polytechnical University of Kraków
Kraków, Poland

and

Department of Applied Mathematics
AGH University of Science and Technology
Kraków, Poland
e-mail: pryk.anat@cybergal.com

Geometric Methods in Physics. XXXV Workshop 2016
Trends in Mathematics, 95–102

An Algebraic Background for Hierarchies of PDE in Dimension $(2|1)$

Claude Roger

Abstract. In $d = 2$ with variables (x, t), the superalgebraic trick of adding a supplementary odd variable allows the construction of a "square root of time", an operator D satisfying $D^2 = \partial/\partial t$ in superspace of dimension $(2|1)$. We already used that trick to obtain a Miura transform in dimension$(2|1)$ for non-stationary Schrödinger type operators [6]. We shall discuss here the construction of an algebra of pseudodifferential symbols in dimension $(2|1)$; that algebra generalizes the one for $d = 1$, used in construction of hierarchies from isospectral deformations of stationary Schrödinger type operators.

Mathematics Subject Classification (2010). 17 B65, 34 L30, 35 Q53, 81 T60.

Keywords. Schrödinger operators, superalgebras, supersymmetry, hierarchies of PDE.

1. Reminders on the classic $d = 1$ case

This is the famous Korteweg–de Vries equation and hierarchy, together with its very rich related analytical, geometrical and algebraic structures; they are described in an extensive literature, for example the treatise of Dickey [2], which we shall refer to. It begins with stationary Schrödinger type operators in $d = 1$ with variable x, like $L = \partial^2 + u(x)$, then one considers its isospectral deformations, by conjugacy in the space of differential operators or pseudo differential symbols, as follows: $L \to L_t = U(t)LU(t)^{-1}$. The latter formula gives infinitesimally a Lax type equation $\dot{L} = [L, M]$, where as usual, the dot stands for time derivative, so $\dot{L} = \dot{u}$. For suitable values of operator M, denoted $M = M_p$ for integer p, one gets a hierarchy of equations $\dot{L} = [L, M_p]$. If moreover, the operators M_p are mutually commuting $[M_p, M_q] = 0$, one deduces an infinite family of conserved quantities for those equations, then said to be completely integrable system with an infinite number of degrees of freedom.

Those equations are Hamiltonian for the Poisson structure of the dual of Virasoro algebra (cf. [3]) of vector fields $u(x)\partial$, with the well-known bracket $[u(x)\partial, v(x)\partial] = (u(x)v'(x) - v(x)u'(x))\partial$, and central term; from that Lie algebra one constructs the associative algebra OD of differential operators of the form $u_0 + u_1\partial + \cdots + u_n\partial^n$, and finally, by formally inverting ∂ into ∂^{-1}, one gets the division algebra of pseudo-differential symbols ΨD. A symbol $D \in \Psi D$ has the form

$$D = \sum_{n=-\infty}^{n=N} a_n \partial^n.$$

Associative algebra ΨD encodes all computations necessary for construction of operators M_p, and also the right Poisson structures for those Hamiltonian equations, the dual of Virasoro algebra and its higher-order generalizations known as $\mathcal{W}$-algebras. Let us recall that the key formula for algebraic calculations in ΨD is the following:

$$\partial^{-1} a = \sum_{n=0}^{\infty} (-1)^n \partial^n(a) \partial^{-1-n}.$$

That algebra allows computation of square root of $L = \partial^2 + u(x)$, and further its successive half-integer powers, giving hierarchies and conserved quantities. As well, Miura's transform factorizes $\partial^2 + u(x) = (\partial + v(x)) \circ (\partial - v(x))$ and turns out to be a rather useful tool.

2. Non-stationary Schrödinger operators

Now, we are in the $d = 2$ case with variables (x, t), let us set for short $\frac{\partial}{\partial x} = \partial$ $\frac{\partial}{\partial t} = \partial_t$; we shall try to generalize the constructions described in part 1 above to non-stationary operators of Schrödinger type with potential: $\mathfrak{Schr} = \partial^2 - \partial_t + u(x, t)$.

Remarks.

1. One may think of the space parametrized by (x, t) as spacetime with space dimension 1, the variable t being the physical time, but it is not necessarily the case.
2. Those operators are, strictly speaking, heat operators, but Wick rotation $t \to it$ transforms them into the actual Schrödinger operator.
3. Hierarchies of PDE in dimension 2 and more have been on the agenda since some time, with various point of view, cf. for example [9].

In the present paper, we shall use a supersymmetric trick: we enlarge the space with an odd dimension and work in the superspace of dimension $(2|1)$, parametrized with even variables (x, t) and one odd variable θ. The even variables can be polynomial, analytic or differentiable, we can have as well $x \in \mathbb{R}$ or $x \in S^1$, and the same for t. We shall not make things more precise for the moment, and all necessary technical preliminaries on superalgebra will be developed in the next part.

We shall now consider the odd differential operator $D_\theta = \theta\partial_t + \partial_\theta$, which satisfies $D_\theta^2 = \partial_t$; so D_θ represents the promised "square root of time", and we can write non stationary Schrödinger operator as a difference of two squares $\mathfrak{Schr} = \partial^2 - D_\theta^2 + u(x,t)$.

3. Some techniques of graded algebra

We shall recall in this part the most basic definitions and formulas for graded algebra, a detailed introduction can be found in [1]. A graded (or super-) algebra is an associative algebra which admits a graduation, following $\mathcal{A} = \oplus_{p\in\mathbb{Z}} A^p$; an element $a \in A^p$ is said to be of degree p, denoted by $|a| = p$. The associative multiplication is graded, i.e.,

$$|ab| = |a| + |b|,$$

and supercommutative (or graded commutative) which means:

$$ab = (-1)^{|a||b|}ba.$$

So, a and b anticommute iff they are both of odd degree, otherwise they commute.

The typical examples are the free algebras on even generators x_i and odd generators θ_α, written as

$$\mathcal{A} = k[x_i] \otimes \Lambda(\theta_\alpha),$$

where as usual the symbol Λ denotes exterior algebra, and $k[x_i]$ the polynomial algebra in indeterminates x_i, with coefficients in a field of vanishing characteristic k; in most examples $k = \mathbb{C}$. One can also give a more global description of those superalgebras: consider two finite-dimensional vector spaces E_0 and E_1 and assume that elements of E_0 (resp.E_1) are even (resp. odd), then the free supercommutative algebra on $E_0 \oplus E_1$ will be

$$\mathcal{A} = S^*(E_0) \oplus \Lambda^*(E_1).$$

In most cases only modulo 2 degree is taken in account, being the only relevant part for sign formulas.

A map f between graded spaces has a degree $|f|$ naturally defined by $|f(a)| = |f| + |a|$, and we shall consider derivations of graded algebras. A map $\delta : \mathcal{A} \to \mathcal{A}$ is a derivation if for any $a, b \in \mathcal{A}$, one has:

$$\delta(ab) = \delta(a)b + (-1)^{|a||\delta|}a\delta(b).$$

So one has odd and even derivations. Now we can consider the graded commutator of derivations following the formula:

$$[\delta_1, \delta_2] = \delta_1 \circ \delta_2 - (-1)^{|\delta_1||\delta_2|}\delta_2 \circ \delta_1.$$

This bracket defines a Lie superalgebra structure on the space of derivations (see [1] for precise formalism about Lie superalgebras); the geometric interpretation of derivations as tangent vector fields allows to consider the space of derivations $\mathrm{Der}(\mathcal{A})$ as the Lie superalgebra of tangent vector fields on the underlying super manifold whose space of functions is $\mathcal{A}$ [7]. Finally, we shall make use for any graded algebra, of the following involution: $a \to \bar{a} = (-1)^{|a|}a$.

4. The division algebra of pseudodifferential symbols in $d = (2|1)$

Our differential and pseudo differential operators will act on the superalgebra $\mathcal{A}$ of functions on superspace of dimension $(2|1)$. We shall use as generators the following two operators: $\mathcal{D} = \partial + iD_\theta$ and $\bar{\mathcal{D}} = \partial - iD_\theta$, where i is present mainly for technical reasons. Computations are a bit delicate, since

$$\mathcal{D}(ab) = \mathcal{D}(a)b + a\mathcal{D}(b)$$

if a is even, but

$$\mathcal{D}(ab) = \mathcal{D}(a)b + a\bar{\mathcal{D}}(b)$$

if a is odd. In terms of composition of operators, an element $a \in \mathcal{A}$ being viewed as an operator of order zero, one has

$$\mathcal{D} \circ a = a \circ \mathcal{D} + \mathcal{D}(a)$$

if $|a| = 0$ (mod.2), and

$$\mathcal{D} \circ a = a \circ \bar{\mathcal{D}} + \mathcal{D}(a)$$

if $|a| = 1$ (mod.2). This is why we cannot consider operators in powers of $\mathcal{D}$ only, we must add the conjugate $\bar{\mathcal{D}}$ in order to get a closed algebra.

Those elements generate the associative graded algebra of differential operators on $\mathcal{A}$, denoted as $OD(\mathcal{D}, \bar{\mathcal{D}})$. Its generic element has the following form:

$$\sum_{k\geq 0, l\geq 0}^{\text{finite}} a_{k,l}\mathcal{D}^k\bar{\mathcal{D}}^l.$$

We can change the generators from $(\mathcal{D}, \bar{\mathcal{D}})$ to (∂, D_θ), and so

$$OD(\mathcal{D}, \bar{\mathcal{D}}) = OD(\partial, D_\theta).$$

Straightforward computations give easily:

$$\begin{aligned} \mathcal{D}^2 &= \partial^2 - \partial_t + 2i\partial D_\theta \\ \bar{\mathcal{D}}^2 &= \partial^2 - \partial_t - 2i\partial D_\theta \end{aligned}$$

so

$$\frac{\mathcal{D}^2 + \bar{\mathcal{D}}^2}{2} = \partial^2 - \partial_t = \mathfrak{Schr}_0,$$

the non-stationary Schrödinger operator with zero potential. This is why algebra $OD(\mathcal{D}, \bar{\mathcal{D}})$ is relevant for our problem! Moreover, one has $\mathcal{D}\bar{\mathcal{D}} = \bar{\mathcal{D}}\mathcal{D} = \partial^2 + \partial_t$, and omitting the i coefficient, we could have obtained $\partial^2 - \partial_t$, like in [7], where that formula was used for generalization of Miura transform.

We are now ready to construct the algebra of pseudo differential symbols $\Psi D(\mathcal{D}, \bar{\mathcal{D}})$ by localization of $OD(\mathcal{D}, \bar{\mathcal{D}})$; let us invert formally $\mathcal{D}$ and $\bar{\mathcal{D}}$ in $\mathcal{D}^{-1}$ and $\bar{\mathcal{D}}^{-1}$ respectively, so a generic pseudodifferential symbol will have the form:

$$\sum_{k=-\infty,l=-\infty}^{\text{finite}} a_{k,l}\mathcal{D}^k\bar{\mathcal{D}}^l.$$

As in the $d = 1$ case, the difficulty is to establish a coherent formula for the composition $\mathcal{D}^{-1} \circ a$; we shall prove by recurrence on the degree that:

$$\mathcal{D}^{-1} \circ a = \sum_{i,j=0;i+j>0}^{+\infty} a_{i,j}\mathcal{D}^{-i}\bar{\mathcal{D}}^{-j}.$$

(Here we omitted symbol $\circ$ when obvious.) As in the $d = 1$ case the formulas will be obtained from $\mathcal{D}\circ\mathcal{D}^{-1}\circ a = a$; we shall use decomposition of any element $a \in \mathcal{A}$ according parity, $a = a^0 + a^1$. We shall proceed by recurrence, from:

$$a = \mathcal{D}\left(\sum_{i,j=0;i+j>0}^{+\infty} a_{i,j}\mathcal{D}^{-i}\bar{\mathcal{D}}^{-j}\right),$$

so:

$$\mathcal{D}(a_{i,j}\mathcal{D}^{-i}\bar{\mathcal{D}}^{-j}) = \mathcal{D}(a_{i,j})\mathcal{D}^{-i}\bar{\mathcal{D}}^{-j} + a^0_{i,j}\mathcal{D}^{-i+1}\bar{\mathcal{D}}^{-j} + a^1_{i,j}\mathcal{D}^{-i}\bar{\mathcal{D}}^{-j+1}.$$

One then readily deduce from above the recurrence formula

$$\mathcal{D}(a_{i,j}) = -a^0_{i+1,j} - a^1_{i,j+1}$$

The recurrence is easily introduced: if $a = a^0 + a^1$, then $a_{1,0} = a^0_{1,0} = a^0$ and $a_{0,1} = a^1_{0,1} = a^1$; so one gets a unique solution to the equation above but the general formulas for $a_{i,j}$ are not obvious to be made explicit; for example one has $a_{i+1,0} = (-1)^i\partial^i a^0$, and also $a_{0,j+1} = (-1)^j\partial^j a^1$; we shall give also some samples in low degree: $a_{1,1} = -iD_\theta a$, $a_{1,2} = i\partial D_\theta a - \partial_t a^1$ and $a_{2,1} = i\partial D_\theta a - \partial_t a^0$. Computations are exactly parallel when one computes $\bar{\mathcal{D}}^{-1} \circ a$.

Remark. In [8] we considered algebras of differential operators and pseudodifferential symbols in dimension(1|1) with variables (x, θ), variable t being a loop space coordinate; in [4] the authors give a detailed study of the algebra of differential operators in dimension (1|1), they award its paternity to Manin and Radul [5] in their work about SuSy (supersymmetric) extension of KP hierarchy. From $D^2_\theta = \partial_t$, one can consider $OD(\mathcal{D}, \bar{\mathcal{D}}) = OD(\partial, D_\theta)$ as a quadratic extension of $OD(\partial, \partial_1)$ which is simply the algebra of differential operators in $d = 2$.; but those algebras are neither commutative nor anticommutative. Let us stress also the fact that $OD(\mathcal{D}, \bar{\mathcal{D}}) = OD(\partial, D_\theta)$ doesn't imply that $\Psi D(\mathcal{D}, \bar{\mathcal{D}}) = \Psi D(\partial, D_\theta)$, since different choice of generators to be inverted change the global algebraic structure.

5. Schrödinger type operators in dimension $(2|1)$

We shall consider operators of the type $\mathcal{D}^2 + \mathcal{U}$, where potential $\mathcal{U} = u + i\theta\phi$ has an odd and even part. Moreover, $\mathcal{D}^2 = \partial^2 - \partial_t + 2i\partial D_\theta = \mathfrak{Schr}_0 + 2i\partial D_\theta$. We get an unwanted supplementary term, it can be considered as the "price to pay" for generalization to $d = 2$, like in the case of Miura transform [6].

The wave functions must be looked for among super functions as $F = f + i\theta\alpha$ and spectral values have odd and even part too: $\Lambda = \lambda + i\theta\xi$. One has

$$\mathcal{D}^2(F) = \mathfrak{Schr}(f) + i\theta\mathfrak{Schr}(\alpha) + 2i\partial D_\theta(F);$$

then one readily computes $D_\theta(F) = \theta\partial_t f + i\alpha$. Then:

$$(\mathcal{D}^2 + \mathcal{U})(F) = \mathfrak{Schr}(f) + uf - 2\partial\alpha + i\theta(\mathfrak{Schr}(\alpha) + u\alpha + 2\partial\partial_t f + \phi f)$$

Finally, equation

$$(\mathcal{D}^2 + \mathcal{U})(F) = \Lambda F$$

induces the following system:

$$\mathfrak{Schr}(f) + uf - 2\partial\alpha = \lambda f$$
$$\mathfrak{Schr}(\alpha) + u\alpha + 2\partial\partial_t f + \phi f = \lambda\alpha + \xi f.$$

In the particular case when the wave function F is purely even, so $F = f$, one obtains a coupled system in f:

$$\mathfrak{Schr}(f) + uf = \lambda f$$
$$2\partial\partial_t f + \phi f = \xi f.$$

The second equation is of Klein–Gordon type, so the system couples hyperbolic and parabolic system.

6. KdV like equations in $d = 2$

We shall obtain evolution equations w.r.t. a real parameter s on a pair of functions (u, ϕ) in variables (x, t, s) appearing as components of a super function $\mathcal{U} = u + i\theta\phi$ as above. We now develop a formalism parallel to the classical $d = 1$ case [2], and thus consider the equation:

$$4\,\mathcal{U}_s = \mathcal{D}^3 + 6\,\mathcal{U}\,\mathcal{D}(\mathcal{U}),$$

One deduces from it the following system of evolution PDE:

$$u_s = \partial^3 u - 3\partial\partial_t u + 6u\partial u - 6u\phi - 3\partial^2\phi + \partial_t\phi$$
$$\phi_s = \partial^3\phi - 3\partial\partial_t\phi + 3\partial^2\partial_t u - \partial_t^2 u + 6\phi\partial u + 6u\partial\phi - 6\phi^2 + 6u\partial_t u$$

Let us now consider the case when the potential $\mathcal{U}$ is purely even, when $\phi = 0$. The second equation yields:

$$3\partial^2\partial_t u - \partial_t^2 u + 6u\partial_t u = 0,$$

so:

$$\partial_t(3\partial^2 u - \partial_t u + 3u^2) = 0.$$

One deduces immediately:

$$3\partial^2 u - \partial_t u + 3u^2 = f(x), \tag{$*$}$$

where f is an arbitrary function independent of t. The first equation gives:

$$u_s = \partial^3 u - 3\partial\partial_t u + 6u\partial u.$$

Now by derivation of equation ($*$), one obtains

$$3\partial^3 u - \partial\partial_t u + 6u\partial u = \partial f(x)$$

and we can cancel the non linear term, getting the linear dispersive equation with sources:

$$u_s = -2\partial^3 u - 2\partial\partial_t u + \partial f(x). \tag{$**$}$$

Another option is to cancel the term with ∂_t between ($*$) ans the first equation, and one gets an equation in (x, s) without t as follows:

$$u_s = -8\partial^3 u - 12u\partial u + 3\partial f(x);$$

after change of scale $s \to -s$ and $x \to 2x$, one obtains finally:

$$u_s = \partial^3 u + 6u\partial u - \frac{3}{2}\partial f(x),$$

which is exactly Korteweg–de Vries equation with source.

Now, if $\mathcal{U}$ is purely odd, i.e., $u = 0$, the above system reduces to:

$$3\partial^2\phi = \partial_t\phi$$

$$\phi_s = \partial^3\phi - 3\partial\partial_t\phi - 6\phi^2.$$

Finally, cancellation of terms in ∂_t will give the following equation in (x, s):

$$\phi_s + 8\partial^3\phi + 6\phi^2 = 0.$$

Any solution of that equation in (x, s) can be prolonged to a general solution in (x, s, t), simply by using a heat kernel.

References

[1] Pierre Deligne and John W. Morgan, *Notes on supersymmetry (following Joseph Bernstein)*, Quantum fields and strings: a course for mathematicians, Vol. 1, 2 (Princeton, NJ, 1996/1997), Amer. Math. Soc., Providence, RI, 1999, pp. 41–97.

[2] L.A. Dickey, *Soliton equations and Hamiltonian systems*, second ed., Advanced Series in Mathematical Physics, vol. 26, World Scientific Publishing Co., Inc., River Edge, NJ, 2003. MR 1964513

[3] Laurent Guieu and Claude Roger, *L'algèbre et le groupe de Virasoro*, Les Publications CRM, Montreal, QC, 2007, Aspects géométriques et algébriques, généralisations. [Geometric and algebraic aspects, generalizations], with an appendix by Vlad Sergiescu. MR 2362888

[4] Simon Li, Ekaterina Shemyakova, and Theodore Voronov, *Differential operators on the superline, Berezinians, and Darboux transformations*, arXiv:1605.07286v2 [math.DG]. MR 3462720

[5] Yu.I. Manin and A.O. Radul, *A supersymmetric extension of the Kadomtsev–Petviashvili hierarchy*, Comm. Math. Phys. **98** (1985), no. 1, 65–77. MR 785261

[6] Claude Roger, *A matrix Miura transform*, J. Geom. Phys. **74** (2013), 87–90. MR 3118574

[7] ———, *Algebras generated by odd derivations*, J. Geom. Symmetry Phys. **40** (2015), 53–59. MR 3496358

[8] Claude Roger and Jérémie Unterberger, *A Hamiltonian action of the Schrödinger–Virasoro algebra on a space of periodic time-dependent Schrödinger operators in* $(1+1)$*-dimensions*, J. Nonlinear Math. Phys. **17** (2010), no. 3, 257–279. MR 2733203

[9] Artur Sergyeyev and Błażej M. Szablikowski, *Central extensions of cotangent universal hierarchy:* $(2+1)$*-dimensional bi-Hamiltonian systems*, Phys. Lett. A **372** (2008), no. 47, 7016–7023. MR 2468982

Claude Roger
Institut Camille Jordan
Université de Lyon, Université Lyon I
43 boulevard du 11 novembre 1918
F-69622 Villeurbanne Cedex, France
e-mail: roger@math.univ-lyon1.fr

Geometric Methods in Physics. XXXV Workshop 2016
Trends in Mathematics, 103–109

Lagrangian Manifolds and Maslov Indices Corresponding to the Spectral Series of the Schrödinger Operators with Delta-potentials

Andrey I. Shafarevich

Abstract. We study semi-classical eigenvalues of a Schrödinger operator with delta-potential on 2D or 3D symmetric manifold. We describe Lagrangian manifolds, corresponding to such eigenvalues and compute the asymptotics of eigenvalues for different values of the parameter, defining the operator. We describe also the effect of the jump of the Maslov index while passing through the critical value of this parameter. These results were obtained in a number of joint papers with T. Filatova, T. Ratiu and A. Suleimanova.

Mathematics Subject Classification (2010). 47F05, 53D25.

Keywords. Operators with delta-potentials, Lagrangian manifolds, Maslov indices.

1. Introduction

This contribution reviews and completes our joint papers with T. Filatova, T. Ratiu and A. Suleimanova [1–3].

Many physical and mathematical works treat Schrödinger operators with delta-potentials (point potentials, zero-range potentials). The model of point potentials can be used to describe short-range impurities, admixtures, defects, and similar phenomena in diverse systems. One of the first works in which the zero-range potentials were used to study the band spectrum of periodic systems was the paper [4], where a model of nonrelativistic electron moving in a rigid crystalline lattice was considered. Since then, the model has become very popular, especially in atomic and nuclear physics.

A rigorous mathematical justification of the method of delta-potentials was given in [5], where it was suggested to use Krein's formula to describe the resolvents

This work was completed with the support of the Russian Scientific Foundation (grant 16-11-10069).

of operators with point perturbations. For an extensive bibliography of the works devoted to applications of the method of point potential, see the monographs [6, 7].

In the present paper, we describe the spectral series of the Schrödinger operator with delta-potential of the form $H = -\frac{h^2}{2}\Delta + \beta\delta_{x_0}(x)$, $\beta \in \mathbb{R}$, in the semiclassical limit as $h \to 0$ on a two- or three-dimensional compact surface admitting a special symmetry. For a large class of equations with smooth coefficients, the semiclassical theory was developed by Maslov (see, e.g., [8]); in particular, this theory implies the following result. Let N be a Riemannian manifold and $V : N \to \mathbb{R}$ a smooth function (the potential). If the Hamiltonian system in T^*N defined by the Hamiltonian $(1/2)|p|^2 + V$ is completely integrable, then the corresponding Liouville tori Λ define semiclassical spectral series of the operator $H = -\frac{h^2}{2}\Delta + V(x)$ (here $x \in N$ and (x, p) stand for the standard coordinates on T^*N). Namely, the asymptotic behavior as $h \to 0$ of the eigenvalues of H is calculated from the Bohr–Sommerfeld–Maslov conditions

$$\frac{1}{2\pi h}\int_\gamma (p, dx) + \frac{1}{4})\mu(\gamma) = m \in \mathbb{Z}, \tag{1}$$

where γ is an arbitrary cycle on Λ, μ stands for the Maslov index, and $m = O(1/h)$. The formal asymptotic behavior of the eigenfunctions (quasimodes) is of the form $\psi = K_\Lambda(1)$, where K_Λ stands for the Maslov canonical operator on the torus Λ satisfying quantization conditions.

In general, the construction of the canonical operator cannot be applied to operators with delta-potentials; at present, the geometry of the corresponding classical problem remains only slightly investigated. Below we describe the invariant Lagrangian manifolds corresponding to the spectral series of the above operator with delta-potential and obtain quantization conditions determining the asymptotic behavior of the eigenvalues. In general, these conditions are nonstandard; both for large and for small values of the coefficient α, which defines the operator (see Sect. 2.2) they pass to equations of the form (1) with diverse values of the Maslov index μ; possibly, this indicates the presence of a more complicated geometric objects associated with the semiclassical theory of operators with singular coefficients.

2. Setting of the problem

2.1. Spectral problem

Consider the spectral problem

$$\Big(-\frac{h^2}{2}\Delta + \beta\delta_{x_0}(x)\Big)\Psi = E\Psi, \quad x \in N, \quad \beta \in \mathbb{R}, \tag{2}$$

where $\delta_{x_0}(x)$ stands for the Dirac delta function concentrated at the point x_0, in the semiclassical limit as $h \to 0$ on a two-dimensional manifold in $\mathbb{R}^3$ or on a three-dimensional manifold in $\mathbb{R}^4$,

$$N = (f(z)\cos\varphi, f(z)\sin\varphi, z),$$

or

$$N = (f(z)\cos\theta\cos\varphi, f(z)\cos\theta\sin\varphi, f(z)\sin\theta, z),$$

where $z \in [z_0, z_1]$, $0 \le \varphi \le 2\pi$, and $-\frac{\pi}{2} \le \theta \le \frac{\pi}{2}$. We impose the following conditions on the function $f(z)$:

- $f(z_0) = f(z_1) = 0$, $f(z) > 0$ for $z \in (z_0, z_1)$;
- $f(z) = \sqrt{(z_1 - z)(z - z_0)}\omega(z)$, where $\omega(z)$ is a polynomial.

Under these assumptions, the surface N is an analytic manifold diffeomorphic to a sphere; the points x_0 and x_1 corresponding to the values $z = z_0, z_1$ of the parameter z are the poles of this surface (and the delta function is concentrated at one of the poles).

Remark 1. The second condition can be weakened. Seemingly, it is sufficient to assume that f is analytic in a neighborhood of the closed interval $[z_0, z_1]$, except for the points z_0 and z_1 at which f has a root singularity.

Below we present a formal definition of the operator with δ-potential on the surface N.

2.2. Formal definition of the operator *H*

The operator

$$H = -\frac{h^2}{2}\Delta + \beta\delta_{x_0}(x), \quad x \in N, \quad \beta \in \mathbb{R},$$

in the space $L_2(N)$ is defined by the construction of self-adjoint extensions (see [5]). Namely, H is constructed in such a way that the following conditions hold.

- The operator H is self-adjoint.
- On the functions vanishing at the point x_0, H coincides with the operator $H_0 = -\frac{h^2}{2}\Delta$, where Δ stands for the Laplace–Beltrami operator.

To be more precise, consider a self-adjoint operator H_0 with the domain $D(H_0) = W_2^2(N)$, where $W_2^2(N)$ stands for the Sobolev space of second order. Restricting the operator H_0 to functions $\psi(x)$ such that $\psi(x_0) = 0$, we obtain a symmetric operator $H_0|_{\psi(x_0)=0}$.

Definition 2. By the operator $H = -\frac{h^2}{2}\Delta + \beta\delta_{x_0}(x)$ we mean the self-adjoint extension of the operator $H_0|_{\psi(x_0)=0}$.

Remark 3. All extensions of this kind are parametrized by a single real parameter α (see boundary conditions below). In particular, for $\alpha = 0$, we obtain $H = H_0$.

Every extension of this kind is defined by a boundary condition at the point x_0; to be more precise, the domain of the operator H consists of the functions of the form $\psi = \psi_0 + c_1 G(x, x_0; i) + c_2 G(x, x_0, -i)$, where $\psi_0 \in W_2^2(N)$, $\psi_0(x_0) = 0$, and $G(x, y, \lambda)$ stands for Green's function of the operator Δ, i.e., the integral kernel of the resolvent, $(\Delta - \lambda)^{-1} f = \int_M G(x, y; z) f(y)\Omega$ (Ω stands for the volume form on N).

The functions of the above form have a singularity at the point x_0; namely, the following expansions hold:

$$\psi(x) = \frac{a}{2\pi}\log d(x,x_0)^{-1} + b + o(1), \quad \dim N = 2,$$
$$\psi(x) = -\frac{a}{4\pi}d(x,x_0)^{-1} + b + o(1), \qquad \dim N = 3,$$

where $a, b \in \mathbb{C}$ and $d(x, x_0)$ stands for the geodesic distance between x and x_0 on N. The domain of the extension H corresponding to the parameter α consists of the functions satisfying the boundary condition

$$a = \frac{2\alpha}{h^2}b.$$

3. Formulation of the result

3.1. Description of the Lagrangian manifold

The semiclassical asymptotic behavior of the eigenvalues of the operator H is evaluated from the quantization condition on the Lagrangian manifold which we shall now describe. Let $(x,p) \in T^*N$, where T^*N stands for the cotangent bundle of N, $x \in N$, and p a vector cotangent to N (the momentum). Consider the Hamiltonian system (the geodesic flow)

$$\dot{x} = \frac{\partial \mathcal{H}}{\partial p}, \qquad \dot{p} = -\frac{\partial \mathcal{H}}{\partial x}, \tag{3}$$

where $\mathcal{H} = |p|^2/2$, and consider the trajectories of the system, $x = X(\omega, t)$, $p = P(\omega,t)$, $\omega \in S_{\sqrt{2E}}$, $t \in \mathbb{R}$, given by the initial conditions

$$x(0) = x_0, \quad p(0) = \omega, \qquad \omega \in S_{\sqrt{2E}}, \quad |\omega| = \sqrt{2E}. \tag{4}$$

Here $S_{\sqrt{2E}}$ stands for the sphere or the circle of radius $\sqrt{2E}$ in the cotangent space at the point x_0. Thus, trajectories of the Hamiltonian system are issued from the point x_0 (at which the delta-potential is concentrated) along the surface N with the momentum "running" around the sphere $S_{\sqrt{2E}}$ (i.e., $p \in \Lambda_0$, where $\Lambda_0 = \{x = x_0, |p| = \sqrt{2E}\}$). These trajectories are contained in the manifold $\Lambda = \bigcup_t g_t\Lambda_0$ (g_t stands for the Hamiltonian phase flow) diffeomorphic to T^2 if $\dim N = 2$ and to $S^2 \times S^1$ if $\dim N = 3$ (see Fig. 1, left). It describes the classical motions corresponding to the quantum problem. The projections of the trajectories to N are geodesics.

We denote by γ the cycle on the manifold Λ formed by the closed trajectory of the Hamiltonian system (3) with the initial conditions (4) (see Fig. 1, right).

The projection of Λ to the x-space is arranged as follows: for every point x of N, except for x_0 and x_1, there are two points of Λ of the form (x,p) and $(x,-p)$ that are projected to x. A sphere or a circle is projected to each of the points x_0 and x_1.

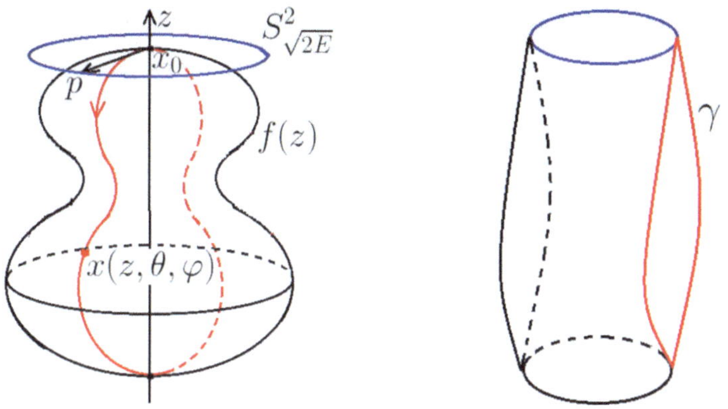

FIGURE 1. Classical trajectories on the surface N corresponding to the quantum problem (left). Lagrangian manifold and the cycle γ (right).

3.2. Formulation of the result: the quantization condition

The quantization condition on the manifold Λ with respect to the cycle γ is just the desired equation for the spectrum of the problem. More precisely, the following assertion holds.

Theorem 4. *Let $Ch^{-\epsilon} < \frac{\alpha}{h^n} < Ch^{\epsilon}$, $n = \dim N$ for some sufficiently small $\epsilon > 0$. Let there be a number $E = O(1)$ satisfying the quantization condition*

$$\tan\left(\frac{1}{2h}\oint_\gamma (p, dx)\right) = \frac{2}{\pi}\left(\log\left(\frac{\sqrt{2E}}{h}\right) + \frac{\pi h^2}{\alpha} + c\right), \quad n = 2,$$

$$\tan\left(\frac{1}{2h}\oint_\gamma (p, dx) + O(h)\right) = \frac{2h^3}{\sqrt{2E}\alpha}, \quad n = 3,$$

where γ stands for the cycle indicated above (the closed trajectory) on the Lagrangian manifold Λ, c stands for the Euler constant. Then there is an eigenvalue E_0 of the operator H such that $|E - E_0| = o(h)$ as $h \to 0$.

Remark 5. The explicit formula for the integral in the left-hand side has the following form $\oint_\gamma (p, dx) = \int_{z_0}^{z_1} \sqrt{2E(f'^2 + 1)}dz$.

Remark 6. The asymptotic behavior of the eigenfunction, outside an arbitrarily small neighborhood of x_0 independent of h, is of the form $K_{\tilde{\Lambda}}(1)$, where K stands for the Maslov canonical operator and $\tilde{\Lambda}$ is a noncompact Lagrangian manifold obtained from Λ by deleting the sphere or the circle projected to the point x_0 (this manifold is obviously homeomorphic to the cylinder $S^{n-1} \times \mathbb{R}$).

3.3. Critical values of α and the jump of the Maslov index

Consider the limit cases of the quantization condition described in the theorem. Let

$$\frac{\alpha \log 1/h}{h^2} \to 0, \quad \dim N = 2, \quad or \quad \frac{\alpha}{h^3} \to 0, \quad \dim N = 3;$$

then the quantization conditions up to small terms acquire the standard form

$$\frac{1}{2\pi h} \oint_\gamma (p, dx) + \frac{1}{2} = k \in \mathbb{Z}$$

(note that the Maslov index of the cycle γ is equal to two). Suppose now that

$$\frac{\alpha \log 1/h}{h^2} \to \infty, \quad \dim N = 2, \quad or \quad \frac{\alpha}{h^3} \to \infty, \quad \dim N = 3;$$

then we have

$$\frac{1}{2\pi h} \oint_\gamma (p, dx) + \frac{1}{2} = k, \quad n = 2, \quad or \quad \frac{1}{2\pi h} \oint_\gamma (p, dx) = k, \quad n = 3. \tag{5}$$

These equations also have the form of the Bohr–Sommerfeld–Maslov condition; however, in 3D case, the "Maslov index" of the cycle γ is equal to zero. Thus, when passing through the critical value $\alpha = O(h^3)$, we face a jump of the integral-valued invariant, which coincides with the Maslov index in the case of a smooth potential; the presence of the delta function leads to the change of this invariant by 2. Note that the jump does not take place in two-dimensional case. This possibly indicates the existence of some topological construction (still unclear to us) which generalizes the Maslov canonical operator to the case of singular coefficients.

On the "classical level" the difference between two- and three-dimensional cases can be explained as follows. The existence of the strong delta-potential leads to the reflection of classical particles at the point x_0; this reflection is described by the change of the sign of the momentum p. So one has to consider the map $g : \Lambda_0 \to \Lambda_0$, $g(p) = -p$ (we remind that Λ_0 is a circle or a sphere in $T^*_{x_0}N$). The degree of this map depends on the dimension: it equals 1 for a circle and -1 for a sphere. So the equations (5) can be written uniformly as follows

$$\frac{1}{2\pi h} \oint_\gamma (p, dx) + \frac{1}{4}(\mu + (\deg g - 1)) = k \in \mathbb{Z},$$

where $\deg g$ stands for the degree of g.

Remark 7. Note that the critical value of α is $O(h^3)$ for $n = 3$ and $O(h^2/\log(1/h))$ for $n = 3$. This additional logarithm appears in different problems connected with delta-potentials (see, e.g., [6]).

Acknowledgment

The author thanks V.P. Maslov and L.D. Faddeev for fruitful discussions. This work was supported by the Russian Scientific Foundation (grant 16-11-10069)

References

[1] T.A. Filatova and A.I. Shafarevich, "Semiclassical Spectral Series of the Schrödinger Operator with a Delta Potential on a Straight Line and on a Sphere," Teoret. Mat. Fiz. **164** (2), 1064–1080 (2010)

[2] T. Ratiu, T.A. Filatova, and A.I. Shafarevich, "Noncompact Lagrangian Manifolds Corresponding to the Spectral Series of the Schrödinger Operator with Delta-Potential on a Surface of Revolution," Dokl. Math. **86** (2), 694–696 (2012)

[3] T. Ratiu, A.A. Suleimanova, and A.I. Shafarevich, "Spectral Series of the Schrödinger Operator with Delta-Potential on a Three-Dimensional Spherically Symmetric Manifold", Russian Journal of Mathematical Physics, **20** (3), 326–335 (2013).

[4] R. de L. Kronig and W.G. Penney, "Quantum Mechanics of Electrons in Crystal Lattices," Proc. R. Soc. Lond. Ser. A bf130, 499–513 (1931)

[5] F.A. Berezin and L.D. Faddeev, "Remark on the Schrödinger Equation with Singular Potential," Dokl. Akad. Nauk SSSR **137**, 1011–1014 (1961)

[6] S. Albeverio, F. Gesztesy, and R. Høegh-Krohn, H. Holden, *Solvable Models in Quantum Mechanics* (Providence: AMS Chelsea Publishing, 2005)

[7] S. Albeverio and P. Kurasov, *Singular Perturbations of Differential Operators* (Cambridge: Cambridge University Press, 2000)

[8] V.P. Maslov and M.V. Fedoryuk, *Semi-Classical Approximation in Quantum Mechanics*, Reidel, Dordrecht, 1981.

Andrey I. Shafarevich
Moscow State University
1, Leninskie Gory
119992, Moscow, Russia
e-mail: shafarev@yahoo.com

Geometric Methods in Physics. XXXV Workshop 2016
Trends in Mathematics, 111–117

Electronic Properties of Graphene Nanoribbons in a Uniform Magnetic Field

Jan Smotlacha and Richard Pincak

Abstract. The electronic spectra of the zigzag and armchair graphene nanoribbons can be influenced by the additional effects like the reconstruction of the edge, the vacancy defects, magnetic field etc. Here, the combination of the influence of the vacancy defects and of the uniform magnetic field on the electronic spectrum was investigated. For this purpose, the usual Schrödinger equation was replaced by the Harper equations which contain the influence of the magnetic field. The results show the fractal structure of the dependence of the energy levels on the magnetic field.

Mathematics Subject Classification (2010). Primary 47A75; Secondary 47B15.
Keywords. Electronic spectrum, magnetic field, fractals.

1. Properties of zigzag and armchair nanoribbons

The calculation of the electronic spectra of the graphene nanostructures follows from their molecular structure which is based on the hexagonal carbon lattice. In the case of the planar graphene, on the base of the translational symmetry, this lattice can be divided into 2 inequivalent sublattices denoted A and B (Fig. 1). In the case of the zigzag and armchair graphene nanoribbons (Fig. 2), depending on the width of the corresponding nanostructure, the number n of the corresponding inequivalent sublattices is considerably larger and we denote them $A_1, \dots, A_n$.

The calculation of the electronic spectra starts on the solution of the Schrödinger equation for an electron bounded in the superposition of the Coulomb potentials coming from the atomic sites [1],

$$\hat{H}\psi = E\psi, \qquad \psi = C_{A_1}\psi_{A_1} + \cdots + C_{A_n}\psi_{A_n}. \tag{1}$$

Here, $A_1, \dots, A_n$ represent the atomic sites from inequivalent sublattices. In the tight-binding approximation, we suppose the solution of the form

$$\psi_{A_i} = \sum_{A_i} \exp[i\vec{k} \cdot \vec{r}_{A_i}] X(\vec{r} - \vec{r}_{A_i}), \tag{2}$$

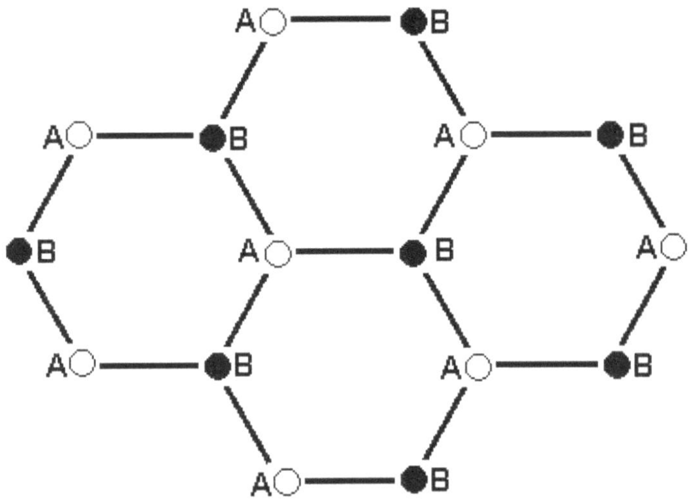

FIGURE 1. The molecular structure of the planar graphene.

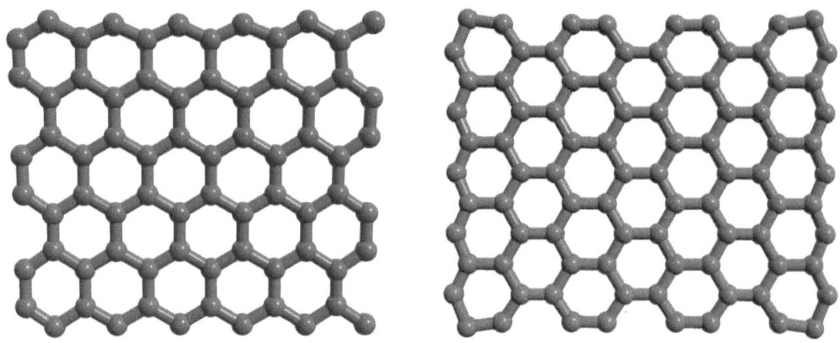

FIGURE 2. Zigzag (left) and armchair (right) nanoribbons. For both cases, the size of the horizontal dimension is considered infinite.

where $X(\vec{r})$ is the atomic orbital function. In the next calculations, we suppose the zero overlap, i.e., for $i \neq j$,

$$\int X(\vec{r} - \vec{r}_{A_i})X(\vec{r} - \vec{r}_{A_j})\mathrm{d}\vec{r} = 0. \tag{3}$$

Using this assumption and the notation

$$\begin{aligned} H_{ij} &= \int \psi^*_{A_i} H \psi_{A_j} \mathrm{d}\overrightarrow{r}, \\ S &= \int \psi^*_{A_i} \psi_{A_i} \mathrm{d}\overrightarrow{r} = \int \psi^*_{A_j} \psi_{A_j} \mathrm{d}\overrightarrow{r}, \quad i,j \in \{1,\ldots,n\}, \end{aligned} \tag{4}$$

we get the matrix equation

$$\begin{pmatrix} H_{A_1A_1} & H_{A_1A_2} & \cdots & \cdots & H_{A_1A_n} \\ H_{A_2A_1} & H_{A_2A_2} & \cdots & \cdots & H_{A_2A_n} \\ \cdots & \cdots & \cdots & \cdots & \cdots \\ \cdots & \cdots & \cdots & \cdots & \cdots \\ H_{A_nA_1} & H_{A_nA_2} & \cdots & \cdots & H_{A_nA_n} \end{pmatrix} \begin{pmatrix} C_{A_1} \\ C_{A_2} \\ \cdots \\ \cdots \\ C_{A_n} \end{pmatrix} = ES \begin{pmatrix} C_{A_1} \\ C_{A_2} \\ \cdots \\ \cdots \\ C_{A_n} \end{pmatrix}, \tag{5}$$

where $H_{A_1A_1} = \cdots = H_{A_nA_n}$. The electronic spectrum we calculate as the zero points of corresponding characteristic polynomial. If we consider the nearest-neighbor approximation, the result has the form outlined in Figure 3 [2].

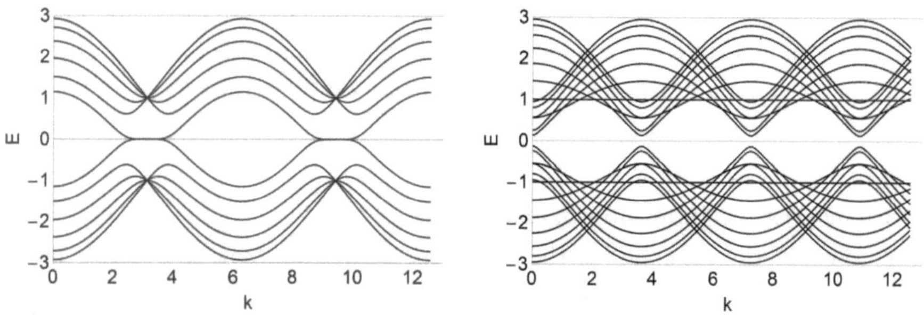

FIGURE 3. Electronic spectrum of zigzag (left) and armchair (right) nanoribbons in the nearest-neighbor approximation.

2. Influence of the magnetic field

Now, we consider the influence of the uniform magnetic field, its direction is perpendicular to the molecular surface. Then, to describe the behaviour of the electron influenced by this magnetic field and by the superposition of the corresponding Coulomb potentials, we use the Harper equations instead of the Schrödinger equation [3, 4]:

$$E\psi_i = \sum_j te^{i\gamma_{ij}}\psi_j, \tag{6}$$

where the indices j correspond to the nearest neighbors of the ith atomic site, t is the nearest-neighbor hopping integral and γ_{ij} is the magnetic phase factor. It is proportional to f, the magnetic flux going through the hexagon:

$$f = \Phi/\Phi_0 = 3\sqrt{3}Ba^2/(2\Phi_0), \tag{7}$$

where B is the value of the magnetic field, a is the length of the atomic bond and $\Phi_0 = hc/e$ is the unit flux expressed with help of the basic physical constants. We are interested in the cases when $f = p/q$ with p and q being mutual primes. For the calculation of the electronic spectrum, we use the procedure which is an analogy of the case without magnetic field: on the base of (6), we compose the matrix equation and by finding the zero points of the corresponding characteristic polynomial, we find the energy levels. For different values of the magnetic flux, the electronic spectrum for the case of the zigzag nanoribbon is sketched in Figure 4 [5].

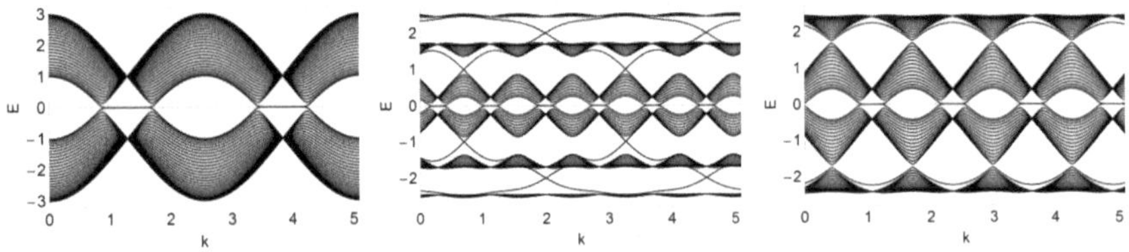

FIGURE 4. Electronic spectrum of zigzag nanoribbons for different values of the magnetic field given by the magnetic flux [10]: $f = 0$ (left), $f = 1/3$ (middle) and $f = 1/2$ (right).

2.1. Next-nearest neighbor approximation

We can consider the influence of the next-nearest neighbors in the atomic lattice as well. For this case, the equation (6) will be changed into the form

$$E\psi_i = \sum_{j,k}(te^{i\gamma_{ij}}\psi_j + t'e^{i\gamma_{ik}}\psi_k), \tag{8}$$

where the notation for the nearest neighbors is known from (6). Analogously, the indices k correspond to the next-nearest neighbors in the atomic lattice, t' is the next-nearest-neighbor hopping integral and γ_{ik} is the magnetic phase factor which is proportional to the magnetic flux f as well. The corresponding matrix equation and the characteristic polynomial can be composed and the electronic spectrum can be calculated in the same way as in the previous case. The resulting electronic spectrum for the cases of zigzag and armchair nanoribbons and different values of the magnetic field we can see in Figures 5 and 6.

If we do a comparison of the cases of zero magnetic field with the same case for the nearest-neighbor approximation (Fig. 4 left), we see that the symmetry of the spectrum related to the horizontal axis is corrupted, but this feature can be suppressed by non-zero magnetic field (Figs. 5 and 6 right).

2.2. Electronic spectrum as a function of the magnetic field

The graphs of the electronic spectrum in Figures 4, 5 and 6 show an interesting feature: from the geometrical point of view, in each of these figures, the graphs have the same (or very similar) shape, but their size is different. This size is inversely proportional to the value of q in the expression for the magnetic flux $f = p/q$. This feature can be viewed as the self-similarity which is closely related to the fractal geometry. This hypothesis we can verify, if we do a plot of the dependence of the energy levels on the magnetic flux. The result we see in Figure 7 [4, 5].

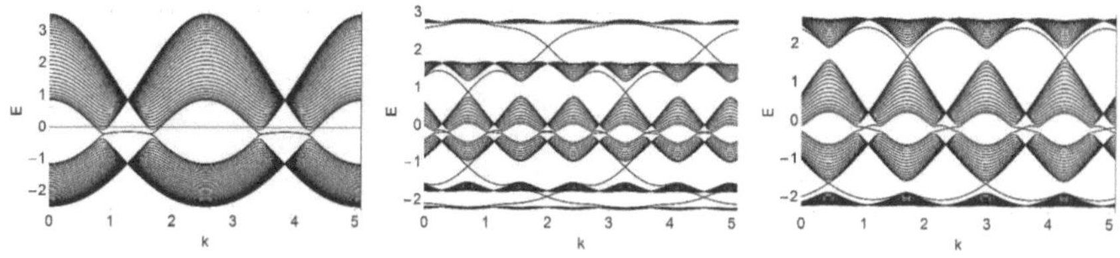

FIGURE 5. Electronic spectrum of the zigzag nanoribbons for different values of the magnetic field given by the magnetic flux: $f = 0$ (left), $f = 1/3$ (middle) and $f = 1/2$ (right).

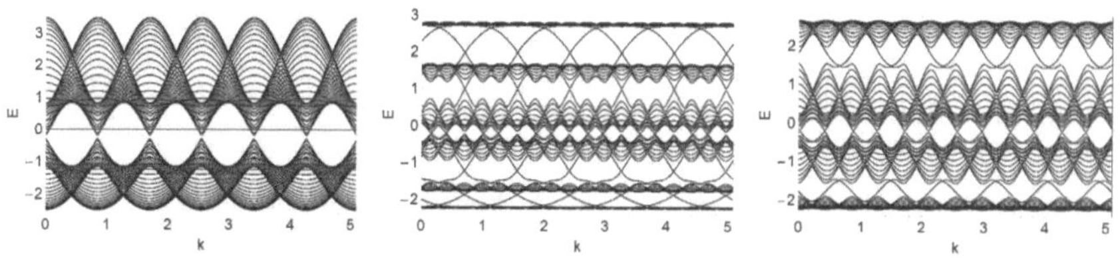

FIGURE 6. Electronic spectrum of the armchair nanoribbons for different values of the magnetic field given by the magnetic flux: $f = 0$ (left), $f = 1/3$ (middle) and $f = 1/2$ (right).

The sketched graphs are really fractal structures. The kind of fractal structure

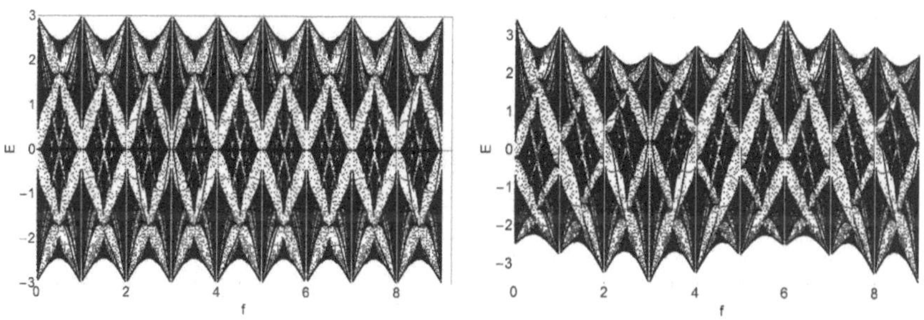

FIGURE 7. Electronic spectrum of the graphene nanoribbons depending on the magnetic flux for different approximations: nearest-neighbor interaction (left) and next-nearest-neighbor interaction (right).

which is represented by them was called the Hofstadter butterfly [6]. Moreover, the dependence on the magnetic flux shows a periodical character. The period depends on the chosen kind of approximation: for the nearest-neighbor approximation, the period for the magnetic flux is 1 (Fig. 7 left), while it is 6 for the next-nearest-neighbor approximation (Fig. 7 right).

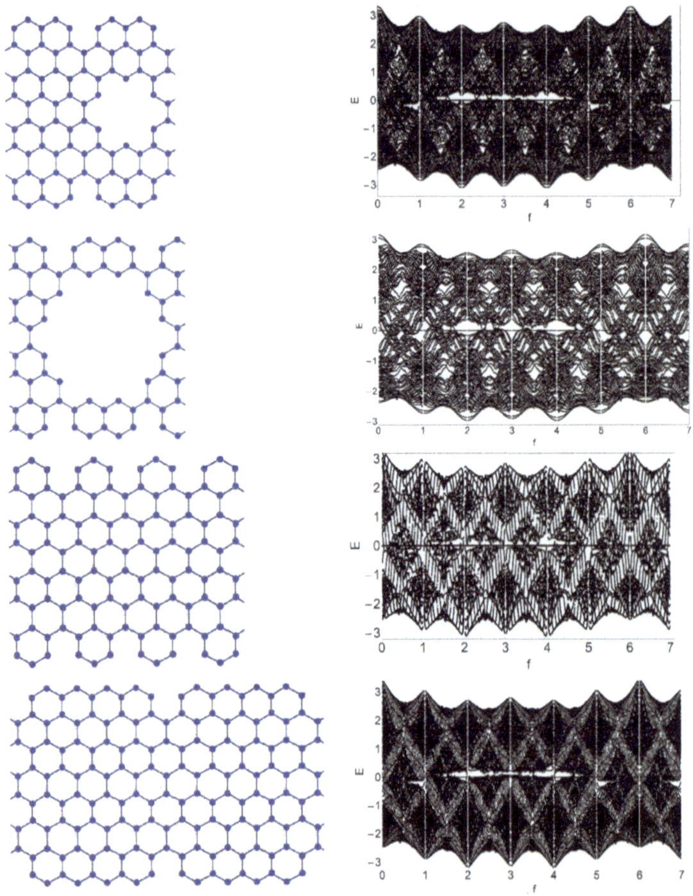

FIGURE 8. Electronic spectra of different zigzag nanoribbons which include vacancies. In the left part, the corresponding molecular surfaces are sketched.

3. Zigzag nanoribbons with atomic vacancies

We can investigate the changes of the electronic spectrum for different variations of graphene nanoribbons. Some examples are shown in Figure 8, where the electronic spectrum was depicted for 4 different forms of zigzag nanoribbons which include vacancies. The outlined graphs show an interesting property of the magnetic field: it can significantly widen the width of the HOMO-LUMO gap.

References

[1] P.R. Wallace, Phys. Rev. **71** (1947) 622.
[2] K. Wakabayashi, K. Sasaki, T. Nakanishi and T. Enoki, Sci. Technol. Adv. Mater. 11 (2010) 054504.
[3] P.G. Harper, Proc. Phys. Soc. London Sect. A 68, 874 (1955).
[4] J. Liu, Z. Ma, A.R. Wright, C. Zhang, Journal of Applied Physics, 103 (10), (2008).
[5] K. Wakabayashi, M. Fujita, H. Ajiki, and M. Sigrist, Phys. Rev. B 59, 8271 (1999).
[6] D.R. Hofstadter, Phys. Rev. B 14, 2239 (1976).

Jan Smotlacha
Bogoliubov Laboratory of Theoretical Physics
Joint Institute for Nuclear Research
141980 Dubna, Moscow region, Russia
e-mail: `smota@centrum.cz`

Richard Pincak
Institute of Experimental Physics
Slovak Academy of Sciences
Watsonova 47,043 53 Kosice, Slovak Republic
e-mail: `pincak@saske.sk`

Geometric Methods in Physics. XXXV Workshop 2016
Trends in Mathematics, 119–127

Formal Normal Forms for Germs of Vector Fields with Quadratic Leading Part. The Rational First Integral Case

Ewa Stróżyna

Abstract. We complete classification of germs of plane vector fields with quadratic leading part initiated in [1]. There were two cases completely analyzed, a simplest one and a most complex one. Here we study the remaining cases. In the proofs we use a new method introduced in the work [2] concerning the Bogdanov–Takens singularity.

Mathematics Subject Classification (2010). Primary 05C38, 15A15; Secondary 05A15, 15A18.

Keywords. Singularity of plane vector field, formal orbital normal form, non-orbital normal form.

1. Introduction

The problem of classification of germs of vector fields in $(\mathbb{C}^n, 0)$, or in $(\mathbb{R}^n, 0)$, is very natural and important. In fact, here are considered two classification problems: the usual one, when one applies local changes of coordinates, and the orbital one, when one additionally applies a reparametrisation of time (i.e., when one is interested in classification of local phase portraits).

A standard approach to this problem is the following. Usually the considered vector fields are of the form

$$V(x) = V_0(x) + \cdots$$

where V_0 is a polynomial quasi-homogeneous vector field (with respect to some grading in the space $\mathbb{C}[[x]]$ of formal power series). The changes $x = h(y)$ of variables are generated by formal vector fields $Z(x)$, $h = \exp Z$; these Z's are

Supported by Polish NCN Grant No 2012/05/B/ST1/03195.

also subject to the same quasi-homogeneous grading. The linear in Z part of the transformed vector field is

$$\mathrm{ad}_V Z = [V, Z] = \mathrm{ad}_{V_0} Z + \cdots .$$

The operator $Z \longmapsto \mathrm{ad}_{V_0} Z$ is the so-called first level homological operator. The first level normal form is defined by a choice of a space complementary to the image of ad_{V_0}. The latter task is split into finite-dimensional algebraic problems, restricting the operator ad_{V_0} to spaces of polynomial vector fields of fixed quasi-homogeneous degree.

In this paper we consider the complex plane vector fields with zero linear part

$$\dot{x} = \alpha x^2 + \beta xy + \gamma y^2 + \cdots , \quad \dot{y} = \delta x^2 + \zeta xy + \eta y^2 + \cdots \tag{1}$$

and classify them with respect to application of formal diffeomorphisms. This classification essentially depends on the homogeneous quadratic parts

$$\boldsymbol{V}_0 = \left(\alpha x^2 + \beta xy + \gamma y^2\right) \partial_x + \left(\delta x^2 + \zeta xy + \eta y^2\right) \partial_y, \tag{2}$$

with respect to the linear changes of the coordinates. This classification was performed in [1, Section 2.2]. From that classification one can see that the reduction process of the higher-order terms in Eq. (1) is done recurrently with respect to some definite quasi-homogeneous grading (in the space of power series in two variables). In most cases that grading is standard, defined by the standard **Euler vector field**

$$\boldsymbol{E} = x\partial_x + y\partial_y, \tag{3}$$

but it can be nonstandard. In this paper we deal only with the cases when only the standard grading is in use.

The division of $\boldsymbol{V}_0$'s into different cases is determined by forms of the so-called **Principal First Integral** (**PFI**) of $\boldsymbol{V}_0$. Generally, this first integral is

$$F = x^a y^b (y - x)^c , \tag{4}$$

but in some limit cases logarithmic summands can appear. Important are also so-called **Inverse Integrating Multipliers** (**IIMs**), which should be polynomial.

In [1] the case with polynomial principal first integral, i.e., with $a = p$, $b = q$, $c = r$ relatively prime positive integers, was studied completely. Also in [1] the 'road map' to treat other cases was sketched. Here we complete that task.

Let us say few words about the method used in our reduction; it was invented by the author with H. Zoladek in [2] (and used in [1]). We want to reduce a vector field of the form $\boldsymbol{V}_0 + \boldsymbol{W}$ (like in Eq. (1)) to some normal form by application of a diffeomorphism $\exp \boldsymbol{Z}$ generated by a vector field $\boldsymbol{Z}$. Recall that the linear in $\boldsymbol{Z}$ part of the action of $\exp \boldsymbol{Z}$ on $\boldsymbol{V}_0$ equals the commutator $-\mathrm{ad}_{\boldsymbol{V}_0} \boldsymbol{Z}$ plus higher-order terms. We divide the perturbations $\boldsymbol{W}$ into two parts: 'transversal' to $\boldsymbol{V}_0$ and 'tangential' to $\boldsymbol{V}_0$; also the vector fields $\boldsymbol{Z}$ are subject to such division. We measure the transversal to $\boldsymbol{V}_0$ part by the bi-vector fields $\boldsymbol{V}_0 \wedge \boldsymbol{W} = h(x, y) \cdot \partial_x \wedge \partial_y$, i.e., by one function h. The tangential to $\boldsymbol{V}_0$ part is of the form $g(x, y)\boldsymbol{V}_0$, hence it is also measured by one function g. The homological operator $\mathrm{ad}_{\boldsymbol{V}_0}$ is split into two '1-dimensional' homological operators:

$$f \longmapsto C(\boldsymbol{V}_0)f := \boldsymbol{V}_0(f), \quad f \longmapsto D(\boldsymbol{V}_0)f := \boldsymbol{V}_0(f) - \operatorname{div} \boldsymbol{V}_0 \cdot f. \tag{5}$$

In this way we realize the so-called first level reduction and obtain a first level normal form. In the second level reduction we use the homological operators $C(\boldsymbol{V}_0 + \boldsymbol{V}_1)$ and $D(\boldsymbol{V}_0 + \boldsymbol{V}_1)$ associated with two lowest degree terms from the first level normal form. In most complex cases we need four such steps.

2. Homological equations

2.1. Koszul complexes and homological operators

We deal with vector fields of the form $\boldsymbol{V} = \boldsymbol{V}_0 + \cdots$, where for $\boldsymbol{V}$ (and $\boldsymbol{V}_0$) we define some linear operators. Let

$$\mathcal{F} = \mathbb{C}[[x,y]], \quad \mathcal{Z} = \{\boldsymbol{Z} = z_1(x,y)\partial_x + z_2(x,y)\partial_y : z_i \in \mathbb{C}[[x,y]]\}$$

be the spaces of formal power series and formal vector fields. By $\mathcal{F}_d$ and $\mathcal{Z}_d$ we denote the spaces of functions and vector fields of degree d, where we put $\deg \partial_x = \deg \partial_y = -1$. We note the following identity:

$$[\boldsymbol{E}, \boldsymbol{Z}] = \deg \boldsymbol{Z} \cdot \boldsymbol{Z} \tag{6}$$

for a homogeneous vector field $\boldsymbol{Z}$.

We put

$$\begin{aligned} \mathrm{ad}_{\boldsymbol{V}} \boldsymbol{Z} &= [\boldsymbol{V}, \boldsymbol{Z}], \\ A(\boldsymbol{V})f &= f \cdot \boldsymbol{V}, \\ B(\boldsymbol{V})\boldsymbol{Z} &= \boldsymbol{V} \wedge \boldsymbol{Z} / \partial_x \wedge \partial_y, \\ C(\boldsymbol{V})f &= \boldsymbol{V}(f) = \partial f / \partial \boldsymbol{V}, \\ D(\boldsymbol{V})f &= \boldsymbol{V}(f) - \operatorname{div}(\boldsymbol{V}) \cdot f. \end{aligned} \tag{7}$$

The operators $C(\boldsymbol{V})$, $\mathrm{ad}_{\boldsymbol{V}}$ and $D(\boldsymbol{V})$ are called the **homological operators**. It turns out that the following diagram, with rows given by so-called **Koszul complexes**,

$$\begin{array}{ccccccccc} 0 & \longrightarrow & \mathcal{F} & \stackrel{A(\boldsymbol{V})}{\longrightarrow} & \mathcal{Z} & \stackrel{B(\boldsymbol{V})}{\longrightarrow} & \mathcal{F} & \longrightarrow & 0 \\ & & \downarrow C(\boldsymbol{V}) & & \downarrow \mathrm{ad}_{\boldsymbol{V}} & & \downarrow D(\boldsymbol{V}) & & \\ 0 & \longrightarrow & \mathcal{F} & \stackrel{A(\boldsymbol{V})}{\longrightarrow} & \mathcal{Z} & \stackrel{B(\boldsymbol{V})}{\longrightarrow} & \mathcal{F} & \longrightarrow & 0 \end{array} \tag{8}$$

is commutative.

It is easy to see that $\ker C(\boldsymbol{V})$ consists of **First Integrals** (**FIs**) of $\boldsymbol{V}$ and that $\ker D(\boldsymbol{V})$ consists of **Inverse Integrating Multipliers** (**IIMs**) of $\boldsymbol{V}$, i.e., of functions M such that $\operatorname{div} M^{-1}\boldsymbol{V} \equiv 0$.

2.2. The case with isolated singularity

When the origin is an isolated critical point of $\boldsymbol{V}_0$ then the above Koszul complexes are exact and we can split the reduction problem to the analysis of the operators $C(\boldsymbol{V})$ and $D(\boldsymbol{V})$. For this we have to resolve the singularity of the homogeneous vector field $\boldsymbol{V}_0$.

Recall that this resolution is a holomorphic map $\pi : (\mathbb{S}, E) \longmapsto \left(\mathbb{C}^2, 0\right)$ which is one-to-one outside the exceptional divisor $E \simeq \mathbb{CP}^1 = \pi^{-1}(0)$. We get a holomorphic foliation in the complex surface $\mathbb{S}$ such that in the general (non-dicritical) case the divisor E is invariant with three singular points (counted with multiplicities). The above singular points on E correspond to invariant lines of $\boldsymbol{V}_0$.

In our analysis of homological operators we use the blowing-up coordinates:

$$(x, u) = (x, y/x)\,.$$

These are coordinates in one chart of the surface $\mathbb{S}$ (and u is a coordinate along E); in another chart the coordinates are $(y, v) = (y, x/y)\,.$

Firstly, we recall that a homogeneous polynomial $f(x, y)$ of degree d takes the form

$$f = x^d \tilde{f}(u) \tag{9}$$

for a polynomial $\tilde{f}$. We have also

$$\boldsymbol{V}_0 = xa(u)\partial_u - x^2 b(u)\partial_x, \quad \operatorname{div} \boldsymbol{V}_0 = xc(u) \tag{10}$$

for some polynomials a, b, c.

The homological equations

$$C(\boldsymbol{V}_0)f = g, \quad D(\boldsymbol{V}_0)f = g \tag{11}$$

(for $f \in \mathcal{F}_d$ with given $g = x^{d+1}\tilde{g}(u) \in \mathcal{F}_{d+1}$) take the form

$$\begin{aligned} a(u)\frac{\mathrm{d}\,\tilde{f}}{\mathrm{d}\,u} &= db(u)\tilde{f} + \tilde{g}, \\ a(u)\frac{\mathrm{d}\,\tilde{f}}{\mathrm{d}\,u} &= [db(u) - c(u)]\,\tilde{f} + \tilde{g}. \end{aligned} \tag{12}$$

Generally, $a(u) = \text{const} \cdot u(u-1)$ and the solutions to the latter equations are of the form

$$\begin{aligned} \tilde{f}(u) &= \text{const} \cdot u^{\alpha}\,(u-1)^{\beta} \int^u \tau^{-\alpha-1}\,(\tau-1)^{-\beta-1}\,\tilde{g}(\tau)\,\mathrm{d}\,\tau, \\ \tilde{f}(u) &= \text{const} \cdot u^{\gamma}\,(u-1)^{\delta} \int^u \tau^{-\gamma-1}\,(\tau-1)^{-\delta-1}\,\tilde{g}(\tau)\,\mathrm{d}\,\tau \end{aligned} \tag{13}$$

(for some exponents $\alpha.\beta, \gamma, \delta$). The integrals in Eqs. (13) define known **Schwarz–Christoffel functions** (**SC functions**), or incomplete Schwarz–Christoffel integrals.

Recall that the solutions to Eqs. (11) should be polynomial; otherwise, the corresponding polynomial g lies outside $\operatorname{Im} C(\boldsymbol{V}_0)$ (or $\operatorname{Im} D(\boldsymbol{V}_0)$). Therefore we should localize obstacles to functions (13) to be polynomials. These obstacles are the **periods** of the SC functions defined as follows.

If $\alpha, \beta \notin \mathbb{Z}$ (respectively, $\gamma, \delta \notin \mathbb{Z}$) then the periods are defined by the following complete SC integrals:

$$\begin{aligned} \Omega_C(g) &= \mathrm{P.V.} \int_0^1 \omega_C(g), \quad \omega_C = u^{-\alpha-1} (u-1)^{-\beta-1} \tilde{g}(u)\, \mathrm{d}\, u, \\ \Omega_D(g) &= \mathrm{P.V.} \int_0^1 \omega_D(g), \quad \omega_D = u^{-\gamma-1} (u-1)^{-\delta-1} \tilde{g}(u)\, \mathrm{d}\, u. \end{aligned} \tag{14}$$

Often $\alpha, \ldots, \delta > 0$ and the above integrals diverge, so one should take some regularization; the principal value symbol P.V. means such regularization. A natural regularization uses an analytic continuation of the Euler Beta function, as a function of parameters. Anyway, we get the following explicit description of the images of our homological operators in the cases when they are of codimension 1 in $\mathcal{F}_{d+1}$:

$$\operatorname{Im} C(\boldsymbol{V}_0) = \{\Omega_C = 0\}, \quad \operatorname{Im} D(\boldsymbol{V}_0) = \{\Omega_D = 0\}. \tag{15}$$

In special situations the periods are defined in special ways. Sometimes some of these images are of codimension 2 and they are defined by vanishing of two periods, e.g., residua of $\omega_{C,D}$ at $u = 0$ and $u = 1$. We will encounter such cases.

We finish this subsection with the following identity which will be used in the sequel: we have $\boldsymbol{V}_0 \wedge \boldsymbol{E} = sxy(y-x)\partial_x \wedge \partial_y$, $s = a + b + c$, for $\boldsymbol{V}_0$ with the first integral (4), i.e.,

$$B(\boldsymbol{V}_0)\boldsymbol{E} = sxy(y-x). \tag{16}$$

Usually the orbital normal form is $\boldsymbol{V}_0 + \Phi(x,y)\boldsymbol{E}$ and the whole normal form is $(1 + \Psi(x,y))\cdot(\boldsymbol{V}_0 + \Phi\boldsymbol{E})$ for special choice of the formal series Φ and Ψ. Therefore Eq. (16) indicates that the right-hand side of second of Eqs. (11) should equal $su(u-1)\tilde{\Phi}(u)$. There are no such restrictions for the right-hand side of the first of Eqs. (11).

3. The rational PFI with 1-factor IIM case

The principal first integral is

$$F = \frac{xy}{(y-x)^r}, \ r \geq 3;$$

thus we have the vector field

$$\boldsymbol{V}_0 = x\,[(b+c)y - bx]\,\partial_x + y\,[(a+c)x - ay]\,\partial_y.$$

with $a = b = 1$, $c = -r$. When $r \neq 4$ we have **Subcase 1** otherwise we have **Subcase 2**.

3.1. The first level analysis

Lemma 1. *We have* $\ker C_d(\boldsymbol{V}_0) = 0$ *for any* d *and* $\ker D_d(\boldsymbol{V}_0) = 0$ *if* $d \neq r+1$ *and* $\ker D_{r+1}(\boldsymbol{V}_0) = \mathbb{C} \cdot M$, *where*

$$M = (2-r)\,(y-x)^{r+1}. \tag{17}$$

Proof. The first statement is obvious. Next, we have $\boldsymbol{X}_F = (y-x)^{-r-1}\boldsymbol{V}_0$ with

$$\boldsymbol{V}_0 = x\left((1-r)\,y - x\right)\partial_x + y\left((1-r)\,x - y\right)\partial_y. \qquad \square$$

Lemma 2. *The IIM* (17) *equals* $B(\boldsymbol{V}_0)\boldsymbol{T}$ *with*

$$\begin{aligned} \boldsymbol{T} &= F^{-1}\boldsymbol{E} + \frac{x+y}{(y-x)^2}F^{-1}\boldsymbol{V}_0 \\ &= (y-x)^{r-2}\left\{((2-r)\,y-(2+r)\,x)\partial_x + ((2-r)\,x-(2+r)\,y)\,\partial_y)\right\}. \end{aligned} \tag{18}$$

Moreover, we have

$$\mathrm{ad}_{\boldsymbol{V}_0}\boldsymbol{T} = 2(4-r)\,(y-x)^{r-2}\,\boldsymbol{V}_0 \tag{19}$$

(*which is nonzero for* $r \neq 4$).

Proof. The fact that $\boldsymbol{V}_0 \wedge F^{-1}\boldsymbol{E} = M\cdot\partial_x\wedge\partial_y$ follows from Eq. (16) with $s=2-r$; of course, $B(\boldsymbol{V}_0)\cdot h\boldsymbol{V}_0 = 0$. But the vector field $F^{-1}\boldsymbol{E}$ has poles along the lines $x=0$ and $y=0$. We remove these poles by adding the term proportional to $\boldsymbol{V}_0$.

Finally, the derivation of Eq. (19) uses Eq. (6), $\boldsymbol{V}_0\left((x+y)/(y-x)^2\right) = \left(x^2+y^2+2\,(3-r)\,xy\right)/\,(y-x)^3$ and $\boldsymbol{V}_0\left(F^{-1}\right)=0$. $\square$

In the first level analysis of the homological operators we use only the operators associated with $\boldsymbol{V}=\boldsymbol{V}_0$. Firstly we localize the subspaces $\mathcal{N}(C_d)$ and $\mathcal{N}(D_d)$ complementary to $\operatorname{Im} C_d$ and $\operatorname{Im} D_d$. Recall that $\dim\mathcal{N}(C_d)=1$ for any d and $\dim\mathcal{N}(D_d)=1$ for $d\neq r+1$ and $=2$ otherwise.

With $a=b=1$, $c=-r$ and $s=2-r$ we have

$$\alpha = -d\frac{1}{r-2},\ \beta = d\frac{r}{r-2},\ \gamma = -(d-3)\frac{1}{r-2}+1,\ \delta = (d-3)\frac{r}{r-2}+1,$$

$$\alpha+\beta = d + d\frac{1}{r-2},\quad \gamma+\delta = d-1+(d-3)\frac{1}{r-2}.$$

We put

$$\begin{aligned} g^C &= x^{d+1}, \\ g^D &= x^{d-1}y(y-x) \text{ if } d\neq r+1, \\ g_0^D &= y^{r+1}\left((1-r)\,x-y\right),\ \ g_1^D = x^{r+1}\left((1-r)\,y-x\right), \end{aligned} \tag{20}$$

if $d=r+1$, as a potential basis for $\mathcal{N}(C_d)$ and $\mathcal{N}(D_d)$. Note that

$$g_0^D = B(\boldsymbol{V}_0)y^r\partial_x,\ g_1^D = B(\boldsymbol{V}_0)x^r\partial_y.$$

We get the form $\omega_C(g^C) = \frac{\mathrm{d}\,u}{u^{\alpha+1}(u-1)^{\beta+1}}$; for $d/\,(r-2)$ non-integer, its period $\Omega_C\left(g^C\right) = \text{const}\cdot B(\alpha,\beta)\neq 0$. If $d/\,(r-2)=m\in\mathbb{Z}$ then we get the function

$$\tilde{f}^C = \frac{1}{2-r}\frac{(u-1)^{mr}}{u^m}\int^u \frac{\tau^{m-1}\,\mathrm{d}\,\tau}{(\tau-1)^{mr+1}}$$

The residuum of $\omega_C(g^C)$ at $u=1$ vanishes; therefore the correct period is

$$\Omega_C\left(g^C\right) = \int_\infty^0 \omega_C\left(g^C\right)\neq 0.$$

If $(d-3)/(r-2) \notin \mathbb{Z}$ then the corresponding period $\Omega_D(g^D) \neq 0$. If $d-3 = m(r-2)$, $m \in \mathbb{Z}$, then we arrive at the function

$$\tilde{f}^D = \frac{1}{2-r} \frac{(u-1)^{mr+1}}{u^{m-1}} \int^u \frac{\tau^{m-2}}{(\tau-1)^{mr+2}} \, \mathrm{d}\tau.$$

For $m > 1$ the above argument with the function $\tilde{f}^C$ works, the unique period $\Omega_D(g^D) = \int_\infty^0 \omega_D(g^D) \neq 0$.

But for $m = 1$, i.e., $d = r+1$, we have two generators, g_0^D and g_1^D, and we define two periods $\Omega_D^{0,1}(g_j^D) = \mathrm{Res}_{u=0,1}\, \omega_D(g_j^D)$, $j = 0, 1$. We have

$$\omega_D(g_0^D) = \frac{(1-r)u - 1}{u(u-1)^{r+2}} \, \mathrm{d}u, \quad \omega_D(g_1^D) = \frac{u^r(u+r-1)}{(u-1)^{r+2}} \, \mathrm{d}u.$$

and we define the **period matrix**

$$\begin{pmatrix} \Omega_D^0(g_0^D) & \Omega_D^0(g_1^D) \\ \Omega_D^1(g_0^D) & \Omega_D^1(g_1^D) \end{pmatrix}. \tag{21}$$

Since $\Omega_D^0(g_0^D) = (-1)^{r-1} \neq 0$, $\Omega_D^1(g_1^D) = 1 \neq 0$ and $\Omega_D^0(g_1^D) = 0$, this matrix takes triangular form, with nonzero entries at the diagonal, and hence is nondegenerate.

But it is not the end of the first level analysis. We have not yet used the kernel of $D_{r+1}(\boldsymbol{V}_0)$ generated by $(y-x)^{r+1}$, via the vector field $\boldsymbol{T}$ from Lemma 2. If the orbital normal form differs from $\boldsymbol{V}_0$ then we can use this $\boldsymbol{T}$ to cancel higher-order terms from the orbital normal form. But, when the orbital normal form is $\boldsymbol{V}_0$, then we get the term

$$(4-r)(y-x)^{r-2}\boldsymbol{V}_0 \neq 0$$

for $r \neq 4$; for $r - 4$ we get nothing. It turns out that the function $g = (y-x)^{r-2}$ lies outside $\mathrm{Im}\, C_d(\boldsymbol{V}_0)$, $d = r-3$.

Indeed, the corresponding period $\Omega_C(g) = \mathrm{P.V.} \int_0^1 u^{-\alpha-1}(u-1)^{-\beta-1} du$, with $\alpha = -(r-3)/(r-2)$, $\beta = r(r-3)/(r-2)$, $\alpha + \beta \notin \mathbb{Z}$, is nonzero.

Therefore we can present now:

the first level normal forms in the **Rational PFI with 1-Factor IIM Case**:

$$\begin{gathered} (1+\psi(x))(\boldsymbol{V}_0 + \mathbf{U} + \varphi(x)\boldsymbol{E}), \\ \boldsymbol{U} = ay^r\partial_x + bx^r\partial_y, \\ \mathcal{I}(\varphi) = \mathbb{Z}_{\geq 2}, \ \ \mathcal{I}(\psi) = \mathbb{Z}_{\geq 1}; \end{gathered} \tag{22}$$

where $\mathcal{I}(\varphi)$ and $\mathcal{I}(\psi)$ are the indices' set for the series $\varphi = \sum_{i \in \mathcal{I}(\varphi)} a_i x^i$ and $\psi = \sum_{i \in \mathcal{I}(\psi)} b_i x^i$, or

$$\begin{gathered} (1+\psi(x))\boldsymbol{V}_0, \\ \mathcal{I}(\psi) = \mathbb{Z}_{\geq 1} \setminus \{r-2\} \ \ (\textbf{Subcase 1}), \\ \mathcal{I}(\psi) = \mathbb{Z}_{\geq 1} \ \ (\textbf{Subcase 2}) \end{gathered} \tag{23}$$

(*the latter form is complete*).

3.2. The second level analysis

In this section we study the homological operators associated with vector fields of the form

$$\boldsymbol{V} = \boldsymbol{V}_0 + \boldsymbol{V}_1,$$

where $\boldsymbol{V}_1$ is a homogeneous vector field of lowest degree $\deg \boldsymbol{V}_1 > 1$ which was not reduced in the first level analysis. We have to consider several possibilities:

$$\boldsymbol{V}_1 = ax^k \boldsymbol{E} \quad \text{or} \quad \boldsymbol{V}_1 = ay^r \partial_x + bx^r \partial_y \tag{24}$$

(from the orbital normal form) or

$$\boldsymbol{V}_1 = cx^l \boldsymbol{V}_0 \tag{25}$$

(associated with the orbital factor). Of course, it is possible that there appear terms (24) and (25) simultaneously, of the same degree or of different degrees. However, we prefer to consider actions of the homological operators associated with them separately, with the orbital normal form priority. Therefore, the case (24) is used when terms of the form (25) are present, moreover, even with degree smaller than the degree of (24). The terms (25) are used when the orbital normal form is $\boldsymbol{V}_0$.

Our analysis is essentially reduced to the operator $D(\boldsymbol{V})$ acting on functions of the form $\xi M + f$, where $\xi \in \mathbb{C}$ and $M = (y - x)^{r+1}$ is the generator of $\ker D_{r+1}(\boldsymbol{V}_0)$, and followed by projection onto a space of homogeneous functions. Therefore we get the following **second level homological operator**:

$$\begin{aligned} \widetilde{D}(\boldsymbol{V}) &: \mathbb{C} \oplus \mathcal{F}_d \longmapsto \mathcal{F}_{d+1}, \\ (\xi, f) &\longmapsto \xi D(\boldsymbol{V}_1) M + D(\boldsymbol{V}_0) f, \end{aligned} \tag{26}$$

where $d = k + r = \deg \boldsymbol{V}_1 + r$. This operator acts between spaces of the same dimension.

The complete normal form in the **Rational PFI with 1-Factor IIM Case** *with* $\boldsymbol{V}_1 = ax^k \boldsymbol{E}$ *is: either*

$$\begin{gathered} (1 + \psi(x)) \left(\boldsymbol{V}_0 + \boldsymbol{V}_1 + bx^r \partial_y + \varphi(x) \boldsymbol{E} \right), \quad k < r - 1, \\ \mathcal{I}(\psi) = \mathbb{Z}_{\geq 1}, \quad \mathcal{I}(\varphi) = \mathbb{Z}_{>k} \setminus \{r - 1, k + r - 2\} \end{gathered} \tag{27}$$

or

$$\begin{gathered} (1 + \psi(x)) \left(\boldsymbol{V}_0 + \boldsymbol{V}_1 + \varphi(x) \boldsymbol{E} \right), \quad k > r - 1, \\ \mathcal{I}(\psi) = \mathbb{Z}_{\geq 1}, \quad \mathcal{I}(\varphi) = \mathbb{Z}_{>k} \setminus \{k + r - 2\} \end{gathered} \tag{28}$$

The second level normal form in the **Rational PFI with 1-Factor IIM Case** *with* $\boldsymbol{V}_1 = ay^r \partial_x + bx^r \partial_y$ *is: either*

$$\begin{gathered} (1 + \psi(x)) \left(\boldsymbol{V}_0 + \boldsymbol{V}_1 + \varphi(x) \boldsymbol{E} \right), \\ a + (-1)^{r+1} b \neq 0, \\ \mathcal{I}(\varphi) = \mathbb{Z}_{\geq r} \setminus \{2r - 3\}, \quad \mathcal{I}(\psi) = \mathbb{Z}_{\geq 1}, \end{gathered} \tag{29}$$

(this form is complete) or

$$(1+\psi(x))\,(\boldsymbol{V}_0+\boldsymbol{V}_1+\varphi(x)\boldsymbol{E})\,,$$
$$\boldsymbol{V}_1=a(y^r\partial_x+(-1)^rx^r\partial_y),\tag{30}$$
$$\mathcal{I}(\varphi)=\mathbb{Z}_{\geq r},\ \ \mathcal{I}(\psi)=\mathbb{Z}_{\geq 1},$$

or

$$(1+\psi(x))\,(\boldsymbol{V}_0+\boldsymbol{V}_1)\,,$$
$$\boldsymbol{V}_1=a(y^r\partial_x+(-1)^rx^r\partial_y),\quad \mathcal{I}(\psi)=\mathbb{Z}_{\geq 1}\setminus\{r-2\}, (\textbf{Subcase 1})\tag{31}$$

(this form is complete) or

$$(1+\psi(x))\,(\boldsymbol{V}_0+\boldsymbol{V}_1)$$
$$\boldsymbol{V}_1=a(y^4\partial_x+x^4\partial_y),\quad \mathcal{I}(\psi)=\mathbb{Z}_{\geq 1},\ (\textbf{Subcase 2}).\tag{32}$$

3.3. Third level

We consider homological operators associated with the vector fields $\boldsymbol{V}=\boldsymbol{V}_0+\boldsymbol{V}_1+\boldsymbol{V}_2$ such that $\boldsymbol{V}_0+\boldsymbol{V}_1$ has nontrivial IIM, i.e.,

$$\boldsymbol{V}_1=a(y^r\partial_x+(-1)^rx^r\partial_y).\tag{33}$$

In fact we are left with two subcases.

The complete normal form in the **Rational PFI with 1-Factor IIM Case** *with* $\boldsymbol{V}_1=a(y^r\partial_x+(-1)^rx^r\partial_y)$ *and* $\boldsymbol{V}_2=cx^l\boldsymbol{E}$ *is:*

$$(1+\psi(x))\,(\boldsymbol{V}_0+\boldsymbol{V}_1+\boldsymbol{V}_2+\varphi(x)\boldsymbol{E})\,,$$
$$\mathcal{I}(\varphi)=\mathbb{Z}_{>l}\setminus\{l+r-2\}\,,\ \ \mathcal{I}(\psi)=\mathbb{Z}_{\geq 1}.\tag{34}$$

The complete normal form in the **Subcase 2** *of the* **Rational PFI with 1-Factor IIM Case** *with* $\boldsymbol{V}_1=a(y^4\partial_x+x^4\partial_y)$ *and* $\boldsymbol{V}_2=cx^j\boldsymbol{V}_0$ *is: either*

$$\left(1+cx^j+\psi(x)\right)(\boldsymbol{V}_0+\boldsymbol{V}_1)\,,\quad \mathcal{I}(\psi)=\mathbb{Z}_{>j}\setminus\{j-2\},\tag{35}$$

or

$$\boldsymbol{V}_0+\boldsymbol{V}_1.\tag{36}$$

References

[1] E. Stróżyna, *Normal forms for germs of vector fields with quadratic leading part. The polynomial first integral case.* J. Differential Equations **259** (2015), 6718–6748.

[2] E. Stróżyna and H. Żołądek, *The complete formal normal form for the Bogdanov-Takens singularity*, Moscow Mathematical J. **15** (2015), 141–178.

Ewa Stróżyna
Faculty of Mathematics and Information Science
Warsaw University of Technology
Koszykowa 75
00-662 Warsaw, Poland
e-mail: E.Strozyna@mini.pw.edu.pl

Geometric Methods in Physics. XXXV Workshop 2016
Trends in Mathematics, 129–144

The Poncelet Theorems in Interpretation of Rafał Kołodziej

Henryk Żołądek

I dedicate this article to the memory of Rafał Kołodziej

Abstract. We present some results of R. Kołodziej related with Poncelet's theorems and several proofs of the Great Poncelet Theorem.

Mathematics Subject Classification (2010). primary 37A05; secondary 37E10, 14H52.

Keywords. Billard, Poncelet Theorems, elliptic curve.

1. Introduction

The principal aim of this paper is to present an interesting construction of an invariant measure for certain transformation of an ellipse. This construction is due Rafał Kołodziej (1956–2011) and has turned out useful in some classical geometric problems related with investigations of the nineteenth century mathematician Jean-Victor Poncelet.

The transformation is associated with a pair of ellipses Γ and Δ such that Δ lies inside Γ. From a point $p = p_0 \in \Gamma$ we can draw two straight lines tangent to Δ; we choose one of them $L = L(p)$ in the direction compatible with an orientation of Γ. Then the second point p_1 of intersection of L with Γ is the image of the transformation

$$T : \Gamma \longmapsto \Gamma,$$

$p_1 = T(p)$. We shall call it the **Poncelet map.**

Iterating this map we get a broken line $p_0 p_1 p_2 \ldots$ such that $p_2 = T(p_1)$, $p_3 = T(p_2)$, etc. It is inscribed in Γ and described on Δ.

The following result is well known (see [1]).

Supported by Polish OPUS grant No 2012/05/B/ST1/03195.

Theorem 1 (Small Poncelet Theorem). *In the case when the ellipses are confocal the above-defined broken line is a trajectory of the billiard in* Γ.

Kołodziej [9] computed the rotation number of the map T in the case of billiard in an ellipse (see Section 2.3).

More interesting is the following

Theorem 2 (Great Poncelet Theorem). *If, for a one starting point* p_0 *the above-broken line is closed, i.e., is equal* $p_0p_1p_2\dots p_{n-1}$ *and* $T^n(p_0) = p_0$, *then the* n^{th} *iteration of the whole map is the identity,* $T^n(p) = p$ *for any* $p \in \Gamma$. *In other words the polygon* $p_0p_1p_2\dots p_{n-1}$ *can be moved in such a way that it remains inscribed in* Γ *and described on* Δ[1].

Nowadays the latter theorem has many proofs and it has acquired the surname 'the Poncelet porism'. For recent results related with Theorem 2 we recommend the reader the review paper [5] by V. Dragović and M. Radnović.

In the next section we present the Kołodziej proof and in Section 3 we present several other interesting proofs of this theorem.

2. Constructions of Kołodziej

2.1. The case with two circles

Assume that Γ and Δ are circles.

Let pq and $p'q'$ be straight segments with endpoints at the circle Γ and tangent to Δ; moreover, p' is infinitesimally close to p. The point r of intersection of these segments is close to the tangency points of the segments pq and $p'q'$ to the circle Δ (see Figure 1).

[1] Jean-Victor Poncelet (1788–1867) discovered this theorem during the period of his imprisonment in Russia (1813–1814) after the Napoleonic campaign.

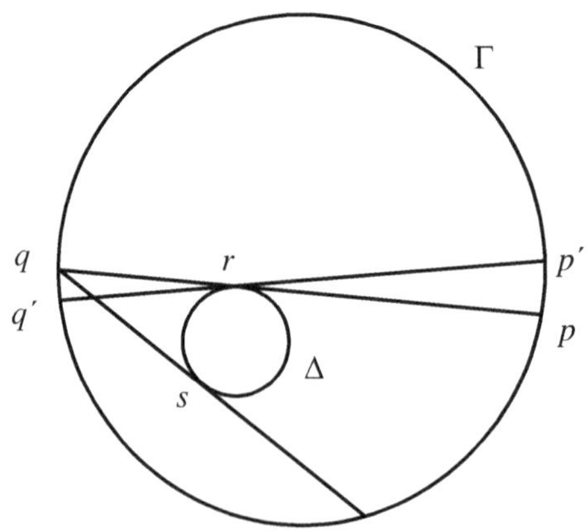

FIGURE 1. The Poncelet map for two circles.

Note that the sides pp' and $p'r$ of the triangle $\Delta pp'r$ and the sides $q'q$ and qr of the triangle $\Delta q'qr$ are based on the same arc joining p and q' in Γ. Therefore these triangles are similar, because they have also the same angles at the vertex r.

This yields the relation

$$\frac{|pp'|}{|pr|} = \frac{|qq'|}{|q'r|}. \tag{1}$$

In the limit case $p' \to p$ the lengths of the segments pp' and qq' can be replaced with the lengths of the corresponding arcs in the circle Γ. Moreover, we can assume $|q'r| = |qr|$. Here we have used the property that Γ is a circle.

The second observation uses the assumption that Δ is a circle. It says that

$$|qr| = |qs|,$$

where s is the point of tangency for the other line drawn from the point q and tangent to Δ.

This implies the following

Lemma 1 (Kołodziej). *The measure*

$$\mu = \frac{d\ell(p)}{|pr|},$$

where $d\ell$ is the Lebesgue measure on the circle Γ and r is the tangency point of the line $L(p)$ with the circle Δ, is invariant with respect to the map T, i.e., $|pp'| / |pr| \approx |qq'| / |qs|$.

Proof of the Great Poncelet Theorem for circles. Note that the density

$$\rho(p) = 1/ |pr|$$

of the measure μ is separated from zero and from infinity. The property of admitting such a measure by a map implies its conjugation with the rotation map.

Indeed, choosing a point $p_* \in \Gamma$ we can define the homeomorphism

$$H(p) = \int_{p_*}^{p} \rho d\ell = \mu\left([p_*, p]\right)$$

between Γ and $\mathbb{R} \bmod \Lambda$, where $\Lambda = \mu(\Gamma)$ is the 'mass' of Γ. Then the identities

$$H(T(p)) = \mu\left([p_*, T(p_*)]\right) + \mu\left(T([p_*, p])\right) = \lambda + H(p)$$

demonstrate that the map

$$T_1 = H \circ T \circ H^{-1}(x) = x + \lambda$$

is a shift by $\lambda = \mu\left([p_*, T(p_*)]\right)$; modulo Λ it is a rotation.

If the number λ/Λ is irrational the rotation T_1 is non-periodic, and if $\lambda/\Lambda \in \mathbb{Q}$ then the rotation T_1 (and the map T) are periodic. The second eventuality takes place under the hypothesis of the Great Poncelet Theorem. $\square$

The above proof can also be found in the book of A. Boyarsky and P. Góra [2].

2.2. Reduction to the case with two spheres

A real conic curve is a circle if and only if its complex projective version intersects the line at infinity in two concrete points: $[1 : i : 0]$ and $[1 : -i : 0]$, $i = \sqrt{-1}$. Indeed, the equation for a circle $(x-a)^2+(y-b)^2 = r^2$, rewritten in the homogeneous coordinates $[x_1 : x_2 : x_3]$ in $\mathbb{CP}^2$, takes the form $(x_1 - ax_3)^2 + (x_2 - bx_3)^2 = r^2 x_3^2$ and the line at infinity L_∞ is given by the equation $x_3 = 0$.

Two ellipses Γ and Δ, as above, have four points of intersection in $\mathbb{CP}^2$. Moreover, these points are not real and are divided into two pairs of mutually conjugate points.

Let us choose one such pair. The line passing through these two points is real in the sense that its defining equation has real coefficients. Therefore we can apply a suitable projective transformation such that it sends this line to the line at infinity; this transformation is real, i.e., is defined by a coset $[A] \in PSL(3,\mathbb{R})$ of a real matrix A.

The two points of the intersection $\Gamma \cap \Delta \cap L_\infty$ are of the form $[1 : \alpha \pm i\beta : 0]$, i.e., the equations of the both ellipses are of the form

$$(y - \alpha x)^2 + \beta^2 x^2 + \cdots = 0.$$

It is clear that, after some linear change, they become reduced to two circles.

Note also that the projective transformations send lines to lines. Therefore the Poncelet problem with ellipses is transformed to an analogous problem with circle.

Remark 1. In the Kołodziej paper [9] a similar construction is realized but only in the case of confocal ellipses[2].

First, one obtains a situation when the exterior ellipse Γ is a circle concentric with the internal ellipse Δ (we forget about the foci).

Next, one embeds the plane $\mathbb{R}^2$ (with Γ and Δ) into the 3-dimensional space $\mathbb{R}^3$. One takes the sphere $\mathbb{S} = \mathbb{S}^2 \subset \mathbb{R}^3$ with the equator along Γ. One considers planes $\mathbb{P}$ tangent to $\mathbb{S}$ and projections $\pi = \pi_{\mathbb{P}}$ onto $\mathbb{P}$ from the points $t = t_{\mathbb{P}} \in \mathbb{S}$ antipodal to the tangency points of the planes with the sphere.

The image of the circle $\Gamma = \mathbb{S} \cap \mathbb{R}^2$ is always the circle $\pi(\Gamma) \subset \mathbb{P}$, because $\pi|_{\mathbb{S}}$ is a stereographic projection. When we vary the plane $\mathbb{P}$ such that the point t moves along the meridian of the sphere corresponding to direction of the longer axis of the ellipse Δ, then we encounter a situation when $\pi(\Delta)$ becomes a circle. To see this one has to compare the situation when t is a south pole of the sphere with the situation when t lies near the equator Γ. (In [9] one can find an interpretation of the above projective transformation in terms of a suitable Lorenz map.)

In fact, it is not very difficult to modify Kołodziej's construction above to the general situation, i.e., when the circle Γ and the ellipse Δ are not concentric.

[2] Kołodziej in [9] did not consider the general case. His principal aim was to compute the rotation number for the billiard map in an ellipse. It was the problem proposed to him by his master thesis supervisor Maciej Wojtkowski; Kołodziej solved it in an excellent way. Also in [9] nothing is said about the Great Poncelet Theorem.

Its core is to get a situation when the ellipses become concentric. M. Wojtkowski proposed the following approach.

Let the equations $(Gx, x) = 0$ and $(Dx, x) = 0$, $x \in \mathbb{R}^3$, define the ellipses Γ and Δ in the homogeneous coordinates. If we were able to diagonalize simultaneously these quadratic forms then we would be done. It is well known that such diagonalization exists when one of the forms is definite (positively or negatively): then we reduce that form to $\pm(x, x)$ and the second one is diagonalizable in standard way (via an orthonormal eigenbasis). Unfortunately, none of our forms is definite. But their difference $((G - D)x, x)$ is such a form, because the ellipses Γ and Δ do not intersect in the real domain. So, we diagonalize simultaneously the forms $((G - D)x, x)$ and (Gx, x), then the form (Dx, x) will be diagonal too.

Remark 2. The idea of using an invariant measure to prove the Great Poncelet Theorem has appeared also in works of other authors. It is worth to mention the paper [8] by J. King from 1994 (much later than Kołodziej's work [9]).

We find there Kołodziej's measure from Lemma 1 in the case of two circles. But King uses only affine transformations and is not able to reduce simultaneously two ellipses to circles. For this reason his construction is performed in two steps.

First, he assumes that Γ is a circle and he gets Eq. (1). Hence the condition of invariance of the measure $\mu = \rho(p)\mathrm{d}\ell(p)$ takes the following form:

$$\frac{\rho(q)}{\rho(p)} = \frac{|qr|}{|pr|}$$

(see Figure 1). Next, he applies an affine map which reduces the ellipse Δ to a circle $\widetilde{\Delta}$. The images of the corresponding points p, q, r, s from Figure 1 will be denoted by $\tilde{p}, \tilde{q}, \tilde{r}, \tilde{s}$. It turns out that $|\tilde{q}\tilde{r}| / |\tilde{p}\tilde{r}| = |qr| / |pr|$. But we have also $|\tilde{q}\tilde{r}| = |\tilde{q}\tilde{s}|$. Thus our measure takes the form

$$\mu(p) = \mathrm{d}\ell(p) / |\tilde{p}\tilde{r}| .$$

In King's paper one can find informations about earlier works (of Jacobi, Bertrand, Schonberg) where the idea of invariant measure was raised up but not much convincingly.

2.3. Billiard in an ellipse and another reduction to circles

Recall that a billiard trajectory in a domain $\Omega \subset \mathbb{R}^2$ with boundary $\Gamma = \partial\Omega$ consists of straight segments in Ω with endpoints in Γ satisfying the condition: the angle of incidence equals the angle of reflection. Each segment is characterized by the starting point $p \in \Gamma$ and the angle $\theta \in [-\pi/2, \pi/2]$ between the segment and the normal to Γ in p.

With a billiard one associates the so-called **billiard map**

$$S : \Gamma \times [-\pi/2, \pi/2] \longmapsto \Gamma \times [-\pi/2, \pi/2] ,$$

defined by the final point of a segment and by the angle of the next segment. It is well known that the billiard map preserves the following measure (see [12]):[3]

$$\lambda = \cos\theta \mathrm{d}\ell(p)\mathrm{d}\theta. \tag{2}$$

Assume that the ellipses Γ and Δ are confocal with the foci a and b. As we have mentioned in Introduction, the Poncelet broken line is a billiard trajectory in the domain bounded by Γ. We can say more. The Small Poncelet Theorem says that the billiard map in an ellipse is integrable[4]. This means that the orbits of the map S lie on level curves of some function $H(p,\theta)$, a **first integral**.

Any such level is defined by fixing an ellipse Δ confocal with Γ. The ellipse $\Delta = \Delta_\beta$ is precisely determined by its eccentricity β =(distance between foci)/(length of main axis). Formally we can write

$$H(p,\theta) = \beta$$

(It is reasonable to take into account also the case when $\Delta = \Delta_\beta$ is a hyperbola confocal with Γ, see [9].)

Let us come back to our confocal ellipses Γ and Δ. Denote by α and β their eccentricities (the ratios between $|ab|$ and the lengths of the main axes). Let $\Phi : \mathbb{R}^2 \diagdown b \longmapsto \mathbb{R}^2 \diagdown b$ be the inversion map with respect to center in the focus b.

Consider four circles: A with center in the focus a and with radius $|ab|/\alpha$ (the length of the longer axis of Γ), B with center in a and with radius $|ab|/\beta$ and their inversions $C = \Phi(A)$ and $D = \Phi(B)$ (see Figure 2).

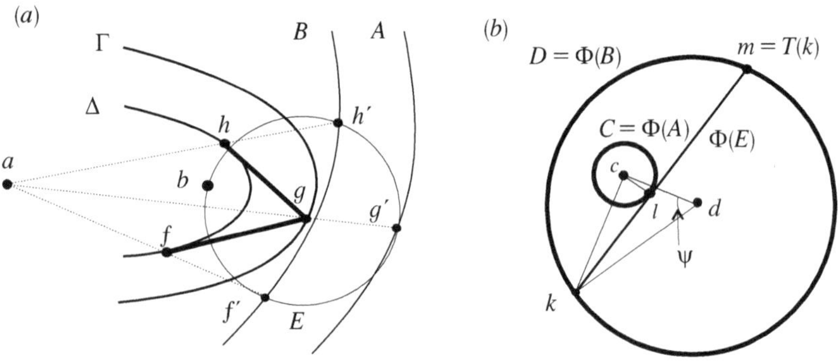

FIGURE 2. Kołodziej's construction.

[3]This measure is derived from the property of invariance of the 2-dimensional Lebesgue measure for the billiard flow.

Consider a stream of particles, with constant density 1, constant velocity 1 and falling locally at an angle $\pi/2 - \theta$ onto Γ. Then in unit of time on the interval $I \subset \Gamma$ with length ds falls $\cos\theta$ds particles of the stream. The same number of particles is reflected from I and then falls on the next interval $I_1 \subset \Gamma$ at the angle $\pi/2 - \theta_1$.

[4]For a proof we refer to the monograph [12] of I. Kornfeld, Y. Sinai and S. Fomin. Another, 'mechanical' proof can be found in the book [10] of V. Kozlov and D. Treshchev.

Lemma 2 (Kołodziej). *With any broken line fgh, such that $f, h \in \Delta$, $g \in \Gamma$ and the lines fg and gh are tangent to Δ, we can associate a segment km which has endpoints in D and is tangent to C.*

This means that the billiard map S in Γ restricted to the level curve $\{H = \beta\}$ is conjugated to the Poncelet map associated with the circles C and D.

Proof. Let f' and h' be the projections of f and h on the circle B from center at the focus a and let g' be the analogous projection of the point g on the circle A (Figure 2(a)). It turns out that the circle E through f', g' and h' passes also through the focus b (see [9]). Hence the image $\Phi(E)$ of this circle is the straight line along the segment km, where $k = \Phi(f'), m = \Phi(h') \in D$ and the point $l = \Phi(g') \in km \cap C$ (Figure 2(b)). □

Below we cite (without proofs) some formulas from [9]. First, the lengths of the radii of the circles C and D equal $r_C = |ab|^{-1} \alpha/(1-\alpha^2)$ and $r_D = |ab|^{-1} \beta/(1-\beta^2)$ respectively. Let c and d be the centers of these circles. It turns out that $|cd| = |ab|^{-1} \left(\beta^2 - \alpha^2\right)/(1-\alpha^2)(1-\beta^2)$. Finally, from trigonometric formulas for the triangles Δklc and Δkcd one gets the following expression for the length of the segment kl with $k \in C$ and tangent to D in l :

$$|kl|^2 = \text{const} \cdot \left(1 - \kappa^2 \sin^2\left(\left(\pi - \psi\right)/2\right)\right), \quad \kappa = 2\sqrt{\beta}/(1+\beta), \tag{3}$$

where $\psi = \measuredangle kdc$ and the constant does not depend on the angle ψ (see Figure 2(b)).

On the other hand, from Lemma 1 we know that $1/|kl|$ is the density of the measure invariant for the Poncelet map T (which is conjugated to the billiard map $S|_{\{H=\beta\}}$). Moreover, the Lebesgue measure on the circle D is proportional to $d\psi$. Therefore the probabilistic measure invariant with respect to the map T equals

$$\nu(\psi) = \frac{\mathrm{d}\psi}{\sqrt{1-\kappa^2\sin^2\left(\left(\pi-\psi\right)/2\right)}} \Bigg/ \int_0^{2\pi} \frac{\mathrm{d}\psi}{\sqrt{1-\kappa^2\sin^2\left(\left(\pi-\psi\right)/2\right)}}.$$

Choosing the initial angle $\psi = \varphi$ corresponding to the situation where the segment km is parallel to the segment cd (Figure 2(b)) we find the final angle, equal $\pi - \varphi$. The integral $\int_\varphi^{\pi-\varphi} \mathrm{d}\nu(\psi)$ equals the rotation number of the map T, as well as of the map $S|_{\{H=\beta\}}$. Evaluation of this integral leas to the following result.

Theorem 3 (Kołodziej). *The above-mentioned rotation number equals*

$$\frac{1}{2} - \frac{F(\varphi/2,\kappa)}{F(\pi/2,\kappa)},$$

where $F(\chi,\kappa) = \int_0^\chi d\psi/\sqrt{1-\kappa^2\sin^2\psi}$ is an elliptic integral

$$\sin\varphi = \alpha\left(1-\beta^2\right)/\beta(1-\alpha^2)$$

and κ is defined in Eq. (3).

Second proof of the Great Poncelet Theorem (GPT). . Taking into account existence of the invariant measure (2) and of the first integral we are able to give another proof of the GPT (probably coming from G. Birkhoff).

Let us treat the measure λ as a differential 2-form: $\lambda = \cos\theta \mathrm{d}x \wedge \mathrm{d}\theta$ where x is the length parameter on Γ. Then we can define the so-called **Gelfand–Leray form**

$$\eta = \lambda / \mathrm{d}H,$$

which satisfies the condition $\eta \wedge \mathrm{d}H = \lambda$ and is uniquely determined on each level curve of the function H. Because the both objects, the form λ and the function H, are invariant with respect to the billiard map S, also the measure $\eta|_{\{H=\beta\}}$ is invariant for the map $S|_{\{H=\beta\}}$. Therefore we can use the arguments from Section 2.1.

On the other hand, the analysis from Section 2.2 demonstrates that the group of projective transformations of the plane is sufficiently large to transform any pair of ellipses in generic position (4 complex intersection points) to a pair of confocal ellipses. □

3. Other proofs of the Great Poncelet Theorem

3.1. Proof of Tabachnikov

The idea of this proof is close to the proof from Section 2.3.

Proof. For two nested ellipses, Γ and Δ, we define some **pencil of conics**. If

$$F(x, y) = 0 \quad \text{and} \quad G(x, y) = 0$$

are the equations defining Γ and Δ then the equations $\lambda F(x, y) + \mu G(x, y) = 0$ define all curves from the pencil. Each curve from the pencil passes through the four points of the intersection $\Gamma \cap \Delta$; they are nonreal. Sergey Tabachnikov [15] denotes $\Gamma = \Gamma_1$, $\Delta = \Gamma_0$ and introduces one more curve Γ_∞ from the pencil such that Γ_∞ surrounds Γ_1 and Γ_0.

In the domain Ω_∞ bounded by Γ_∞ he introduces the hyperbolic distance $\mathrm{dist}(p, q) = |\log [p, q, r, s]|$, where $[\cdot]$ denotes the cross ratio of four points in the line pq, from which r are s the points of the intersection of the line pq with Γ_∞. This metric provides an isomorphism of the domain Ω_∞ with the hyperbolic plane (the so-called Klein–Beltrami model) and defines a hyperbolic measure in Ω_∞.

Tabachnikov defines a map T of the ring $\Omega_{0,\infty}$ between the curves Γ_0 and Γ_∞ as follows. From a point $p \in \Omega_{0,\infty}$ one draws a straight line $L = L(p)$, tangent to Γ_0 in a point r, and chooses a point $p_1 = T(p) \in L \setminus p$ via the equality $\mathrm{dist}(p_1, r) = \mathrm{dist}(p, r)$. It turns out that the map T has the following properties:

- It preserves the hyperbolic measure.
- Each ellipse in Ω from our pencil is invariant with respect to T. In particular, the map T restricted to Γ_1 is the Poncelet map.

These properties imply the existence of an invariant measure of the Gelfand–Leray type for $T|_{\Gamma_1}$, and this implies the GPT. □

3.2. Proof with use of elliptic curves

This proof is probably the best disseminated. One can find it in Wikipedia, but we shall refer rather to the paper [6].

Proof. We shall consider the complex and projective versions of the curves Γ and Δ. As Riemann surfaces they have zero **genus**, $g = 0$, and are homeomorphic with the two-dimensional sphere. Also it is standard that the space of projective lines tangent to Δ forms the so-called **dual curve** Δ^*, which is also of degree 2 and has zero genus.

Consider the following set

$$E = \{(p, L) : p \in \Gamma,\ L \in \Delta^*,\ p \in L\} \subset \Gamma \times \Delta^*. \tag{4}$$

It is a projective complex algebraic curve. In order to compute its genus we consider the natural projection $\pi : E \longmapsto \Gamma$,

$$(p, L) \longmapsto p.$$

It is a ramified covering of degree $d = 2$, because from a typical point in Γ come out two lines tangent to Δ. Moreover, it has four ramification points r_0, r_1, r_2 and r_3 in E corresponding to intersection points q_0, q_1, q_2 and q_3 of the curves Γ and Δ (we have only one tangent line in q_j to Δ). Of course, the ramification indices of these points equal $\nu_j = 2$. Now we use the classical **Riemann–Hurwitz formula**

$$d \cdot \chi(\Gamma) = \chi(E) + \sum (\nu_j - 1),$$

where $\chi = 2 - 2g$ is the **Euler characteristic**.[5] In our case we get $\chi(E) = 0$, i.e., E is homeomorphic with a torus. Therefore E is the so-called **elliptic curve**.

Besides the above description of an elliptic curve (as a two-fold covering of $\mathbb{CP}^1$ with four ramification points) there exists its another model:

$$E \simeq \mathbb{C}/\Lambda, \tag{5}$$

where $\Lambda = \mathbb{Z}\omega_1 + \mathbb{Z}\omega_2$ is the lattice generated by the **periods** ω_1 and ω_2.

The Poncelet map in E (we denote it by $\mathcal{T}$) can be represented as a composition of two involutions, $\mathcal{T} = \iota_2 \circ \iota_1$. Here the map ι_1 relies upon exchanging the point p in (p, L) with another intersection point of L with the curve Γ, whereas the map ι_2 replaces the line L in (p, L) with another line tangent to Δ and passing through p.

Each of the involutions $\iota_j : E \longmapsto E$ is lifted to a corresponding involution of the universal covering, i.e., to $\mathbb{C}$. But it is easy to see that then we obtain affine and holomorphic diffeomorphisms of $\mathbb{C}$ and that they are of the form $u \longmapsto -u + v_j$. This gives

$$\mathcal{T}(u) = u + w,$$

[5] This formula is obtained using suitable (compatible with π) triangulations of E and Γ.

i.e., the complex elliptic version of the Poncelet transformation is a shift on the torus (as an Abelian group). In the Poncelet case we have $\mathcal{T}^n(u_0) = u_0$ for some point u_0, which leads to $\mathcal{T}^n = id$. The latter identity holds also in the real part of the curve E. □

Probably the first proof of the GPT with use of elliptic functions was given by C. Jacobi (see [6]). Also from papers of P. Griffiths and J. Harris [6, 7] one can learn about works of A. Cayley devoted to conditions onto the curves Γ and Δ guaranteeing that the Poncelet broken line be closed. We shortly describe that result following its approach from [7].

One of standard models of an elliptic curve is a smooth cubic in $\mathbb{CP}^2$ defined by the equation (in the affine part):

$$s^2 = (t - t_1)(t - t_2)(t - t_3), \tag{6}$$

thus E is a Riemann surface of the function $\sqrt{(t - t_1)(t - t_2)(t - t_3)}$. Then the projection π takes the form $(t, s) \longmapsto t$ and the ramification points are $(t_j, 0)$, $j = 1, 2, 3$, and $(\infty, \infty) = [0 : 1 : 0]$. The cycles γ_1 and γ_2 generating the group of 1-dimensional homologies of the curve E are lifts of loops in the plane of t's surrounding the points t_1 and t_2 and t_2 and t_3 respectively.

It turns out that the 1-form

$$\eta = \mathrm{d}t/s$$

is holomorphic and nonzero in the whole curve E (also in the ramification points). Denote $O = [0 : 1 : 0]$. For a point $z \in E$ we consider the following integral

$$u(z) = \int_O^z \eta; \tag{7}$$

it is an *incomplete elliptic integral.* This integral essentially depends on the integration path. More precisely, when we replace a given path by adding to it the loop γ_1 (or γ_2), then to the integral (7) the **period** (*complete elliptic integral*) $\omega_1 = \int_{\gamma_1} \eta$ (or $\omega_2 = \int_{\gamma_2} \eta$) is added. This is the way to obtain the representation (5). The map inverse to (7) is expressed via the doubly periodic Weierstrass function:

$$z = (t, s) = (\mathcal{P}(u), \mathcal{P}'(u)).$$

The function $\mathcal{P}(u)$ has pole in $u = 0 : t \sim u^{-2}$, $s \sim u^{-3}$. The structure of an Abelian group on E comes from the integral (7).

Another way to define the structure of an Abelian group on E uses the Abel theorem (see below). For any line in $\mathbb{CP}^2$ its three points z_1, z_2, z_3 of intersection with E (in the version (6)) obey the equation

$$z_1 + z_2 + z_3 = 0$$

in the group operation sense. Because $[0 : 1 : 0]$ is the only inflection point of the curve, it is the neutral element of the group.

Interesting is the question which cubic polynomial in the right-hand side of Eq. (6) corresponds to the curve (4). The answer is

$$s^2 = \det(tG + D), \tag{8}$$

where G and D are 3×3 matrices which define the curves in homogeneous coordinates: $\Gamma = \{(Gx, x) = 0\}$ and $\Delta = \{(Dx, x) = 0\}$.

For the proof we consider the pencil of conics

$$\Delta_t = \{((tG + D)\,x, x) = 0\}.$$

The values $t = t_j$ for which $\det(tG + D) = 0$ correspond to the situations when the conic Δ_t is reduced to two lines. Recall that we have four base points of the pencil: q_0, q_1, q_2, q_3. From q_0 we draw a line L_t tangent to Δ_t; it intersects Γ in another point $p(t)$. It is easy to see that $p(t_j) = q_j$ (after proper enumeration). Moreover, $\Delta_\infty = \Gamma$, i.e., $p(\infty) = q_0$. Hence the map $t \longmapsto p(t)$ defines a kind of isomorphism between $\mathbb{CP}^1$ and Γ preserving the ramification points of the 2-valued functions in (4) and (8) defining the elliptic curve.

Note that for $t = 0$ we have $\Delta_0 = \Delta$, so the line L_0 is tangent to Δ and $p(0) = T(q_0)$ (T is the Poncelet map). Thus one of the points of the curve (8) above $t = 0$ is the image $w = \mathcal{T}(O)$ of the point $O = \{u = 0\} = \{[0 : 1 : 0]\}$ for the elliptic Poncelet map. More precisely, O corresponds to the pair $(q_0, L(q_0))$ and $w = \mathcal{T}(O)$ corresponds to the pair $(p(0), L(p(0))) = (T(q_0), L(p(0)))$.

The property $\mathcal{T}^n = id$ means that

$$nw = 0 \mod \Lambda. \tag{9}$$

We would like to rewrite this condition in terms of Eq. (8).

Recall the above-mentioned result of Abel.[6]

Theorem 4 (Abel Theorem). *For given $u_i, v_i \in \mathbb{C}$, $i = 1, \ldots, n$, there exists a meromorphic function $f(u)$ with period lattice Λ and with zeroes in u_i and poles in v_j if and only if*

$$u_1 + \cdots + u_n = v_1 + \cdots + v_n \mod \Lambda. \tag{10}$$

Choose $v_j = O$, $j = 1, \ldots, n$, i.e., $v_j = 0 \bmod \Lambda$. Then the function $f(u)$ from the statement of the Abel theorem has pole at $u = 0$ of order n. The space of such functions for $n \geq 2$ is n-dimensional and has basis $f_1, \ldots, f_n$. More precisely, the closure of this space constitutes the space $H^0(\mathcal{O}_E(nO))$ of sections of a corresponding linear bundle $\mathcal{O}_E(nO)$ over E. In the model (6) of the elliptic curve the functions f_j can be chosen as follows: $f_1(s,t) = 1$, $f_2 = t$, $f_3 = s$, $f_4 = t^2$, $f_5 = st, \ldots$ (we compare orders of the pole at $u = 0$).

Condition (10) means that $u_1 + \cdots + u_n = 0 \bmod \Lambda$ and is equivalent to the condition $\det(f_i(u_j)) = 0$, because some nonzero function from $H^0(\mathcal{O}_E(nO))$

[6] Relatively simple proof of this theorem, which uses the Cauchy integral formula and standard theta functions, can be found in the book of C. Clemens [3].

vanishes in the points u_j. Approaching suitably the points u_j to $w = \mathcal{T}(O)$ we find that the later condition becomes a Wronskian type condition

$$W = \begin{vmatrix} f_1 & \dots & f_n \\ \dots & \dots & \dots \\ f_1^{(n-1)} & \dots & f_n^{(n-1)} \end{vmatrix} (w) = 0. \tag{11}$$

This condition is equivalent to condition (9). The basis of the space $H^0(\mathcal{O}_E(nO))$ is chosen among the functions t^j and st^k, as above. Moreover, the derivatives with respect to u can be replaced with derivatives with respect to t and then we can put $t = 0$.

If equation (8) gives the expansion

$$s(t) = \sum A_k t^k,$$

then for $n = 2m + 1$ and $n = 2m$ condition (11) means respectively

$$\begin{vmatrix} A_2 & \dots & A_{m+1} \\ \dots & \dots & \dots \\ A_{m+1} & \dots & A_{2m} \end{vmatrix} = 0 \text{ and } \begin{vmatrix} A_3 & \dots & A_{m+1} \\ \dots & \dots & \dots \\ A_{m+1} & \dots & A_{2m} \end{vmatrix} = 0.$$

This is the Cayley condition for the existence of a Poncelet n-gon.

Remark 3. It would be interesting to find a relation of these results with the complete elliptic integral found by Kołodziej in Theorem 3. There the elliptic curve takes the form $s^2 = \left(1 - \kappa^2 t^2\right)\left(1 - t^2\right)$, where $t = \sin\psi$.

Remark 4. It is worth to mention a generalization of the Poncelet theorem to the spacial case, i.e., for two degree 2 surfaces in $\mathbb{CP}^3$. Instead of a polygon we have a polyhedron and one should properly interpret 'drawing' a plane from a point in one of the surfaces. Due to lack of space we do not state a result, we only refer the reader to the article [6] of Griffiths and Harris.

3.3. Proof of Darboux

Gaston Darboux published this proof in the book [4], but we shall follow the article [14] of V. Prasolov.

We begin with the following

Lemma 3 (Darboux). *Suppose that we have n straight lines L_j in $\mathbb{C}^2$ in generic position given by equations $f_j(x) = 0$. Then any algebraic curve of degree $n - 1$ passing through all $n(n-1)/2$ points of mutual intersections of these lines has the form*

$$F(y) = f_1 \dots f_n \left(\frac{\lambda_1}{f_1} + \dots + \frac{\lambda_n}{f_n}\right) = 0. \tag{12}$$

Proof. Here the coefficients λ_j are determined by comparison of the values of the polynomial G (of degree $n-1$), defining this curve, in the points $q_1, \ldots, q_n$ of intersection of a typical line with our lines:

$$G(q_j) = F(q_j) = \lambda_j \prod_{i \neq j} f_i(q_j).$$

Then the function $(G-F)|_{L_j}$ vanishes in n different points of the line L_j and hence $(G-F)\,|_{L_j} \equiv 0$. The polynomial $G - F$ vanishes on n lines and therefore must be identically zero. □

The main idea of Darboux relies on introduction of special coordinates in the plane associated with the quadric Δ. The coordinates of a point p define a pair of points (parameters) on the curve Δ^*, dual to Δ.

In order to simplify formulas we assume that, in some complex homogeneous coordinates $y = [y_1 : y_2 : y_3]$, the curve Δ is defined as follows:

$$y_1 y_3 = y_2^2 \quad \text{or} \quad y = \left[1 : t : t^2\right] \quad (t \in \mathbb{CP}^1).$$

Since the tangent to Δ in a point $y = \left[1 : t : t^2\right]$ has the form $t^2 y_1 - 2t y_2 + y_3 = 0$, solving the system $t_1^2 y_1 - 2t_1 y_2 + y_3 = t_2^2 y_1 - 2t_2 y_2 + y_3 = 0$ with respect to y, we get the following characterization of the point of intersection of the two tangents to Δ in points corresponding to t_1 and t_2:

$$2y_2 = (t_1 + t_2) \cdot y_1, \quad y_3 = t_1 t_2 \cdot y_1.$$

By definition, (t_1, t_2) is the new coordinate system.

In particular, the equation of the curve Δ takes the form $(t_1 - t_2)^2 = 0$ (we have only one tangent). Moreover, the equation of tangent to Δ in $\left[1 : s : s^2\right]$ is following:

$$(t_1 - s)\,(t_2 - s) = 0. \tag{13}$$

The equation of a conic, e.g., Γ, takes the form

$$\varphi(t_1, t_2) = a t_1^2 t_2^2 + b t_1 t_2 (t_1 + t_2) + c t_1 t_2 + d(t_1 + t_2)^2 + e(t_1 + t_2) + f = 0. \tag{14}$$

Assume now that we have a Poncelet polygon with sides $p_{j-1}p_j$, $j = 1, \ldots, n$, inscribed in Γ, of the form (14), and described on Δ with the tangency points corresponding to parameters $s_1, \ldots, s_n$. Thus the lines L_j along the segments $p_{j-1}p_j$ are given by the equations $(t_1 - s_j)\,(t_2 - s_j) = 0$. According to the Darboux Lemma the curve Σ which contains all the mutual intersections of the lines L_j is given by the equation

$$\sum \frac{\lambda_j}{f_j} = \sum \frac{\lambda_j}{(t_1 - s_j)\,(t_2 - s_j)} = 0.$$

Expressing $1/\,(t_1 - s_j)\,(t_2 - s_j)$ as a difference of inverses of linear functions (with a coefficient) and introducing the new rational function

$$R(t) = \sum \frac{\lambda_j}{t - s_j} = \frac{P(t)}{Q(t)}$$

(where $\deg P = n-1$ and $Q = \prod(t - s_j)$) we obtain the equation $R(t_1) = R(t_2)$ for the curve Σ. It is equivalent to the equation

$$R_\mu(t_1) = R_\mu(t_2), \tag{15}$$

where

$$R_\mu(t) = \frac{P}{Q + \mu P} = \sum \frac{\lambda_j(\mu)}{t - s_j(\mu)}.$$

Eq. (15) also arises from lines $L_j(\mu)$ tangent to Δ in points $s_j(\mu)$.

The fact that Eq. (15) describes the same curve Σ (independent on μ) means that the points of mutual intersections of the lines $L_j(\mu)$ move along fixed curve Σ. In order to prove the GPT we should find a curve of degree $n-1$ which contains the points of pairwise intersections of the lines $L_j = L_j(0)$ and contains Γ as an irreducible component.

For $n = 3$ the situation is obvious: $\Sigma = \Gamma$. For $n = 4$ we have 6 points of mutual intersections: $p_0 = p_4, p_1, p_2, p_3$ and $q_1 = L_1 \cap L_3$ and $q_2 = L_2 \cap L_4$. We put $\Sigma = \Gamma \cup q_1 q_2$.

In the case of larger number of lines we apply a procedure of elimination of some variables s_i. For example, from the equations $\varphi(s_1, s_2) = \varphi(s_2, s_3) = 0$ (describing Γ in Eq. (14)) we eliminate s_2. (We represent the polynomial $\varphi(r_1, r_2) = \varphi^{r_1}(r_2) = \varphi^{r_2}(r_1)$ in the form $\varphi^{r_1}(r_2) = A(r_1)r_2^2 + B(r_1)r_2 + C(r_1)$ with quadratic trinomials A, B, C.) The answer is the resultant $\Phi_2(s_1, s_3)$ of the polynomials φ^{s_1} and φ^{s_3}. This resultant is symmetric and vanishes when $s_1 = s_3$, so $\Phi_2 = \varphi_2 \cdot (s_1 - s_3)^2$. As Φ_2 is of degree 4 with respect to each variable the polynomial $\varphi_2(s_1, s_3)$ has the form (14), i.e., it describes a conic curve. The equation $\varphi_2(t_1, t_2) = 0$ defines a conic Γ_2 with the points $L_1 \cap L_3$, $L_2 \cap L_4, \ldots$.

Next one eliminates s_3 from the equations $\varphi_2(s_1, s_3) = \varphi_1(s_3, s_4) = 0$, where we use the notation $\varphi_1 = \varphi$. One gets the equation $\Phi_3(s_1, s_4) = 0$. But this equation is satisfied also after replacing s_4 with s_2 (as $\varphi_1(s_1, s_2) = 0$). Therefore Φ_3 is divisible by φ_1 and we have a new polynomial $\varphi_3 = \Phi_3/\varphi_1$. It turns out that φ_3 is symmetric (repeat its derivation in the order s_4, s_3, s_2, s_1) and has degrees such that it defines a conic Γ_3. The latter contains the points $L_1 \cap L_4$, $L_2 \cap L_5, \ldots$.

In the induction step one eliminates s_{k+1} from the equations $\varphi_k(s_1, s_{k+1}) = \varphi_1(s_{k+1}, s_{k+2}) = 0$. One gets an equation $\Phi_{k+1}(s_1, s_{k+2}) = 0$ where $\varphi_{k+1} = \Phi_{k+1}/\varphi_{k-1}$ turns out to be a polynomial defining a conic Γ_k.

The last curve is $\Gamma_m = \{\varphi_m(t_1, t_2) = 0\}$ for $n = 2m+1$ or $n = 2m$. In the second case the polynomial φ_m is a square, i.e., Γ_m is a double line.

3.4. Geometric proofs

The original Poncelet proof of his 'closure Theorem' used projective geometry, but it was not much popular.

Very popular is a geometric proof due to A. Hart. It is discussed in the monograph of M. Berger [1, Sect. 16.6], which relies on the book [11] of H. Lebesgue. In fact, in [1] a more general theorem is proved. One considers a series of conics

$\Delta_1, \ldots, \Delta_n$ from one pencil and an n-gon with vertices in Γ and with sides tangent to Δ_j. If this n-gon is closed then it is movable as inscribed in Γ and tangent to Δ_j.

The proof given in [1] is inductive and, unfortunately, highly complicated. We do not present it here.[7]

Acknowledgement

This paper is an English version of the Polish paper "Twierdzenia Ponceleta w interpretacji Rafała Kołodzieja" published in Wiadomości Matematyczne **49** (2013) No 2, 29–45. I would like to thank its editor-in-chief, Tadeusz Nadzieja, for permission to translate my article.

Also I would like to thank Maciej Wojtkowski for interesting discussions on Kołodziej's proof and to Eugene Gutkin[8] for an initiative to write an article about Rafał.

References

[1] M. Berger, *Geometry*, Springer-Verlag, New York, 1987.

[2] A. Boyarsky and P. Góra, *Laws of Chaos: Invariant Measures and Dynamical Systems in One Dimension*, Birkhäuser, Boston, 1997.

[3] C.H. Clemens, *A Scrapebook of Complex Curves Theory*, Plenum Press, New York and London, 1980.

[4] G. Darboux, *Principles de Géométrie Analytique*, Gauthier-Villars, Paris, 1917.

[5] V. Dragović and M. Radnović, *Bicentennial of the great Poncelet theorem* (1813–2013)*: current advances*, Bull. Amer. Math. Soc. **51** (2014), No 3, 373-445.

[6] P. Griffiths and J. Harris, *A Poncelet theorem in space*, Comment. Math. Helvetici **52** (1977), 145–160.

[7] P. Griffiths and J. Harris, *On Caley's explicit solution to Poncelet's porism*, Enseign. Math. **24** (1978), 31–40.

[8] J.L. King, *Three problems in search of a measure*, Amer. Math. Monthly **107** (1994), 609–628.

[9] R. Kołodziej, *The rotation number of some transformation related to billiards in an ellipse*, Studia Math. **81** (1985), 293–302.

[10] V. Kozlov and D. Treshchev, *Billiards. A Genetic Introduction to the Dynamics of Systems with Impacts*, Amer. Math. Soc. 1991.

[11] H. Lebesgue, *Les Coniques*, Gauthier-Villars, Paris, 1955.

[12] I.P. Kornfeld, Ya.G. Sinai, S.F. Fomin, *Ergodic Theory*, Springer Verlag, New York, 1982.

[13] J.-V. Poncelet, *Traité des Propriétés Projectives des Figures*, Mett, Paris, 1822.

[7] I was explaining Kołodziej's proof at a local school in Siedlce (Poland). The audience was impressed. Warsaw geometers admitted that they studied the proof from Berger's monograph in series of seminars.

[8] Gutkin had declared to describe Kołodziej's results about antibilliards. Unfortunately, his unexpected death in June 2013 has modified this project.

[14] V.V. Prasolov, *Dokazatel'stvo teoremy Ponsele po Darbu*, Matemat. Prosveshcheniye **3** (2001), 140–144 (in Russian).
[15] S. Tabachnikov, *Poncelet's theorem and dual billiards*, Enseign. Math. **39** (1993), 189–194.

Henryk Żołądek
Institute of Mathematics, Warsaw University
ul. Banacha 2, 02-097 Warsaw, Poland
e-mail: zoladek@mimuw.edu.pl

Geometric Methods in Physics. XXXV Workshop 2016
Trends in Mathematics, 145–151

Weighted Szegő Kernels

Zbigniew Pasternak-Winiarski and Tomasz Lukasz Zynda

Abstract. In this paper we define a weighted Szegő kernel by putting a measurable almost everywhere positive function μ under the inner product integral and try to answer which conditions it must satisfy in order to give a 'good generalization' of a classical case.

Mathematics Subject Classification (2010). 30H20, 32A25, 32C15, 46E22.

Keywords. Szegő kernel, Bergman kernel, reproducing kernel, Hilbert space of holomorphic functions.

1. Reproducing kernels

Let H be a Hilbert space of complex-valued functions on a non-empty set X with an inner product $\langle f|g\rangle$. We call $K : X \times X \to \mathbb{C}$ a **reproducing kernel** of H, if for every $z \in X$ $\overline{K(z,\cdot)} \in H$ and every $f \in H$

$$f(z) = \langle f, \overline{K(z,\cdot)}\rangle. \tag{1}$$

By the Riesz representation theorem, a Hilbert space can't have more than one reproducing kernel. One can prove that if a Hilbert space has a reproducing kernel, then it is given by

$$K(z,w) = \sum_{i\in I} \varphi_i(z)\overline{\varphi_i(w)}, \tag{2}$$

where $\{\varphi_i\}_{i\in I}$ is an arbitrary complete orthonormal system in H.

2. Weighted Bergman kernels

Let us consider a space $L^2H(\Omega)$ of functions which are both holomorphic and square-integrable with respect to the Lebesgue measure on a domain $\Omega \subset \mathbb{C}$. It is called Bergman space and a reproducing kernel of it is called Bergman kernel.

The Bergman kernel is given by (2). It is analytic in a real sense, holomorphic in its first variable and antiholomorphic in its second variable.

Let now $\mu : \Omega \to \mathbb{R}$ will be a weight, i.e., a function measurable and almost everywhere positive on Ω. We can consider weighted Bergman spaces, i.e., spaces $L^2H(\Omega,\mu)$ of functions which are holomorphic and square-integrable on Ω in the sense

$$\int_{\Omega} |f(z)|^2 \mu(z)\, \mathrm{d}^{2n} z < \infty \tag{3}$$

with an inner product given by

$$\langle f|g\rangle_\mu := \int_{\Omega} f(z)\overline{g(z)}\mu(z)\, \mathrm{d}^{2n} z. \tag{4}$$

We say that a weight μ is an admissible weight, if $L^2H(\Omega,\mu)$ is a closed subspace of $L^2(\Omega,\mu)$ and all functionals of evaluation, i.e., functionals

$$E_z : L^2H(\Omega,\mu) \ni f \rightarrowtail f(z) \in \mathbb{C}^n, \tag{5}$$

for $z \in \Omega$, are continuous. A natural question is to ask which conditions must a weight satisfy to be an admissible weight. Article [1] gives us response to that question:

Theorem 1. *Let μ be a weight. The following conditions are equivalent:*

(i) *μ is an admissible weight;*
(ii) *for any compact set $X \subset \Omega$ there exists a constant $C_X > 0$, such that for any $z \in X$ and each $f \in L^2H(\Omega,\mu)$*

$$|E_z f| \le C_X \parallel f \parallel_\mu . \tag{6}$$

3. Weighted Szegő kernel

Let Ω be a bounded domain in $\mathbb{C}^N$ with the boundary $\partial\Omega$ of class C^2. For $\mu : \partial\Omega \to \mathbb{R}$ measurable and almost everywhere positive by $L^2(\partial\Omega,u)$ we will denote a set of functions $f : \partial\Omega \to \mathbb{C}$ square-integrable in the sense

$$\int_{\partial\Omega} |f(z)|^2 \mu(z)\, \mathrm{d}S < \infty, \tag{7}$$

where we consider a real surface integral of a scalar field f. Set $L^2(\partial\Omega,u)$ with an inner product given by

$$\langle f|g\rangle_\mu := \int_{\partial\Omega} f(z)\overline{g(z)}\mu(z)\, \mathrm{d}S \tag{8}$$

is a Hilbert space. Now let us consider space $A(\Omega)$ of functions $F : \overline{\Omega} \to \mathbb{C}$, such that $A(\Omega) := H(\Omega) \cap C(\overline{\Omega})$.

By $L^2(\partial\Omega,u)$ we will denote the closure of restrictions of elements of A to $\partial\Omega$ in L^2 topology. By the Poisson integral, each element of $L^2(\partial\Omega,u)$ has a unique holomorphic extension to Ω, which we will denote by the same symbol.

For $\mu(z) = 1$ for every z we have a classical case, for which we use terms such as Szegő space and Szegő kernel. The question is, how the space $L^2H(\partial\Omega,\mu)$ changes with the change of μ and which μ are 'good enough' to take?

Definition 1. We say that $\mu : \partial\Omega \to \mathbb{R}$ is a weight, if μ is measurable and almost everywhere positive and that a weight μ is a Szegő admissible weight (S-admissible weight for short), if all functionals of evaluation are continuous for every $z \in \Omega$.

It is not hard to prove that Theorem 1 is also true if we change an admissible weight to Szegő admissible weight. Moreover,

Proposition 1. *Let μ_1, μ_2 be weights on $\partial\Omega$. Then*

$$L^2H(\partial\Omega, \mu_1 + \mu_2) = L^2H(\partial\Omega, \mu_1) \cap L^2H(\partial\Omega, \mu_2), \tag{9}$$

where the equality is understood as the equality of sets.

Proof. Let f be holomorphic on Ω. Then

$$\int_{\partial\Omega} |f(z)|^2(\mu_1(z) + \mu_2(z))\,\mathrm{d}S = \int_{\partial\Omega} |f(z)|^2\mu_1(z)\,\mathrm{d}S + \int_{\partial\Omega} |f(z)|^2\mu_2(z)\,\mathrm{d}S. \tag{10}$$

If the integral on the left-hand side is finite, then both integrals on the right-hand side must be also finite, because all considered integrals are non-negative, so $L^2H(\partial\Omega, \mu_1 + \mu_2) \subseteq L^2H(\partial\Omega, \mu_1) \cap L^2H(\partial\Omega, \mu_2)$. On the other hand, if both integrals on the right side are finite, then the integral on the left side must be finite, so $L^2H(\partial\Omega, \mu_1) \cap L^2H(\partial\Omega, \mu_2) \subseteq L^2H(\partial\Omega, \mu_1+\mu_2)$ and at last $L^2H(\partial\Omega, \mu_1+\mu_2) = L^2H(\partial\Omega, \mu_1) \cap L^2H(\partial\Omega, \mu_2)$. □

Theorem 2. *The following are true:*

(i) *if μ_1 and μ_2 are S-admissible weights, then also $\mu_1 + \mu_2$ is an S-admissible weight and*
(ii) *if μ is an S-admissible weight and α is a real positive number, then also $\alpha\mu$ is an S-admissible weight.*

Proof. (i) Let μ_1, μ_2 be S-admissible weights. Then for any compact set $X \Subset \Omega$ there exist $C_X^1, C_X^2 > 0$, such as that for every $z \in X$ and $f \in L^2H(\partial\Omega, \mu_1) \cap L^2H(\partial\Omega, \mu_2)$

$$|E_z f| \leq C_X^1 \parallel f \parallel_{\mu_1} \quad \text{and} \quad |E_z f| \leq C_X^2 \parallel f \parallel_{\mu_2} .$$

Then

$$\begin{aligned} 2|E_z f| &= |E_z f| + |E_z f| \leq C_X^1 \parallel f \parallel_{\mu_1} + C_X^2 \parallel f \parallel_{\mu_2} \\ &\leq 2\max\{C_X^1, C_X^2\}\max\{\parallel f \parallel_{\mu_1}, \parallel f \parallel_{\mu_2}\} \leq 2\max\{C_X^1, C_X^2\} \parallel f \parallel_{\mu_1+\mu_2}, \end{aligned} \tag{11}$$

which after dividing both sides by 2 gives us inequality (ii) from Theorem 1, so the sum of S-admissible weights is an S-admissible weight.

(ii) If $|E_z f| \leq C_X \sqrt{\int_{\partial\Omega} |f(z)|^2\mu(z)\,\mathrm{d}S}$, then also

$|E_z f| \leq C_X \frac{1}{\sqrt{\alpha}} \sqrt{\int_{\partial\Omega} |f(z)|^2\alpha\mu(z)\,\mathrm{d}S}$, so $\alpha\mu$ is also an S-admissible weight.

□

Theorem 3. *Let μ be a weight. If $\mu \geq c$ almost everywhere, for some $c > 0$, then μ is an S-admissible weight.*

Proof. Taking into account the fact that a function equal to 1 on $\partial\Omega$ is an S-admissible weight, we get for any compact set $X \subset \Omega$

$$\begin{aligned} |f(z)| \leq C_X \sqrt{\int_{\partial\Omega} |f(z)|^2 \cdot 1 \, \mathrm{d}S} &= C_X \frac{1}{\sqrt{c}} \sqrt{\int_{\partial\Omega} |f(z)|^2 \cdot c \, \mathrm{d}S} \\ &\leq C_X \frac{1}{\sqrt{c}} \sqrt{\int_{\partial\Omega} |f(z)|^2 \mu(z) \, \mathrm{d}S}, \end{aligned} \tag{12}$$

which ends the proof. Note that $L^2H(\partial\Omega, \mu) \subseteq L^2H(\partial\Omega, 1)$, because if the integral on the right-hand side is finite, then the first integral in the upper line must be also finite. □

Proposition 2. *If μ_1 is an S-admissible weight and $\mu_1 \leq \mu_2$ almost everywhere, then μ_2 is an S-admissible weight.*

Proof. Taking into account the fact that μ_1 is an S-admissible weight, we get that for any $f \in L^2H(\partial\Omega, \mu_2)$ and any compact set $X \subset \Omega$ there exists $C_X > 0$, such that

$$|E_z f| \leq C_X \sqrt{\int_{\partial\Omega} |f(z)|^2 \mu_1(z) \, \mathrm{d}S} \leq C_X \sqrt{\int_{\partial\Omega} |f(z)|^2 \mu_2(z) \, \mathrm{d}S}. \tag{13}$$

□

In particular, if μ is an S-admissible weight, then also e^μ and μ^μ are S-admissible weights. The second fact is true, because $x^x > x$ almost everywhere on the interval $[0, +\infty[$.

A limit (in any sense) of a sequence of S-admissible weights doesn't have to be an S-admissible weight. For example $\mu_n(z) = \frac{1}{n}$, $n \in \mathbb{N}$ is an S-admissible weight, but a limit of this sequence is 0, which is not an S-admissible weight.

However, in some cases, it is true. For example, if $\mu_0 = \lim_{n\to\infty} \mu_n > c$ for some real positive c almost everywhere or if $\mu > \mu_n$ for at least one n, then μ_0 is also an S-admissible weight.

Theorem 4. *Let μ_1, μ_2 be S-admissible weights and $\mu_2 \geq c_2 > 0$ almost everywhere. Then $\mu_1\mu_2$ is also an S-admissible weight.*

Proof. If $f \in L^2H(\partial\Omega, \mu_1\mu_2)$, then

$$\begin{aligned} |E_z f| \leq C_X \sqrt{\int_{\partial\Omega} |f(z)|^2 \mu_1(z) \, \mathrm{d}S} &= C_X \frac{1}{\sqrt{c_2}} \sqrt{\int_{\partial\Omega} |f(z)|^2 \mu_1(z) c_2 \, \mathrm{d}S} \\ &\leq C_X \frac{1}{\sqrt{c_2}} \sqrt{\int_{\partial\Omega} |f(z)|^2 \mu_1(z) \mu_2(z) \, \mathrm{d}S}. \end{aligned} \tag{14}$$

□

From the above, we get that if μ is a weight, such that $\mu \geq c > 0$ almost everywhere, then $W(\mu)$ and $e^{W(\mu)}$, where $W(\mu)$ is any polynomial of μ positive on the interval $[c, +\infty[$, are S-admissible weights.

Note it may happen that $L^2H(\partial\Omega,\mu_1) \cap L^2H(\partial\Omega,\mu_2) = L^2H(\partial\Omega,\mu_1) = L^2H(\partial\Omega,\mu_2)$, even for $\mu_1 \neq \mu_2$, so it is natural to make a definition

Definition 2. Let μ_1, μ_2 be S-admissible weights. We say that μ_1 is equivalent to μ_2 and write $\mu_1 \equiv \mu_2$, if

$$L^2H(\partial\Omega,\mu_1) = L^2H(\partial\Omega,\mu_2), \tag{15}$$

where the equality is understood as the equality of sets.

It is easy to prove, that '$\equiv$' is an equivalence relation. (Not to be confused with another equivalence relation on the set of measurable functions, in which two functions are equivalent if they are equal almost everywhere.)

Theorem 5. *Let μ_1, μ_2 be S-admissible weights, and m, M be real positive numbers. If*

$$m\mu_1 \leq \mu_2 \leq M\mu_1 \tag{16}$$

almost everywhere, then $\mu_1 \equiv \mu_2$

Of course, if (16) holds, then it is also true that

$$\frac{1}{M}\mu_2 \leq \mu_1 \leq \frac{1}{m}\mu_2. \tag{17}$$

Proof. Let μ_1 and μ_2 be S-admissible weights and let (16) hold. Then, for any holomorphic f it is true, that

$$m\int_{\partial\Omega} |f(z)|^2\mu_1(z)\,\mathrm{d}S \leq \int_{\partial\Omega} |f(z)|^2\mu_2(z)\,\mathrm{d}S \leq M\int_{\partial\Omega} |f(z)|^2\mu_1(z)\,\mathrm{d}S. \tag{18}$$

If $f \in L^2H(\partial\Omega,\mu_2)$, then the integral in the center is finite and because of that, the integral on the left-hand side must also be finite, so $f \in L^2H(\partial\Omega,\mu_1)$. If $f \in L^2H(\partial\Omega,\mu_1)$, then the integral on the right-hand side is finite and because of that, the integral in the center must also be finite, so $f \in L^2H(\partial\Omega,\mu_2)$. Reassuming, $L^2H(\partial\Omega,\mu_1) = L^2H(\partial\Omega,\mu_2)$ and $\mu_1 \equiv \mu_2$. □

In particular:

Corollary 1.

(i) *If μ is an S-admissible weight and α is a real positive number, then $\mu \equiv \alpha\mu$.*

(ii) *If μ is an S-admissible weight, such that there exist real positive numbers m, M, such that*

$$m \leq \mu \leq M, \tag{19}$$

(so μ is bounded from up and down) almost everywhere, then $\mu \equiv 1$.

Theorem 6. *If K is a Szegő kernel of $L^2H(\partial\Omega,\mu)$ and μ is essentially bounded on $\partial\Omega$, then*

$$\int_{\partial\Omega} K(z,w)\mu(w)\,\mathrm{d}S = 1 \tag{20}$$

Proof. By taking the function $f \equiv 1$ in (1), we get identity (20). □

Even if spaces $L_1 := L^2H(\partial\Omega, \mu_1)$ and $L_2 := L^2H(\partial\Omega, \mu_2)$ are equal as sets, they still can have different Szegő kernels. For example

Proposition 3. *If $\mu_2 = \alpha\mu_1$, where $\alpha \in \mathbb{R}_+$, then set L_1 is equal to set L_2 and $K_2(z,w) = \frac{1}{\alpha}K_1(z,w)$, where K_i is Szegő kernel of L_i.*

Proof. Set L_1 is equal to L_2, by Corollary 1 (i). If $\{\varphi_i\}_{i\in I}$ is a complete orthonormal system of L_1, then $\{\psi_i\}_{i\in I}$, where $\psi_i(z) := \frac{1}{\sqrt{\alpha}}\varphi(z)$, is a complete orthonormal system of L_2 and

$$K_2(z,w) = \sum_{i\in I}\psi(z)\overline{\psi(w)} = \sum_{i\in I}\frac{1}{\sqrt{\alpha}}\varphi(z)\frac{1}{\sqrt{\alpha}}\overline{\varphi(w)} \tag{21}$$
$$= \frac{1}{\alpha}\sum_{i\in I}\varphi(z)\overline{\varphi(w)} = \frac{1}{\alpha}K_1(z,w).$$
□

Theorem 7. *Let G be a domain in $\mathbb{C}^n$ for $n \geq 2$ such that $\overline{G} \subset \Omega$ and ∂G is of class C^2. Let μ_G be a weight on ∂G and μ_Ω be an S-admissible weight on $\partial\Omega$. Then the function*

$$\mu(\zeta) := \begin{cases} \mu_\Omega(\zeta), & \zeta \in \partial\Omega, \\ \mu_G(\zeta), & \zeta \in \partial G \end{cases} \tag{22}$$

is an S-admissible weight on $\partial(\Omega \setminus \overline{G})$ and the map $L^2H(\partial\Omega, \mu_\Omega) \ni f \rightarrowtail Tf := f_{|\Omega\setminus\overline{G}} \in L^2H(\partial(\Omega \setminus \overline{G}), \mu)$ is a continuous isomorphism of Hilbert spaces.

Proof. Let X be a compact subset in $\Omega\setminus\overline{G}$. Then $X \subset \Omega$ and there exists $C_X > 0$, such that for any $f \in L^2H(\partial\Omega, \mu_G)$ and any $z \in X$

$$|f(z)| \leq C_X \parallel f \parallel_{\mu_\Omega} \tag{23}$$

On the other hand, if $g \in L^2H(\partial(\Omega\setminus\overline{G}), \mu)$, then by Hartog's prolongation theorem, there exists $\tilde{g}$ holomorphic on Ω, such that $\tilde{g}_{|\Omega\setminus\overline{G}} = g$. It is obvious that

$$\int_{\partial\Omega}|\tilde{g}(\zeta)|^2\mu_\Omega(\zeta)\,\mathrm{d}(\partial\Omega) \leq \int_{\partial\Omega}|\tilde{g}(\zeta)|^2\mu_\Omega(\zeta)\,\mathrm{d}(\partial\Omega) + \int_{\partial G}|\tilde{g}(\zeta)|^2\mu_G(\zeta)\,\mathrm{d}(\partial G) \tag{24}$$
$$= \parallel g \parallel_\mu^2 < \infty.$$

Then

$$\parallel \tilde{g} \parallel_{\mu_\Omega} \leq \parallel g \parallel_\mu \tag{25}$$

and $\tilde{g} \in L^2H(\partial\Omega, \mu_\Omega)$.

For any $z \in X$ we have

$$|g(z)| = |\tilde{g}(z)| \leq C_X \parallel \tilde{g} \parallel_{\mu_G} \leq C_X \parallel g \parallel_\mu . \tag{26}$$

Since g is an arbitrary element of $L^2H(\partial\Omega \setminus \overline{G}, \mu)$, we see that μ is an S-admissible weight for $\Omega \setminus \overline{G}$. Moreover, the prolongation $L^2H(\partial\Omega \setminus \overline{G}), \mu) \ni g \rightarrowtail \tilde{g} \in L^2H(\partial\Omega, \mu_G)$ is uniquely defined and then it is an inverse of T. By (25), T^{-1} is bounded and by Banach's inverse theorem, T is continuous. □

In the case $n = 1$, the theorem is not true. For example, if $\Omega := D := D(0,1) = \{z \in \mathbb{C} : |z| < 1\}$, $G := D(0, \frac{1}{2})$, $\mu_\Omega = 1$ for each z and $\mu_G = 1$ for each z, the function

$$g(z) = \frac{1}{z}, z \in \Omega \setminus G, \tag{27}$$

is an element of $L^2H(\partial(\Omega \setminus \overline{G}), \mu)$, but it has no prolongation to a function $\tilde{g} \in L^2H(\partial\Omega, \mu_\Omega)$.

However, using similar a argument as in the proof of the theorem, we can show that if $n = 1$, then the operator of restriction $T : L^2H(\partial\Omega, \mu_\Omega) \to L^2H(\partial\Omega \setminus \overline{G}, \mu)$ is continuous and a one-to-one map onto its image, and that $T(L^2H(\Omega, \mu_\Omega))$ is a closed subspace of $L^2H(\partial(\Omega \setminus \overline{G}), \mu)$.

4. What to do next?

Z. Pasternak-Winiarski in [2] proved that a weighted Bergman kernel depends analytically on weights. A natural question is if the same holds for Szegő kernel.

References

[1] Z. Pasternak Winiarski, *On weights which admit the reproducing kernel of Bergman type*, Internat. J. Math. and Math. Sci., vol. **15** N. 1. 1992.

[2] Z. Pasternak-Winiarski, *On the Dependence of the Reproducing Kernel on the Weight of Integration*, Journal of Functional Analysis **94**, 110–134, 1990.

Zbigniew Pasternak-Winiarski and Tomasz Lukasz Zynda
Faculty of Mathematics and Information Science
Warsaw University of Technology
Koszykowa 75
00-662 Warsaw, Poland
e-mail: `z.pasternak-winiarski@mini.pw.edu.pl`
`tomasz.lukasz.zynda@gmail.com`

Part II

Integrability & Geometry

Geometric Methods in Physics. XXXV Workshop 2016
Trends in Mathematics, 155–174

On a Conjecture of Hong and Won

Ivan Cheltsov

Abstract. We give an explicit counter-example to a conjecture of Kyusik Hong and Joonyeong Won about α-invariants of polarized smooth del Pezzo surfaces of degree one.

Mathematics Subject Classification (2010). 14J45, 14J26, 32Q26.

Keywords. α-invariant of Tian, del Pezzo surface, K-stability.

1. Introduction

In [11], Tian defined the α-invariant of a smooth Fano variety[1] and proved

Theorem 1 ([11]). *Let X be a smooth Fano variety of dimension n such that $\alpha(X) > \frac{n}{n+1}$. Then X admits a Kähler–Einstein metric.*

In [10], Odaka and Sano proved

Theorem 2. *Let X be a smooth Fano variety of dimension n such that $\alpha(X) > \frac{n}{n+1}$. Then X is K-stable.*

Two-dimensional smooth Fano varieties are also known as smooth del Pezzo surfaces. The possible values of their α-invariants are given by

Theorem 3 ([1, Theorem 1.7]). *Let S be a smooth del Pezzo surface. Then*

$$\alpha(S) = \begin{cases} \frac{1}{3} \text{ if } S \cong \mathbb{F}_1 \text{ or } K_S^2 \in \{7,9\}, \\ \frac{1}{2} \text{ if } S \cong \mathbb{P}^1 \times \mathbb{P}^1 \text{ or } K_S^2 \in \{5,6\}, \\ \frac{2}{3} \text{ if } K_S^2 = 4, \\ \frac{2}{3} \text{ if } S \text{ is a cubic surface in } \mathbb{P}^3 \text{ with an Eckardt point}, \\ \frac{3}{4} \text{ if } S \text{ is a cubic surface in } \mathbb{P}^3 \text{ without Eckardt points}, \\ \frac{3}{4} \text{ if } K_S^2 = 2 \text{ and } |-K_S| \text{ has a tacnodal curve}, \\ \frac{5}{6} \text{ if } K_S^2 = 2 \text{ and } |-K_S| \text{ has no tacnodal curves}, \\ \frac{5}{6} \text{ if } K_S^2 = 1 \text{ and } |-K_S| \text{ has a cuspidal curve}, \\ 1 \text{ if } K_S^2 = 1 \text{ and } |-K_S| \text{ has no cuspidal curves}. \end{cases}$$

[1] All varieties are assumed to be algebraic, projective and defined over $\mathbb{C}$.

Let X be an arbitrary smooth algebraic variety, and let L be an ample $\mathbb{Q}$-divisor on it. Donaldson, Tian and Yau conjectured that the following conditions are equivalent:

- the pair (X, L) is K-polystable,
- the variety X admits a constant scalar curvature Kähler metric in $c_1(L)$.

In [6], this conjecture has been proved in the case when X is a Fano variety and $L = -K_X$.

In [12], Tian defined a new invariant $\alpha(X, L)$ that generalizes the classical α-invariant. If X is a smooth Fano variety, then $\alpha(X) = \alpha(X, -K_X)$. By [3, Theorem A.3], one has

$$\alpha(X, L) = \sup\left\{\lambda \in \mathbb{Q} \;\middle|\; \begin{array}{l}\text{the log pair } (X, \lambda D) \text{ is log canonical}\\ \text{for every effective } \mathbb{Q}\text{-divisor } D \sim_{\mathbb{Q}} L\end{array}\right\} \in \mathbb{R}_{>0}.$$

In [8], Dervan generalized Theorem 2 as follows:

Theorem 4 ([8, Theorem 1.1]). *Suppose that* $-K_X - \frac{n}{n+1}\frac{-K_X\cdot L^{n-1}}{L^n}L$ *is nef, and*

$$\alpha(X, L) > \frac{n}{n+1}\frac{-K_X \cdot L^{n-1}}{L^n}.$$

Then the pair (X, L) *is* K*-stable.*

For smooth del Pezzo surfaces, Theorem 4 gives

Theorem 5 ([2, 9]). *Let* S *be a smooth del Pezzo surface such that* $K_S^2 = 1$ *or* $K_S^2 = 2$. *Let* A *be an ample* $\mathbb{Q}$*-divisor on the surface* S *such that the divisor*

$$-K_S - \frac{2}{3}\frac{-K_S \cdot A}{A^2}A$$

is nef. Then the pair (S, A) *is* K*-stable.*

This result is closely related to

Problem 6 (cf. Theorem 3). *Let* S *be a smooth del Pezzo surface. Compute*

$$\alpha(S, A) \in \mathbb{R}_{>0}$$

for every ample $\mathbb{Q}$*-divisor* A *on the surface* S.

Hong and Won suggested an answer to Problem 6 for del Pezzo surfaces of degree one. This answer is given by their [9, Conjecture 4.3], which is Conjecture 11 in Section 2.

The main result of this paper is

Theorem 7 (cf. Theorem 3). *Let* S *be a smooth del Pezzo surface such that* $K_S^2 = 1$. *Let* C *be an irreducible smooth curve in* S *such that* $C^2 = -1$. *Then there is a unique curve*

$$\widetilde{C} \in \big| -2K_S - C\big|.$$

The curve $\widetilde{C}$ is also irreducible and smooth. One has $\widetilde{C}^2 = -1$ and $1 \leqslant |C \cap \widetilde{C}| \leqslant C \cdot \widetilde{C} = 3$. Let λ be a rational number such that $0 \leqslant \lambda < 1$. Then $-K_S + \lambda C$ is ample and

$$\alpha(S, -K_S + \lambda C) = \begin{cases} \min\left(\alpha(S), \frac{2}{1+2\lambda}\right) & \text{if } |C \cap \widetilde{C}| \geqslant 2, \\ \min\left(\alpha(S), \frac{4}{3+3\lambda}\right) & \text{if } |C \cap \widetilde{C}| = 1. \end{cases}$$

Theorem 7 implies that [9, Conjecture 4.3] is wrong. To be precise, this follows from

Example 8. Let S be a surface in $\mathbb{P}(1,1,2,3)$ that is given by

$$w^2 = z^3 + zx^2 + y^6,$$

where x, y, z, w are coordinates such that $\mathrm{wt}(x) = \mathrm{wt}(y) = 1$, $\mathrm{wt}(z) = 2$ and $\mathrm{wt}(w) = 3$. Then S is a smooth del Pezzo surface and $K_S^2 = 1$. Let C be the curve in X given by

$$z = w - y^3 = 0.$$

Similarly, let $\widetilde{C}$ be the curve in S that is given by $z = w + y^3 = 0$. Then $C + \widetilde{C} \sim -2K_S$. Both curves C and $\widetilde{C}$ are smooth rational curves such that $C^2 = \widetilde{C}^2 = -1$ and $|C \cap \widetilde{C}| = 1$. All singular curves in $|-K_S|$ are nodal. Then $\alpha(S) = 1$ by Theorems 3, so that

$$\alpha(S, -K_S + \lambda C) = \min\left(1, \frac{4}{3+3\lambda}\right)$$

by Theorem 7. But [9, Conjecture 4.3] says that $\alpha(S, -K_S + \lambda C) = \min(1, \frac{2}{1+2\lambda})$.

Theorem 7 has two applications. By Theorem 4, it implies

Corollary 9 ([8, Theorem 1.2]). *Let S be a smooth del Pezzo surface such that $K_S^2 = 1$. Let C be an irreducible smooth curve in S such that $C^2 = -1$. Fix $\lambda \in \mathbb{Q}$ such that*

$$3 - \sqrt{10} \leqslant \lambda \leqslant \frac{\sqrt{10} - 1}{9}.$$

Then the pair $(S, -K_S + \lambda C)$ is K-stable.

By [5, Remark 1.1.3], Theorem 7 implies

Corollary 10. *Let S be a smooth del Pezzo surface. Suppose that $K_S^2 = 1$ and $\alpha(S) = 1$. Let C be an irreducible smooth curve in S such that $C^2 = -1$. Fix $\lambda \in \mathbb{Q}$ such that*

$$-\frac{1}{4} \leqslant \lambda \leqslant \frac{1}{3}.$$

Then S does not contain $(-K_S + \lambda C)$-polar cylinders (see [5, Definition 1.2.1]).

Corollary 9 follows from Theorem 5. Corollary 10 follows from [5, Theorem 2.2.3].

Let us describe the structure of this paper. In Section 2, we describe [9, Conjecture 4.3]. In Section 3, we present several well-known local results about singularities of log pairs. In Section 4, we prove eight local lemmas that are crucial for the proof of Theorem 7. In Section 5, we prove Theorem 7 using Lemmas 23, 24, 25, 26, 27, 28, 29, 30.

2. Conjecture of Hong and Won

Let S be a smooth del Pezzo surface, and let A be an ample $\mathbb{Q}$-divisor on S. Put

$$\mu = \inf\Big\{\lambda \in \mathbb{Q}_{>0}\ \Big|\ \text{the } \mathbb{Q}\text{-divisor } K_S + \lambda A \text{ is pseudo-effective}\Big\} \in \mathbb{Q}_{>0}.$$

Then $K_S + \mu A$ is contained in the boundary of the Mori cone $\overline{\mathbb{NE}}(S)$ of the surface S.

Suppose that $K_S^2 = 1$. Then $\overline{\mathbb{NE}}(S)$ is polyhedral and is generated by (-1)-curves in S. By a (-1)-curve, we mean a smooth irreducible rational curve $E \subset S$ such that $E^2 = -1$.

Let Δ_A be the smallest extremal face of the Mori cone $\overline{\mathbb{NE}}(S)$ that contains $K_S + \mu A$. Let $\phi\colon S \to Z$ be the contraction given by the face Δ_A. Then

- either ϕ is a birational morphism and Z is a smooth del Pezzo surface,
- or ϕ is a conic bundle and $Z \cong \mathbb{P}^1$.

If ϕ is birational and $Z \ncong \mathbb{P}^1 \times \mathbb{P}^1$, we call A a divisor of $\mathbb{P}^2$-type. In this case, we have

$$K_S + \mu A \sim_{\mathbb{Q}} \sum_{i=1}^{8} a_i E_i,$$

where E_1, E_2, E_3, E_4, E_5, E_6, E_7, E_8 are eight disjoint (-1)-curves in our surface S, and a_1, a_2, a_3, a_4, a_5, a_6, a_7, a_8 are non-negative rational numbers such that

$$1 > a_1 \geqslant a_2 \geqslant a_3 \geqslant a_4 \geqslant a_5 \geqslant a_6 \geqslant a_7 \geqslant a_8 \geqslant 0.$$

In this case, we put $s_A = a_2 + a_3 + a_4 + a_5 + a_6 + a_7 + a_8$.

If our ample divisor A is not a divisor of $\mathbb{P}^2$-type, then the surface S contains a smooth irreducible rational curve C such that $C^2 = 0$ and

$$K_S + \mu A \sim_{\mathbb{Q}} \delta C + \sum_{i=1}^{7} a_i E_i,$$

where E_1, E_2, E_3, E_4, E_5, E_6, E_7 are disjoint (-1)-curves in S that are disjoint from C, and δ, a_1, a_2, a_3, a_4, a_5, a_6, a_7 are non-negative rational numbers such that

$$1 > a_1 \geqslant a_2 \geqslant a_3 \geqslant a_4 \geqslant a_5 \geqslant a_6 \geqslant a_7 \geqslant 0.$$

In this case, let $\psi\colon S \to \overline{S}$ be the contraction of the curves E_1, E_2, E_3, E_4, E_5, E_6, E_7, and let $\eta\colon S \to \mathbb{P}^1$ be a conic bundle given by $|C|$. Then either $\overline{S} \cong \mathbb{F}_1$ or

$\overline{S} \cong \mathbb{P}^1 \times \mathbb{P}^1$. In both cases, there exists a commutative diagram

$$\begin{array}{ccc} S & \xrightarrow{\quad\psi\quad} & \overline{S} \\ & \underset{\eta}{\searrow} \quad \underset{\pi}{\swarrow} & \\ & \mathbb{P}^1 & \end{array}$$

where π is a natural projection. Then $\delta > 0 \iff \phi$ is a conic bundle and $\phi = \eta$. Similarly, if ϕ is birational and $Z \cong \mathbb{P}^1 \times \mathbb{P}^1$, then $\delta = 0$, $a_7 > 0$, and $\phi = \psi$. Then

- we call A a divisor of $\mathbb{F}_1$-type in the case when $\overline{S} \cong \mathbb{F}_1$,
- we call A a divisor of $\mathbb{P}^1 \times \mathbb{P}^1$-type in the case when $\overline{S} \cong \mathbb{P}^1 \times \mathbb{P}^1$.

In both cases, we put $s_A = a_2 + a_3 + a_4 + a_5 + a_6 + a_7$.

In order to study $\alpha(S, A)$, we may assume that $\mu = 1$, because

$$\alpha(S, A) = \mu\alpha(S, \mu A)$$

If A is a divisor of $\mathbb{P}^2$-type, let us define a number $\alpha_c(S, A)$ as follows:

- if $s_A > 4$, we put $\alpha_c(S, A) = \frac{1}{2+a_1}$,
- if $4 \geqslant s_A > 1$, we let $\alpha_c(S, A)$ to be

$$\max\left(\frac{2}{2 + 2a_1 + s_A - a_2 - a_3}, \frac{4}{3 + 4a_1 + 2s_A - a_2 - a_3 - a_4}, \frac{3}{2 + 3a_1 + s_A}\right),$$

- if $1 \geqslant s_A$, we put $\alpha_c(S, A) = \min(\frac{2}{1+2a_1+s_A}, 1)$.

Similarly, if A is a divisor of $\mathbb{F}_1$-type, we define $\alpha_c(S, A)$ as follows:

- if $s_A > 4$, we put $\alpha_c(S, A) = \frac{1}{2+a_1+\delta}$,
- if $4 \geqslant s_A > 1$, we let $\alpha_c(S, A)$ to be

$$\max\left(\frac{2}{2 + 2a_1 + s_A - a_2 - a_3 + 2\delta}, \frac{4}{3 + 4a_1 + 2s_A - a_2 - a_3 - a_4 + 4\delta}, \frac{3}{2 + 3a_1 + s_A + 3\delta}\right),$$

- if $1 \geqslant s_A$, we put $\alpha_c(S, A) = \min(\frac{2}{1+2a_1+s_A+2\delta}, 1)$.

Finally, if A is a divisor of $\mathbb{P}^1 \times \mathbb{P}^1$-type, we define $\alpha_c(S, A)$ as follows:

- if $s_A > 4$, we put $\alpha_c(S, A) = \frac{1}{2+a_1+\delta}$,
- if $4 \geqslant s_A > 1$, we let $\alpha_c(S, A)$ to be

$$\max\left(\frac{2}{2 + s_A - a_7 - a_2 - a_3 + 2\delta}, \frac{4}{3 + 2s_A - 2a_7 - a_2 - a_3 - a_4 + 4\delta}, \frac{3}{2 + s_A - a_7 + 3\delta}\right),$$

- if $1 \geqslant s_A$, we put $\alpha_c(S, A) = \min(\frac{2}{1+s_A-a_7+2\delta}, 1)$.

The conjecture of Hong and Won is

Conjecture 11 ([9, Conjecture 4.3]). *If $\alpha(S) = 1$, then $\alpha(S, A) = \alpha_c(S, A)$.*

The main evidence for this conjecture is

Theorem 12 ([9]). *Let D be an effective $\mathbb{Q}$-divisor on the surface S such that $D \sim_{\mathbb{Q}} A$. Then the log pair $(S, \alpha_c(S, A)D)$ is log canonical outside of finitely many points.*

As we already mentioned in Section 1, Example 8 shows that Conjecture 11 is wrong. However, the smooth del Pezzo surface of degree one in Example 8 is rather special. Therefore, Conjecture 11 may hold for *general* smooth del Pezzo surfaces of degree one.

By [5, Remark 1.1.3], it follows from Conjecture 11 that S does not contain A-polar cylinders (see [5, Definition 1.2.1]) when $\alpha(S) = 1$ and a_1 and δ are small enough.

3. Singularities of log pairs

Let S be a smooth surface, and let D be an effective $\mathbb{Q}$-divisor on it. Write

$$D = \sum_{i=1}^{r} a_i C_i$$

where each C_i is an irreducible curve on S, and each a_i is a non-negative rational number. We assume here that all curves $C_1, \ldots, C_r$ are different.

Let $\gamma\colon \mathcal{S} \to S$ be a birational morphism such that the surface $\mathcal{S}$ is smooth as well. It is well known that the morphism γ is a composition of n blow ups of smooth points. Thus, the morphism γ contracts n irreducible curves. Denote these curves by $\Gamma_1, \ldots, \Gamma_n$. For each curve C_i, denote by $\mathcal{C}_i$ its proper transform on the surface $\mathcal{S}$. Then

$$K_{\mathcal{S}} + \sum_{i=1}^{r} a_i \mathcal{C}_i + \sum_{j=1}^{n} b_j \Gamma_j \sim_{\mathbb{Q}} \gamma^*\big(K_S + D\big)$$

for some rational numbers $b_1, \ldots, b_n$. Suppose, in addition, that the divisor

$$\sum_{i=1}^{r} \mathcal{C}_i + \sum_{j=1}^{n} \Gamma_j$$

has simple normal crossing singularities. Fix a point $P \in S$.

Definition 13. The log pair (S, D) is *log canonical* (respectively *Kawamata log terminal*) at the point P if the following two conditions are satisfied:

- $a_i \leqslant 1$ (respectively $a_i < 1$) for every C_i such that $P \in C_i$,
- $b_j \leqslant 1$ (respectively $b_j < 1$) for every Γ_j such that $\pi(\Gamma_j) = P$.

This definition does not depend on the choice of the birational morphism γ.

The log pair (S, D) is said to be *log canonical* (respectively *Kawamata log terminal*) if it is log canonical (respectively, *Kawamata log terminal*) at every point in S.

The following result follows from Definition 13. But it is very handy.

Lemma 14. *Suppose that the singularities of the pair (S, D) are not log canonical at P. Let D' be an effective $\mathbb{Q}$-divisor on S such that (S, D') is log canonical at P and $D' \sim_{\mathbb{Q}} D$. Then there exists an effective $\mathbb{Q}$-divisor D'' on the surface S such that*

$$D'' \sim_{\mathbb{Q}} D,$$

the log pair (S, D'') is not log canonical at P, and $\mathrm{Supp}(D') \not\subseteq \mathrm{Supp}(D'')$.

Proof. Let ϵ be the largest rational number such that $(1+\epsilon)D - \epsilon D'$ is effective. Then

$$(1+\epsilon)D - \epsilon D' \sim_{\mathbb{Q}} D.$$

Put $D'' = (1+\epsilon)D - \epsilon D'$. Then (S, D'') is not log canonical at P, because

$$D = \frac{1}{1+\epsilon}D'' + \frac{\epsilon}{1+\epsilon}D'.$$

Furthermore, we have $\mathrm{Supp}(D') \not\subseteq \mathrm{Supp}(D'')$ by construction. □

Let $f\colon \widetilde{S} \to S$ be a blow up of the point P. Let us denote the f-exceptional curve by F. Denote by $\widetilde{D}$ the proper transform of the divisor D via f. Put $m = \mathrm{mult}_P(D)$.

Theorem 15 ([7, Exercise 6.18]). *If (S, D) is not log canonical at P, then $m > 1$.*

Let C be an irreducible curve in the surface S. Suppose that $P \in C$ and $C \not\subseteq \mathrm{Supp}(D)$. Denote by $\widetilde{C}$ the proper transform of the curve C via f. Fix $a \in \mathbb{Q}$ such that $0 \leqslant a \leqslant 1$. Then $(S, aC + D)$ is not log canonical at P if and only if the log pair

$$\left(\widetilde{S}, a\widetilde{C} + \widetilde{D} + \left(a\,\mathrm{mult}_P\left(C\right) + m - 1\right)F\right) \tag{1}$$

is not log canonical at some point in F. This follows from Definition 13.

Theorem 16 ([7, Exercise 6.31]). *Suppose that C is smooth at P, and $(D\cdot C)_P \leqslant 1$. Then the log pair $(S, aC + D)$ is log canonical at P.*

Corollary 17. *Suppose that the log pair (1) is not log canonical at some point in $F \setminus \widetilde{C}$. Then either $a\,\mathrm{mult}_P(C) + m > 2$ or $m > 1$ (or both).*

Let us give another application of Theorem 16.

Lemma 18. *Suppose that there is a double cover $\pi\colon S \to \mathbb{P}^2$ branched in a curve $R \subset \mathbb{P}^2$. Suppose also that (S, D) is not log canonical at P, and $D \sim_{\mathbb{Q}} \pi^*(\mathcal{O}_{\mathbb{P}^2}(1))$. Then $\pi(P) \in R$.*

Proof. The log pair $(\widetilde{S}, \widetilde{D}+(m-1)F)$ is not log canonical at some point $Q \in F$. Then

$$m+\mathrm{mult}_Q\big(\widetilde{D}\big) > 2 \tag{2}$$

by Theorem 15. Suppose that $\pi(P) \notin R$. Then there is $Z \in |\pi^*(\mathcal{O}_{\mathbb{P}^2}(1))|$ such that

- the curve Z passes through the point P,
- the proper transform of the curve Z on the surface $\widetilde{S}$ contains Q.

Denote by $\widetilde{Z}$ the proper transform of the curve Z on the surface $\widetilde{S}$.

By Lemma 14, we may assume that the support of the $\mathbb{Q}$-divisor D does not contain at least one irreducible component of the curve Z, because (S, Z) is log canonical at P. Thus, if Z is irreducible, then $2-m=\widetilde{Z}\cdot\widetilde{D} \geqslant \mathrm{mult}_Q(\widetilde{D})$, which contradicts (2).

We see that $Z=Z_1+Z_2$, where Z_1 and Z_2 are irreducible smooth rational curves. We may assume that $Z_2 \not\subseteq \mathrm{Supp}(D)$. If $P \in Z_2$, then $1=D\cdot Z_2 \geqslant m > 1$ by Theorem 15. This shows that $P \in Z_1$ and $Z_1 \subseteq \mathrm{Supp}(D)$.

Let d be the degree of the curve R. Then $Z_1^2=Z_2^2=\frac{2-d}{2}$ and $Z_1\cdot Z_2=\frac{d}{2}$.

We may assume that $C_1=Z_1$. Put $\Delta=a_2C_2+\cdots+a_rC_r$. Then $a_1 \leqslant \frac{2}{d}$, since

$$1=Z_2\cdot D=Z_2\cdot\Big(a_1C_1+\Delta\Big)=a_1Z_2\cdot C_1+Z_2\cdot\Delta \geqslant a_1Z_2\cdot C_1=\frac{a_1d}{2}.$$

Denote by $\widetilde{C}_1$ the proper transform of the curve C_1 on the surface $\widetilde{S}$. Then $Q \in \widetilde{C}_1$. Denote by $\widetilde{\Delta}$ the proper transform of the $\mathbb{Q}$-divisor Δ on the surface $\widetilde{S}$. The log pair

$$\left(\widetilde{S}, a_1\widetilde{C}_1+\widetilde{\Delta}+\Big(a_1+\mathrm{mult}_P\left(\Delta\right)-1\Big)F\right)$$

is not log canonical at the point Q by construction. By Theorem 16, we have

$$1+\frac{d-2}{2}a_1-\mathrm{mult}_P\left(\Delta\right)=\widetilde{C}_1\cdot\widetilde{\Delta} \geqslant \big(\widetilde{C}_1\cdot\widetilde{\Delta}\big)_Q > 1-\Big(a_1+\mathrm{mult}_P\left(\Delta\right)-1\Big),$$

so that $a_1 > \frac{2}{d}$. But we already proved that $a_1 \leqslant \frac{2}{d}$. □

Fix a point $Q \in F$. Put $\widetilde{m}=\mathrm{mult}_Q(\widetilde{D})$. Let $g\colon \widehat{S} \to \widetilde{S}$ be a blow up of the point Q. Denote by $\widehat{C}$ and $\widehat{F}$ the proper transforms of the curves $\widetilde{C}$ and F via g, respectively. Similarly, let us denote by $\widehat{D}$ the proper transform of the $\mathbb{Q}$-divisor D on the surface $\widehat{S}$. Denote by G the g-exceptional curve. If the log pair (1) is not log canonical at Q, then

$$\left(\widehat{S}, a\widehat{C}+\widehat{D}+\Big(a\,\mathrm{mult}_P\left(C\right)+m-1\Big)\widehat{F}+\Big(a\,\mathrm{mult}_P\left(C\right)+a\,\mathrm{mult}_Q\big(\widetilde{C}\big)+m+\widetilde{m}-2\Big)G\right) \tag{3}$$

is not log canonical at some point in G.

Lemma 19. *Suppose* $m \leqslant 1$, $a\,\mathrm{mult}_P(C)+m \leqslant 2$ *and* $a\,\mathrm{mult}_P\left(C\right)+a\,\mathrm{mult}_Q(\widetilde{C})+2m \leqslant 3$. *Then* (3) *is log canonical at every point in* $G \setminus \widehat{C}$.

Proof. Suppose that (3) is not log canonical at some point $O \in G$ such that $O \notin \widehat{C}$. If $O \notin \widehat{F}$, then $1 \geqslant m \geqslant \widetilde{m} = \widehat{D} \cdot G \geqslant (\widehat{D} \cdot G)_O > 1$ by Theorem 16. Then $O \in \widehat{F}$. Then

$$m - \widetilde{m} = \left(\widehat{D} \cdot \widehat{F}\right)_O > 1 - \left(a\,\mathrm{mult}_P\left(C\right) + a\,\mathrm{mult}_Q\left(\widetilde{C}\right) + m + \widetilde{m} - 2\right)$$

by Theorem 16. This is impossible, since $a\,\mathrm{mult}_P(C) + a\,\mathrm{mult}_Q(\widetilde{C}) + 2m \leqslant 3$. □

Fix a point $O \in G$. Put $\widehat{m} = \mathrm{mult}_O(\widehat{D})$. Let $h\colon \overline{S} \to \widehat{S}$ be a blow up of the point O. Denote by $\overline{C}$, $\overline{F}$, $\overline{G}$ the proper transforms of the curves $\widehat{C}$, $\widehat{F}$ and G via h, respectively. Similarly, let us denote by $\overline{D}$ the proper transform of the $\mathbb{Q}$-divisor D on the surface $\overline{S}$. Let H be the h-exceptional curve. If $O = G \cap \widehat{F}$ and (3) is not log canonical at O, then

$$\left(\overline{S}, a\overline{C} + \overline{D} + \left(2a\mathrm{mult}_P\left(C\right) + a\mathrm{mult}_Q\left(\widetilde{C}\right) + a\mathrm{mult}_O\left(\widehat{C}\right) + 2m + \widetilde{m} + \widehat{m} - 4\right)H \right.$$
$$\left. + \left(a\mathrm{mult}_P\left(C\right) + m - 1\right)\overline{F} + \left(a\mathrm{mult}_P\left(C\right) + a\mathrm{mult}_Q\left(\widetilde{C}\right) + m + \widetilde{m} - 2\right)\overline{G}\right) \qquad (4)$$

is not log canonical at some point in H.

Lemma 20. *Suppose that* $O = G \cap \widehat{F}$, $m \leqslant 1$, $a\,\mathrm{mult}_P(C) + a\,\mathrm{mult}_Q(\widetilde{C}) + m + \widetilde{m} \leqslant 3$ *and*

$$2a\,\mathrm{mult}_P\left(C\right) + a\,\mathrm{mult}_Q\left(\widetilde{C}\right) + a\,\mathrm{mult}_O\left(\widehat{C}\right) + 4m \leqslant 5.$$

Then the log pair (4) is log canonical at every point in $H \setminus \overline{C}$.

Proof. Suppose that the pair (4) is not log canonical at some point $E \in H$ such that $E \notin \overline{C}$. If $E \notin \overline{F} \cup \overline{G}$, then $m \geqslant \widehat{m} = \overline{D} \cdot H \geqslant (\overline{D} \cdot H)_E > 1$ by Theorem 16. Then $E \in \overline{F} \cup \overline{G}$.

If $E \in \overline{G}$, then $E \notin \overline{F}$, so that Theorem 16 gives

$$\begin{aligned}\widetilde{m} - \widehat{m} &= \left(\overline{D} \cdot \overline{F}\right)_E \\ &> 1 - \left(2a\,\mathrm{mult}_P\left(C\right) + a\,\mathrm{mult}_Q\left(\widetilde{C}\right) + a\,\mathrm{mult}_O\left(\widehat{C}\right) + 2m + \widetilde{m} + \widehat{m} - 4\right),\end{aligned}$$

which is impossible, since $2a\,\mathrm{mult}_P(C) + a\,\mathrm{mult}_Q(\widetilde{C}) + a\,\mathrm{mult}_O(\widehat{C}) + 4m \leqslant 5$ by assumption. Similarly, if $E \in \overline{F}$, then $E \notin \overline{G}$, so that Theorem 16 gives

$$\begin{aligned}m - \widetilde{m} - \widehat{m} &= \left(\overline{D} \cdot \overline{F}\right)_E \\ &> 1 - \left(2a\,\mathrm{mult}_P\left(C\right) + a\,\mathrm{mult}_Q\left(\widetilde{C}\right) + a\,\mathrm{mult}_O\left(\widehat{C}\right) + 2m + \widetilde{m} + \widehat{m} - 4\right),\end{aligned}$$

which is impossible, since $2a\,\mathrm{mult}_P(C) + a\,\mathrm{mult}_Q(\widetilde{C}) + a\,\mathrm{mult}_O(\widehat{C}) + 4m \leqslant 5$. □

Let Z be an irreducible curve in S such that $P \in Z$. Suppose also that $Z \not\subseteq \mathrm{Supp}(D)$. Denote its proper transforms on the surfaces $\widetilde{S}$ and $\widehat{S}$ by the symbols $\widetilde{Z}$ and $\widehat{Z}$, respectively. Fix $b \in \mathbb{Q}$ such that $0 \leqslant b \leqslant 1$. If $(S, aC + bZ + D)$ is not log canonical at P, then

$$\left(\widetilde{S}, a\widetilde{C} + b\widetilde{Z} + \widetilde{D} + \Big(a\,\mathrm{mult}_P\,(C) + b\,\mathrm{mult}_P\,(Z) + m - 1\Big)F\right) \tag{5}$$

is not log canonical at some point in F.

Lemma 21. *Suppose that $m \leqslant 1$ and*

$$a\,\mathrm{mult}_P\,(C) + b\,\mathrm{mult}_P\,(Z) + m \leqslant 2.$$

Then (5) is log canonical at every point in $Q \in F \setminus (\widetilde{C} \cup \widetilde{Z})$.

Proof. Suppose that (5) is not log canonical at some point $Q \in F$ such that $Q \notin \widetilde{C} \cup \widetilde{Z}$. Then $m = \widetilde{D} \cdot F \geqslant (\widetilde{D} \cdot F)_Q > 1$ by Theorem 16. But $m \leqslant 1$ by assumption. □

If the log pair (5) is not log canonical at Q, then the log pair

$$\begin{aligned}\Bigg(&\widehat{S}, a\widehat{C} + b\widehat{Z} + \widehat{D} + \Big(a\,\mathrm{mult}_P\,(C) + b\,\mathrm{mult}_P\,(Z) + m - 1\Big)F \\ &+ \Big(a\,\mathrm{mult}_P\,(C) + a\,\mathrm{mult}_Q\,(\widetilde{C}) + b\,\mathrm{mult}_P\,(Z) + b\,\mathrm{mult}_Q\,(\widetilde{Z}) + m + \widetilde{m} - 2\Big)G\Bigg)\end{aligned} \tag{6}$$

is not log canonical at some point in G.

Lemma 22. *Suppose that $m \leqslant 1$, $a\,\mathrm{mult}_P(C) + b\,\mathrm{mult}_P(Z) + m \leqslant 2$ and*

$$a\,\mathrm{mult}_P\,(C) + a\,\mathrm{mult}_Q\,(\widetilde{C}) + b\,\mathrm{mult}_P\,(Z) + b\,\mathrm{mult}_Q\,(\widetilde{Z}) + 2m \leqslant 3.$$

Then the log pair (6) is log canonical at every point in $G \setminus (\widehat{C} \cup \widehat{Z})$.

Proof. We may assume that the log pair (6) is not log canonical at O and $O \notin \widehat{C} \cup \widehat{Z}$. If $O \notin \widehat{F}$, then $m \geqslant \widetilde{m} = \widehat{D} \cdot G \geqslant (\widehat{D} \cdot G)_O > 1$ by Theorem 16, so that $O \in \widehat{F}$. Then

$$\begin{aligned} m - \widetilde{m} &= \big(\widehat{D} \cdot \widehat{F}\big)_O \\ &> 1 - \Big(a\,\mathrm{mult}_P\,(C) + a\,\mathrm{mult}_Q\,(\widetilde{C}) + b\,\mathrm{mult}_P\,(Z) + b\,\mathrm{mult}_Q\,(\widetilde{Z}) + m + \widetilde{m} - 2\Big),\end{aligned}$$

by Theorem 16, so that

$$a\,\mathrm{mult}_P\,(C) + a\,\mathrm{mult}_Q(\widetilde{C}) + b\,\mathrm{mult}_P(Z) + b\,\mathrm{mult}_Q(\widetilde{Z}) + 2m > 3.$$ □

4. Eight local lemmas

Let us use notations and assumptions of Section 3. Fix $x \in \mathbb{Q}$ such that $0 \leqslant x \leqslant 1$. Put

$$\mathrm{lct}_P\big(S, C\big) = \sup\Big\{\lambda \in \mathbb{Q}\ \Big|\ \text{the log pair } \big(S, \lambda C\big) \text{ is log canonical at } P\Big\} \in \mathbb{Q}_{>0}.$$

Lemma 23. *Suppose that C has an ordinary node or an ordinary cusp at P, $a \leqslant \frac{x}{2}$ and*

$$\big(D \cdot C\big)_P \leqslant \frac{4}{3} + \frac{x}{6} - a.$$

Then the log pair $(S, aC + D)$ is log canonical at P.

Proof. We have $2m \leqslant \mathrm{mult}_P(D)\,\mathrm{mult}_P(C) \leqslant (D{\cdot}C)_P \leqslant \frac{4}{3}+\frac{x}{6}-a$, so that $2m+a \leqslant \frac{4}{3}+\frac{x}{6}$. Then $m \leqslant \frac{3}{4}$ and $m+2a = m+\frac{a}{2}+\frac{3a}{2} \leqslant \frac{\frac{4}{3}+\frac{x}{6}}{2}+\frac{3a}{2} \leqslant \frac{\frac{4}{3}+\frac{x}{6}}{2}+\frac{3x}{4} = \frac{2}{3}+\frac{5}{6}x \leqslant \frac{3}{2}$.

Suppose that $(S, aC + D)$ is not log canonical at P. Let us seek for a contradiction. We may assume that (1) is not log canonical at Q. Then $Q \in \widetilde{C}$ by Corollary 17. Then

$$\big(\widetilde{D} \cdot \widetilde{C}\big)_O > 1 - \big(2a + m - 1\big)\big(\widetilde{C} \cdot F\big)_O \geqslant 1 - 2\big(2a + m - 1\big) = 3 - 4a - 2m.$$

On the other hand, we have $\frac{4}{3} + \frac{x}{6} - a \geqslant (D \cdot C)_P \geqslant 2m + (\widetilde{D} \cdot \widetilde{C})_O$, so that $a > \frac{5}{9} - \frac{x}{18}$. Then $\frac{x}{2} \geqslant a > \frac{5}{9} - \frac{x}{18}$, so that $x > 1$. But $x \leqslant 1$ by assumption. □

Lemma 24. *Suppose that C has an ordinary node or an ordinary cusp at P, and*

$$\big(D \cdot C\big)_P \leqslant \mathrm{lct}_P(S, C) + \frac{x}{2}.$$

Suppose also that $a \leqslant \mathrm{lct}_P(S, C) - \frac{x}{2}$. Then $(S, aC + D)$ is log canonical at P.

Proof. We have $2m \leqslant (D \cdot C)_P$. This gives $2m + a \leqslant 1 + \frac{x}{2}$. Thus, we have $m \leqslant \frac{1+\frac{x}{2}}{2} \leqslant \frac{3}{4}$. Similarly, we get $m+2a = m+\frac{a}{2}+\frac{3a}{2} \leqslant \frac{1+\frac{x}{2}}{2}+\frac{3a}{2} \leqslant \frac{1+\frac{x}{2}}{2}+\frac{3}{2}(1-\frac{x}{2}) = 2 - \frac{x}{2} \leqslant 2$.

Suppose that $(S, aC + D)$ is not log canonical at P. Let us seek for a contradiction. We may assume that the pair (1) is not log canonical at Q. Then $Q \in \widetilde{C}$ by Corollary 17. We may assume that (3) is not log canonical at O. Then $O \in \widehat{C}$ by Lemma 19, since

$$3a + 2m \leqslant 2a + 1 + \frac{x}{2} \leqslant 2 - x + 1 + \frac{x}{2} = 3 - \frac{x}{2} \leqslant 3,$$

because $2m + a \leqslant 1 + \frac{x}{2}$ and $a \leqslant 1 - \frac{x}{2}$. If $O \notin \widehat{F}$, then Theorem 16 gives

$$1 + \frac{x}{2} - a \geqslant \big(D \cdot C\big)_P - 2m - \widetilde{m} \geqslant \big(\widehat{D} \cdot \widehat{C}\big)_O > 1 - \big(3a + m + \widetilde{m} - 2\big),$$

which implies that $2a+\frac{x}{2} > 2+m$. But $2a+\frac{x}{2} \leqslant 2-\frac{x}{2}$, because $a \leqslant \mathrm{lct}_P(S, C)-\frac{x}{2} \leqslant 1 - \frac{x}{2}$. This shows that $O = G \cap \widehat{F} \cap \widehat{C}$. In particular, the curve C has an ordinary cusp at P. By assumption, we have $a \leqslant \frac{5}{6} - \frac{x}{2}$ and $2m + a \leqslant \frac{5}{6} + \frac{x}{2}$. This gives $6a + 4m \leqslant 5 - x \leqslant 5$.

Put $E = H \cap \overline{C}$. Then (4) is not log canonical at E by Lemma 20. Then

$$\left(\overline{D} \cdot \overline{C}\right)_E > 1 - \left(6a + 2m + \widetilde{m} + \widehat{m} - 4\right) = 5 - 6a - 2m - \widetilde{m} - \widehat{m}$$

by Theorem 16. Thus, we have $\frac{5}{6}+\frac{x}{2}-a \geqslant (D\cdot C)_P \geqslant 2m+\widetilde{m}+\widehat{m}+(\overline{D}\cdot\overline{C})_E > 5-6a$. This gives $a > \frac{5}{6} - \frac{x}{10}$. But $a \leqslant \frac{5}{6} - \frac{x}{2}$, which is absurd. □

Lemma 25. *Suppose that C is smooth at P, $a \leqslant \frac{1}{3} + \frac{x}{2}$, $m + a \leqslant 1 + \frac{x}{2}$ and*

$$\left(D \cdot C\right)_P \leqslant 1 - \frac{x}{2} + a.$$

Then the log pair $(S, aC + D)$ is log canonical at P.

Proof. We have $m \leqslant (D \cdot C)_P$, so that $m - a \leqslant 1 - \frac{x}{2}$. Then $m \leqslant 1$, since $m + a \leqslant 1 + \frac{x}{2}$.

Suppose that $(S, aC + D)$ is not log canonical at P. Let us seek for a contradiction. We may assume that the pair (1) is not log canonical at Q. Then $Q \in \widetilde{C}$ by Corollary 17. We may assume that (3) is not log canonical at O. Then $O \in \widehat{C}$ by Lemmas 19. Then

$$\left(\widehat{D} \cdot \widehat{C}\right)_O > 1 - \left(2a + m + \widetilde{m} - 2\right) = 3 - 2a - m - \widetilde{m}$$

by Theorem 16. Then $1 - \frac{x}{2} + a \geqslant (D\cdot C)_P \geqslant m + (\widetilde{D}\cdot\widetilde{C})_Q \geqslant m + \widetilde{m} + (\widehat{D}\cdot\widehat{C})_O > 3 - 2a$. This give $a > \frac{2}{3} + \frac{x}{6}$, which is impossible, since $a \leqslant \frac{1}{3} + \frac{x}{2}$ and $x \leqslant 1$. □

Lemma 26. *Suppose that C is smooth at P, $a \leqslant \frac{8}{9} - \frac{x}{18}$, $m + a \leqslant \frac{4}{3} + \frac{x}{6}$ and*

$$\left(D \cdot C\right)_P \leqslant \frac{x}{2} + a.$$

Then the log pair $(S, aC + D)$ is log canonical at P.

Proof. We have $m \leqslant (D \cdot C)_P$, so that $m - a \leqslant \frac{x}{2}$. Then $m \leqslant \frac{2}{3} + \frac{x}{3} \leqslant 1$, since $m + a \leqslant \frac{4}{3} + \frac{x}{6}$.

Suppose that $(S, aC + D)$ is not log canonical at P. Let us seek for a contradiction. We may assume that the pair (1) is not log canonical at Q. Then $Q \in \widetilde{C}$ by Corollary 17. We may assume that (3) is not log canonical at O. Then $O \in \widehat{C}$ by Lemmas 19. Then

$$\left(\widehat{D} \cdot \widehat{C}\right)_O > 1 - \left(2a + m + \widetilde{m} - 2\right) = 3 - 2a - m - \widetilde{m}$$

by Theorem 16. Then $\frac{x}{2}+a \geqslant (D\cdot C)_P \geqslant m+(\widetilde{D}\cdot\widetilde{C})_Q \geqslant m+\widetilde{m}+(\widehat{D}\cdot\widehat{C})_O > 3-2a$. This gives $a > 1 - \frac{x}{6}$, which is impossible, since $a \leqslant \frac{8}{9} - \frac{x}{18}$ and $x \leqslant 1$. □

Lemma 27. *Suppose that C has an ordinary node or an ordinary cusp at P, $a \leqslant \frac{1+x}{3}$ and*

$$\left(D \cdot C\right)_P \leqslant 2 - 2a.$$

Then the log pair $(S, aC + D)$ is log canonical at P.

Proof. We have $2m \leqslant (D \cdot C)_P \leqslant 2 - 2a$. This gives $m + a \leqslant 1$, so that we have $m \leqslant 1$. Then $m + 2a \leqslant 1 + a \leqslant 1 + \frac{1+x}{3} = \frac{4+x}{3} \leqslant \frac{5}{3}$ and $3a + 2m \leqslant 2 + a \leqslant 2 + \frac{1+x}{3} = \frac{7+x}{3} \leqslant \frac{8}{3}$.

Suppose that $(S, aC + D)$ is not log canonical at P. Let us seek for a contradiction. We may assume that the pair (1) is not log canonical at Q. Then $Q \in \widetilde{C}$ by Corollary 17. We may assume that (3) is not log canonical at O. Then $O \in \widehat{C}$ by Lemma 19.

If $O \notin \widehat{F}$, then $(\widehat{D} \cdot \widehat{C})_O > 3 - 3a - m - \widetilde{m}$ by Theorem 16, so that

$$2 - 2a \geqslant \left(D \cdot C\right)_P \geqslant 2m + \left(\widetilde{D} \cdot \widetilde{C}\right)_Q \geqslant 2m + \widetilde{m} + \left(\widehat{D} \cdot \widehat{C}\right)_O > 3 - 3a,$$

which is absurd. This shows that $O = G \cap \widehat{F} \cap \widehat{C}$. Then

$$\left(\widehat{D} \cdot \widehat{C}\right)_O > 1 - \left(2a + m - 1\right) - \left(3a + m + \widetilde{m} - 2\right) = 4 - 5a - 2m - \widetilde{m}$$

by Theorem 16. Then $2 - 2a \geqslant (D \cdot C)_P \geqslant 2m + \widetilde{m} + (\widehat{D} \cdot \widehat{C})_O > 4 - 5a$, so that $a > \frac{2}{3}$. But $a \leqslant \frac{1+x}{3} \leqslant \frac{2}{3}$ by assumption. This is a contradiction. □

Lemma 28. *Suppose that C has an ordinary node or an ordinary cusp at P, $a \leqslant \frac{2}{3}$ and*

$$\left(D \cdot C\right)_P \leqslant \frac{4}{3} + \frac{2x}{3} - 2a.$$

Then the log pair $(S, aC + D)$ is log canonical at P.

Proof. We have $2m \leqslant (D \cdot C)_P$, so that $m + a \leqslant \frac{2}{3} + \frac{x}{3} \leqslant 1$. Then $m \leqslant 1$ and $m + 2a \leqslant \frac{5}{3}$. Similarly, we see that $3a + 2m \leqslant \frac{4}{3} + \frac{2x}{3} + a \leqslant \frac{4}{3} + \frac{2x}{3} + \frac{2}{3} = 2 + \frac{2x}{3} \leqslant \frac{8}{3} < 3$.

Suppose that $(S, aC + D)$ is not log canonical at P. Let us seek for a contradiction. We may assume that the pair (1) is not log canonical at Q. Then $Q \in \widetilde{C}$ by Corollary 17. We may assume that (3) is not log canonical at O. Then $O \in \widehat{C}$ by Lemma 19.

If $O \notin \widehat{F}$, then $\frac{4}{3} + \frac{2x}{3} - 2a \geqslant (D \cdot C)_P \geqslant 2m + \widetilde{m} + (\widehat{D} \cdot \widehat{C})_O > m + 3 - 3a$ by Theorem 16. Therefore, if $O \notin \widehat{F}$, then $a > \frac{5}{3} - \frac{2x}{3} \geqslant 1$. But $a \leqslant \frac{2}{3}$. This shows that $O = G \cap \widehat{F} \cap \widehat{C}$. Then $(\widehat{D} \cdot \widehat{C})_O > 1 - (2a + m - 1) - (3a + m + \widetilde{m} - 2) = 4 - 5a - 2m - \widetilde{m}$ by Theorem 16. Then $\frac{4}{3} + \frac{2x}{3} - 2a \geqslant (D \cdot C)_P \geqslant 2m + \widetilde{m} + (\widehat{D} \cdot \widehat{C})_O > 4 - 5a$, which gives $a > \frac{2}{3}$. □

Lemma 29. *Suppose that C and Z are smooth at P, $(C \cdot Z)_P \leqslant 2$, and $a + b + m \leqslant 1 + \frac{x}{2}$. Suppose also that $a \leqslant \frac{1+x}{3}$, $b \leqslant \frac{1+x}{3}$, $(D \cdot C)_P \leqslant 1 + a - 2b$ and $(D \cdot Z)_P \leqslant 1 + b - 2a$. Then the log pair $(S, aC + bZ + D)$ is log canonical at P.*

Proof. We have $m \leqslant (D \cdot C)_P \leqslant 1 + a - 2b$ and $m \leqslant (D \cdot Z)_P \leqslant 1 + b - 2a$. Then $m + \frac{a+b}{2} \leqslant 1$.

Suppose that $(S, aC + bZ + D)$ is not log canonical at P. Let us seek for a contradiction. We may assume that (5) is not log canonical at Q. Then $Q \in \widetilde{C} \cup \widetilde{Z}$ by Lemma 21. Without loss of generality, we may assume that $\widetilde{C}$ contains Q. Then

$\widetilde{Z}$ also contains Q. Indeed, if $Q \notin \widetilde{Z}$, then $1+a-2b \geqslant (D\cdot C)_P \geqslant m+(\widetilde{D}\cdot\widetilde{C})_Q > 2-a-b$ by Theorem 16. But $1+b-2a \geqslant 0$. Thus, we have $Q = G\cap\widetilde{C}\cap\widetilde{Z}$, so that $(C\cdot Z)_P = 2$.

We may assume that (6) is not log canonical at O. Then $O \in \widehat{C}\cup\widehat{Z}$ by Lemma 22. In particular, we have $O \notin \widehat{F}$. Without loss of generality, we may assume that $O \in \widehat{C}$. By Theorem 16, we have $1+a-2b-m-\widetilde{m} \geqslant (\widehat{D}\cdot\widehat{C})_O > 1-(2a+2b+m+\widetilde{m}-2)$. This gives $a > \frac{2}{3}$, which is impossible, since $a \leqslant 1+\frac{x}{2} \leqslant \frac{2}{3}$. □

Lemma 30. *Suppose that C and Z are smooth at P, $(C\cdot Z)_P \leqslant 2$, and $a+b+m \leqslant \frac{4}{3}+\frac{x}{6}$. Suppose also that $a \leqslant \frac{2}{3}$, $b \leqslant \frac{2}{3}$, $(D\cdot C)_P \leqslant \frac{2+x}{3}+a-2b$ and $(D\cdot Z)_P \leqslant \frac{2+x}{3}+b-2a$. Then the log pair $(S, aC+bZ+D)$ is log canonical at P.*

Proof. We have $m \leqslant (D\cdot C)_P \leqslant \frac{2+x}{3}+a-2b$ and we have $m \leqslant (D\cdot Z)_P \leqslant \frac{2+x}{3}+b-2a$. Then $m+\frac{a+b}{2} \leqslant \frac{2+x}{3} \leqslant 1$, $m+a+b \leqslant \frac{4}{3}+\frac{x}{6} \leqslant \frac{3}{2}$ and $2a-b \leqslant 1$.

Suppose that $(S, aC+bZ+D)$ is not log canonical at P. Let us seek for a contradiction. We may assume that (5) is not log canonical at Q. Then $Q \in \widetilde{C}\cup\widetilde{Z}$ by Lemma 21. Without loss of generality, we may assume that Q is contained in $\widetilde{C}$. Then $Q \in \widetilde{C}\cap\widetilde{Z}$. Indeed, if $\widetilde{Z}$ does not contain Q, then $\frac{2+x}{3}+a-2b \geqslant m+\left(\widetilde{D}\cdot\widetilde{C}\right)_Q > 2-a-b$ by Theorem 16. The later inequality immediately leads to a contradiction, since $2a-b \leqslant 1$.

We may assume that (6) is not log canonical at O. Then $O \in \widehat{C}\cup\widehat{Z}$ by Lemmas 22. In particular, we have $O \notin \widehat{F}$. Without loss of generality, we may assume that $O \in \widehat{C}$. Then $\frac{2+x}{3}+a-2b-m-\widetilde{m} \geqslant (\widehat{D}\cdot\widehat{C})_O > 1-(2a+2b+m+\widetilde{m}-2)$ by Theorem 16. This gives $a > \frac{7-x}{9}$, which is impossible, since $a \leqslant \frac{2}{3}$. □

5. The proof of the main result

Let S be a smooth del Pezzo surface such that $K_S^2 = 1$. Then $|-2K_S|$ is base point free. It is well known that the linear system $|-2K_S|$ gives a double cover $S \to \mathbb{P}(1,1,2)$. This double cover induces an involution $\tau \in \mathrm{Aut}(S)$.

Let C be an irreducible curve in S such that $C^2 = -1$. Then $-K_S\cdot C = 1$ and $C \cong \mathbb{P}^1$. Put $\widetilde{C} = \tau(C)$. Then $\widetilde{C}^2 = K_S\cdot\widetilde{C} = -1$ and $\widetilde{C} \cong \mathbb{P}^1$. Moreover, we have $C+\widetilde{C} \sim -2K_S$. Furthermore, the irreducible curve $\widetilde{C}$ is uniquely determined by this rational equivalence. Since $C\cdot(C+\widetilde{C}) = -2K_S\cdot C = 2$ and $C^2 = -1$, we have $C\cdot\widetilde{C} = 3$, so that $1 \leqslant |C\cap\widetilde{C}| \leqslant 3$.

Fix $\lambda \in \mathbb{Q}$. Then $-K_S+\lambda C$ is ample $\iff$ $-\frac{1}{3} < \lambda < 1$. Indeed, we have

$$-K_S+\lambda C \sim_{\mathbb{Q}} \frac{1}{2}(C+\widetilde{C})+\lambda C = \Big(\frac{1}{2}+\lambda\Big)C+\frac{1}{2}\widetilde{C} \sim_{\mathbb{Q}} (1+2\lambda)\Big(-K_S-\frac{\lambda}{1+2\lambda}\widetilde{C}\Big). \tag{7}$$

One the other hand, we have $(-K_S+\lambda C)\cdot C = 1-\lambda$ and $(-K_S+\lambda C)\cdot\widetilde{C} = 1-3\lambda$.

Note that Theorem 7 and (7) imply

Corollary 31. *Suppose that* $-\frac{1}{3} < \lambda < 1$. *If* $|C \cap \widetilde{C}| \geqslant 2$, *then*

$$\alpha(S, -K_S + \lambda C) = \begin{cases} \min\left(\frac{\alpha(X)}{1+2\lambda}, 2\right) & \text{if } -\frac{1}{3} < \lambda < 0, \\ \min\left(\alpha(X), \frac{2}{1+2\lambda}\right) & \text{if } 0 \leqslant \lambda < 1. \end{cases}$$

Similarly, if $|C \cap \widetilde{C}| = 1$, *then*

$$\alpha(S, -K_S + \lambda C) = \begin{cases} \min\left(\frac{\alpha(X)}{1+2\lambda}, \frac{4}{3+3\lambda}\right) & \text{if } -\frac{1}{3} < \lambda < 0, \\ \min\left(\alpha(X), \frac{4}{3+3\lambda}\right) & \text{if } 0 \leqslant \lambda < 1. \end{cases}$$

Now let us prove Theorem 7. Suppose that $0 \leqslant \lambda < 1$. Put

$$\mu = \begin{cases} \min\left(\alpha(S), \frac{2}{1+2\lambda}\right) & \text{when } |C \cap \widetilde{C}| \geqslant 2, \\ \min\left(\alpha(S), \frac{4}{3+3\lambda}\right) & \text{when } |C \cap \widetilde{C}| = 1. \end{cases} \tag{8}$$

Lemma 32. *One has* $\alpha(S, -K_S + \lambda C) \leqslant \mu$.

Proof. Since we have $(\frac{1}{2}+\lambda)C + \frac{1}{2}\widetilde{C} \sim_{\mathbb{Q}} -K_S + \lambda C$, we see that $\alpha(S, -K_S+\lambda C) \leqslant \frac{2}{1+2\lambda}$. Similarly, we see that $\alpha(S, -K_S + \lambda C) \leqslant \alpha(S)$. If $|C \cap \widetilde{C}| = 1$, then the log pair

$$\left(S, \frac{2+4\lambda}{3+3\lambda}C + \frac{2}{4+3\lambda}\widetilde{C}\right)$$

is not Kawamata log terminal at the point $C \cap \widetilde{C}$, so that $\alpha(S, -K_S + \lambda C) \leqslant \frac{4}{3+3\lambda}$. □

Thus, to complete the proof of Theorem 7, we have to show that $\alpha(S, -K_S + \lambda C) \geqslant \mu$. Suppose that $\alpha(S, -K_S + \lambda C) < \mu$. Let us seek for a contradiction.

Since $\alpha(S, -K_S + \lambda C) < \mu$, there exists an effective $\mathbb{Q}$-divisor D on S such that

$$D \sim_{\mathbb{Q}} -K_S + \lambda C,$$

and $(S, \mu D)$ is not log canonical at some point $P \in S$.

By Lemma 14 and (7), we may assume that $\mathrm{Supp}(D)$ does not contain C or $\widetilde{C}$. Indeed, one can check that the log pair $(S, \mu(\frac{1}{2} + \lambda)C + \frac{\mu}{2}\widetilde{C})$ is log canonical at P.

Let $\mathcal{C}$ be a curve in the pencil $|-K_S|$ that passes through P. Then $\mathcal{C} + \lambda C \sim -K_S + \lambda C$. Moreover, the curve $\mathcal{C}$ is irreducible, and the log pair $(S, \mu\mathcal{C} + \mu\lambda C)$ is log canonical at P. Thus, we may assume that $\mathrm{Supp}(D)$ does not contain C or $\mathcal{C}$ by Lemma 14.

Lemma 33. *The curve* $\mathcal{C}$ *is smooth at the point* P.

Proof. Suppose that $\mathcal{C}$ is singular at P. If $\mathcal{C} \not\subseteq \mathrm{Supp}(D)$, then Theorem 15 gives

$$1+\lambda=\mathcal{C}\cdot\Big(-K_S+\lambda C\Big)=\mathcal{C}\cdot D\geqslant \mathrm{mult}_P\big(\mathcal{C}\big)\,\mathrm{mult}_P\big(D\big)\geqslant 2\,\mathrm{mult}_P\big(D\big)>\frac{2}{\mu},$$

which is impossible by (8). Thus, we have $\mathcal{C}\subseteq \mathrm{Supp}(D)$. Then $C\not\subseteq \mathrm{Supp}(D)$.

Write $D=\epsilon\mathcal{C}+\Delta$, where ϵ is a positive rational number, and Δ is an effective $\mathbb{Q}$-divisor on the surface S whose support does not contain the curves $\mathcal{C}$ and C. Then

$$1-\lambda=C\cdot\Big(-K_S+\lambda C\Big)=C\cdot D=C\cdot\Big(\epsilon\mathcal{C}+\Delta\Big)=\epsilon+C\cdot\Delta\geqslant\epsilon,$$

so that $\epsilon\leqslant 1-\lambda$. Similarly, we have

$$1+\lambda-\epsilon=\mathcal{C}\cdot\Delta\geqslant\big(\mathcal{C}\cdot\Delta\big)_P. \tag{9}$$

We claim that $\lambda\leqslant\frac{1}{2}$. Indeed, suppose that $\lambda>\frac{1}{2}$. Then it follows from (9) that

$$\big(\Delta\cdot\mathcal{C}\big)_P\leqslant 1+\lambda-\epsilon=\frac{1+2\lambda}{2}\left(\frac{4}{3}+\frac{\frac{4-4\lambda}{1+2\lambda}}{6}-\frac{2}{1+2\lambda}\epsilon\right).$$

Thus, we can apply Lemma 23 to the log pair $(S,\frac{2}{1+2\lambda}D)$ with $x=\frac{4-4\lambda}{1+2\lambda}$ and $a=\frac{2}{1+2\lambda}\epsilon$. This implies that $(S,\frac{2}{1+2\lambda}D)$ is log canonical at P, which is impossible, because $\mu\leqslant\frac{2}{1+2\lambda}$.

If $\mathcal{C}$ has a node at P, then we can apply Lemma 24 to (S,D) with $x=2\lambda$ and $a=\epsilon$. This implies that (S,D) is log canonical, which is absurd, since $\mu\leqslant 1$.

Therefore, the curve $\mathcal{C}$ has an ordinary cusp at P and $\lambda\leqslant\frac{1}{2}$. Then $\mu\leqslant\alpha(S)=\frac{5}{6}$. Thus, we can apply Lemma 23 to the log pair $(S,\frac{5}{6}D)$ with $x=\frac{5}{3}\lambda$ and $a=\frac{5}{6}\epsilon$, since

$$\big(\Delta\cdot\mathcal{C}\big)_P\leqslant\frac{6}{5}\left(\frac{5}{6}+\frac{5}{6}\lambda-\frac{5}{6}\epsilon\right).$$

This implies that $(S,\frac{5}{6}D)$ is log canonical at P, which is impossible, since $\mu\leqslant\frac{5}{6}$. □

The next step in the proof of Theorem 7 is

Lemma 34. *The point P is not contained in the curve C.*

Proof. Suppose that $P\in C$. Let us seek for a contradiction. If $C\not\subseteq\mathrm{Supp}(D)$, then

$$1-\lambda=C\cdot\Big(-K_S+\lambda C\Big)=C\cdot D\geqslant\mathrm{mult}_P\big(C\big)\,\mathrm{mult}_P\big(D\big)\geqslant\mathrm{mult}_P\big(D\big)>\frac{1}{\mu}$$

by Theorem 15. But (8) implies that $\mu>\frac{1}{1-\lambda}$, which is impossible, because $\mu\leqslant 1$. Therefore, we must have $C\subseteq\mathrm{Supp}(D)$. Then $\mathcal{C}\not\subseteq\mathrm{Supp}(D)$ and also $\widetilde{C}\not\subseteq\mathrm{Supp}(D)$.

Write $D=\epsilon C+\Delta$, where ϵ is a positive rational number, and Δ is an effective divisor whose support does not contain $\mathcal{C}$, C and $\widetilde{C}$. Then $1+\lambda-\epsilon=$

$\mathcal{C} \cdot \Delta \geqslant \mathrm{mult}_P(\Delta)$. Similarly, we have $1+3\lambda-3\epsilon = \widetilde{C} \cdot \Delta \geqslant 0$. Finally, we have $1-\lambda+\epsilon = C \cdot \Delta \geqslant (C \cdot \Delta)_P$.

If $\lambda \leqslant \frac{1}{2}$, we can apply Lemma 25 to the log pair (S, D) with $x = 2\lambda$ and $a = \epsilon$. This implies that (S, D) is log canonical, which is impossible since $\mu \leqslant 1$.

Therefore, we have $\lambda > \frac{1}{2}$. Since $\epsilon \leqslant \frac{1}{3} + \lambda$, we have $\frac{2}{1+2\lambda}\epsilon \leqslant \frac{2}{1+2\lambda}(\frac{1}{3}+\lambda) = \frac{8}{9} - \frac{\frac{4-4\lambda}{1+2\lambda}}{18}$. Since $\epsilon + \mathrm{mult}_P(\Delta) \leqslant 1+\lambda$, we have $\frac{2}{1+2\lambda}\epsilon + \frac{2}{1+2\lambda}\,\mathrm{mult}_P(\Delta) \leqslant \frac{2}{1+2\lambda}(1+\lambda) = \frac{4}{3} + \frac{\frac{4-4\lambda}{1+2\lambda}}{6}$. But

$$\left(\Delta \cdot C\right)_P \leqslant 1-\lambda+\epsilon = \frac{1+2\lambda}{2}\left(\frac{\frac{4-4\lambda}{1+2\lambda}}{2} + \frac{2}{1+2\lambda}\epsilon\right).$$

Thus, we can apply Lemma 26 to the log pair $(S, \frac{2}{1+2\lambda}D)$ with $x = \frac{4-4\lambda}{1+2\lambda}$ and $a = \frac{2}{1+2\lambda}\epsilon$. This implies that $(S, \frac{2}{1+2\lambda}D)$ is log canonical at P, which is impossible, since $\mu \leqslant \frac{2}{1+2\lambda}$. □

Let $h\colon S \to \overline{S}$ be the contraction of the curve C. Put $\overline{D} = h(D)$. Then $\overline{D} \sim_{\mathbb{Q}} -K_{\overline{S}}$. Moreover, it follows from Lemma 34 that $(\overline{S}, \mu\overline{D})$ is not log canonical at the point $h(P)$.

By construction, the surface $\overline{S}$ is a smooth del Pezzo surface such that $K_{\overline{S}}^2 = K_S^2 + 1 = 2$. Then $|-K_{\overline{S}}|$ gives a double cover $\pi\colon \overline{S} \to \mathbb{P}^2$ branched in a smooth quartic curve $R_4 \subset \mathbb{P}^2$. By Lemma 18, there exists a unique curve $\overline{Z} \in |-K_{\overline{S}}|$ such that $\overline{Z}$ is singular at $h(P)$. Moreover, the log pair $(\overline{S}, \overline{Z})$ is not log canonical at the point $h(P)$ by [4, Theorem 1.12]. Note that $\pi(\overline{Z})$ is the line in $\mathbb{P}^2$ that is tangent to the curve R_4 at the point $\pi \circ h(P)$.

Let Z be the proper transform of the curve $\overline{Z}$ on the surface S. Then $h(C) \not\in \overline{Z}$. Indeed, if $h(C)$ is contained in $\overline{Z}$, then $Z \sim -K_S$, which is impossible by Lemma 33. Thus, we see that $C \cap Z = \varnothing$. Then $Z \sim -K_S + C$.

Lemma 35. *The curve Z is reducible.*

Proof. Suppose that Z is irreducible. Then Z has an ordinary node or ordinary cusp at P. In fact, if $Z \not\subseteq \mathrm{Supp}(D)$, then $2 = Z \cdot D > \frac{2}{\mu}$ by Theorem 15, which contradicts to (8). Therefore, we have $Z \subseteq \mathrm{Supp}(D)$. Put $\widetilde{Z} = \tau(Z)$. Then $Z + \widetilde{Z} \sim -4K_S$ and

$$\frac{3\lambda+1}{4}Z + \frac{1-\lambda}{4}\widetilde{Z} \sim_{\mathbb{Q}} \frac{1-\lambda}{4}\left(Z + \widetilde{Z}\right) + \lambda Z \sim_{\mathbb{Q}} -K_S + \lambda C.$$

Furthermore, one can show (using Definition 13) that the log pair

$$\left(S, \mu\frac{3\lambda+1}{4}Z + \mu\frac{1-\lambda}{4}\widetilde{Z}\right)$$

is log canonical at P. Hence, we may assume that $\widetilde{Z} \not\subseteq \mathrm{Supp}(D)$ by Lemma 14.

Write $D = \epsilon Z + \Delta$, where ϵ is a positive rational number, and Δ is an effective $\mathbb{Q}$-divisor on the surface S whose support does not contain Z and $\widetilde{Z}$. Then $2 + 4\lambda - 6\epsilon = \widetilde{Z} \cdot \Delta \geqslant 0$. Thus, we have $\epsilon \leqslant \frac{1+2\lambda}{3}$. Finally, we have

$$2 - 2\epsilon = Z \cdot \Delta \geqslant \big(Z \cdot \Delta\big)_P.$$

Therefore, if $\lambda \leqslant \frac{1}{2}$, then we can apply Lemma 27 to (S, D) with $x = 2\lambda$ and $a = \epsilon$. This implies that (S, D) is log canonical at P. But $\mu \leqslant 1$. Thus, we have $\lambda > \frac{1}{2}$.

We have $\mu \leqslant \frac{2}{1+2\lambda}$. Then $(S, \frac{2}{1+2\lambda}D)$ is not log canonical at P. We have $\frac{2}{1+2\lambda}\epsilon \leqslant \frac{2}{3}$. Thus, we can apply Lemma 28 to $(S, \frac{2}{1+2\lambda}D)$ with $x = \frac{4-4\lambda}{1+2\lambda}$ and $a = \frac{2}{1+2\lambda}\epsilon$, because

$$\big(\Delta \cdot Z\big)_P \leqslant \frac{1+2\lambda}{2}\left(\frac{4}{3} + \frac{2\frac{4-4\lambda}{1+2\lambda}}{3} - 2\frac{2}{1+2\lambda}\epsilon\right) = 2 - 2\epsilon.$$

This implies that $(S, \frac{2}{1+2\lambda}D)$ is log canonical at P, which is absurd, since $\mu \leqslant \frac{2}{1+2\lambda}$. □

Since Z is reducible, $Z = Z_1 + Z_2$, where Z_1 and Z_2 are smooth irreducible curves. Then $Z_1^2 = Z_2^2 = -1$ and $Z_1 \cdot Z_2 = 2$. Moreover, we have $P \in Z_1 \cap Z_2$ and $(Z_1 \cdot Z_2)_P \leqslant 2$. Furthermore, we have $Z_1 \cap C = \varnothing$ and $Z_2 \cap C = \varnothing$.

We have $Z_1 \subseteq \mathrm{Supp}(D)$ and $Z_2 \subseteq \mathrm{Supp}(D)$. Indeed, if $Z_1 \nsubseteq \mathrm{Supp}(D)$, then

$$1 = Z_1 \cdot \Big(-K_S + \lambda C\Big) = Z_1 \cdot D \geqslant \mathrm{mult}_P\big(Z_1\big)\,\mathrm{mult}_P\big(D\big) \geqslant \mathrm{mult}_P\big(D\big) > \frac{1}{\mu} \geqslant 1$$

by Theorem 15. This shows that $Z_1 \subseteq \mathrm{Supp}(D)$. Similarly, we have $Z_2 \subseteq \mathrm{Supp}(D)$. But

$$(1-\lambda)\mathcal{C} + \lambda\big(Z_1 + Z_2\big) \sim_{\mathbb{Q}} -K_S + \lambda C.$$

On the other hand, the log pair $(S, \mu(1-\lambda)\mathcal{C} + \mu\lambda(Z_1 + Z_2))$ is log canonical at P. Therefore, we may assume that $\mathcal{C} \nsubseteq \mathrm{Supp}(D)$ by Lemma 14.

Put $\widetilde{Z}_1 = \tau(Z_1)$ and put $\widetilde{Z}_2 = \tau(Z_2)$. Then $Z_1 + \widetilde{Z}_1 \sim -2K_S$ and $Z_2 + \widetilde{Z}_2 \sim -2K_S$. This gives $\mathcal{C} \cdot Z_1 = \mathcal{C} \cdot Z_2 = 1$, $Z_1 \cdot \widetilde{Z}_1 = Z_2 \cdot \widetilde{Z}_2 = 3$, $Z_1 \cdot \widetilde{Z}_2 = Z_2 \cdot \widetilde{Z}_1 = 0$, $\widetilde{Z}_1 \cdot C = \widetilde{Z}_2 \cdot C = 2$. Moreover, we have $Z_1 + Z_2 \sim -K_S + C$. Then

$$\frac{1+\lambda}{2}Z_1 + \lambda Z_2 + \frac{1-\lambda}{2}\widetilde{Z}_1 \sim_{\mathbb{Q}} \frac{1-\lambda}{2}\Big(Z_1 + \widetilde{Z}_1\Big) + \lambda\Big(Z_1 + Z_2\Big) \sim_{\mathbb{Q}} -K_S + \lambda C$$

Note that $P \notin \widetilde{Z}_1$, because $P \in Z_2$ and $\widetilde{Z}_1 \cdot Z_2 = 0$. Using this, we see that the log pair

$$\left(S, \mu\frac{1+\lambda}{2}Z_1 + \mu\lambda Z_2 + \mu\frac{1-\lambda}{2}\widetilde{Z}_1\right)$$

is log canonical at the point P. Hence, we may assume that $\widetilde{Z}_1 \nsubseteq \mathrm{Supp}(D)$ by Lemma 14. Similarly, we may assume that $\widetilde{Z}_2 \nsubseteq \mathrm{Supp}(D)$ using Lemma 14 one more time.

Now let us write $D = \epsilon_1 Z_1 + \epsilon_2 Z_2 + \Delta$, where ϵ_1 and ϵ_2 are positive rational numbers, and Δ is an effective divisor whose support does not contain Z_1 and Z_2. Then

$$1 + \lambda - \epsilon_1 - \epsilon_2 = \mathcal{C} \cdot \Delta \geqslant \mathrm{mult}_P\left(\Delta\right).$$

This gives $\epsilon_1 + \epsilon_2 + \mathrm{mult}_P(\Delta) \leqslant 1 + \lambda$. We also have $\epsilon_1 \leqslant \frac{1+2\lambda}{3}$, since

$$1 + 2\lambda - 3\epsilon_1 = \widetilde{Z}_1 \cdot \Delta \geqslant 0.$$

Similarly, see that $\epsilon_2 \leqslant \frac{1+2\lambda}{3}$. Moreover, we have

$$1 + \epsilon_1 - 2\epsilon_2 = Z_1 \cdot \Delta \geqslant \left(Z_1 \cdot \Delta\right)_P.$$

Finally, we have

$$1 + \epsilon_2 - 2\epsilon_1 = Z_2 \cdot \Delta \geqslant \left(Z_2 \cdot \Delta\right)_P.$$

Thus, if $\lambda \leqslant \frac{1}{2}$, then we can apply Lemma 29 to (S, D) with $x = 2\lambda$, $a = \epsilon_1$ and $b = \epsilon_1$. This implies that (S, D) is log canonical at P, which is absurd. Hence, we have $\lambda > \frac{1}{2}$.

Since $\lambda > \frac{1}{2}$, we have $\mu \leqslant \frac{2}{1+2\lambda}$. Then the log pair $(S, \frac{2}{1+2\lambda}D)$ is not log canonical at P. On the other hand, we have $\frac{2}{1+2\lambda}\epsilon_1 \leqslant \frac{2}{3}$ and $\frac{2}{1+2\lambda}\epsilon_2 \leqslant \frac{2}{3}$. We also have

$$\begin{aligned}\frac{2}{1+2\lambda}\epsilon_1 + \frac{2}{1+2\lambda}\epsilon_2 + \frac{2}{1+2\lambda}\,\mathrm{mult}_P(\Delta) &\leqslant \frac{2}{1+2\lambda}(1+\lambda)\\ &= \frac{2}{1+2\lambda} + \lambda\frac{2}{1+2\lambda} = \frac{4}{3} + \frac{\frac{4-4\lambda}{1+2\lambda}}{6},\end{aligned}$$

Moreover, we have

$$\left(\Delta \cdot Z_1\right)_P \leqslant 1 + \epsilon_1 - 2\epsilon_2 = \frac{1+2\lambda}{2}\left(\frac{2}{3} + \frac{\frac{4-4\lambda}{1+2\lambda}}{3} + \frac{2}{1+2\lambda}\epsilon_1 - 2\frac{2}{1+2\lambda}\epsilon_2\right).$$

Furthermore, we also have

$$\left(\Delta \cdot Z_2\right)_P \leqslant 1 + \epsilon_1 - 2\epsilon_2 = \frac{1+2\lambda}{2}\left(\frac{2}{3} + \frac{\frac{4-4\lambda}{1+2\lambda}}{3} + \frac{2}{1+2\lambda}\epsilon_2 - 2\frac{2}{1+2\lambda}\epsilon_1\right).$$

Thus, we can apply Lemma 30 to $(S, \frac{2}{1+2\lambda}D)$ with $x = \frac{4-4\lambda}{1+2\lambda}$, $a = \frac{2}{1+2\lambda}\epsilon_1$ and $b = \frac{2}{1+2\lambda}\epsilon_2$. This implies that $(S, \frac{2}{1+2\lambda}D)$ is log canonical at P, which is absurd.

The obtained contradiction completes the proof of Theorem 7.

Acknowledgement

The main result of this paper was proved in the summer 2016. This happened during the 35th workshop on Geometric Methods in Physics in Białowieża. We would like to thank the organizers of this workshop for creative working conditions. We also would like to thank the referee for helpful comments.

References

[1] I. Cheltsov, *Log canonical thresholds of del Pezzo surfaces*, Geometric and Functional Analysis **18** (2008), 1118–1144.

[2] I. Cheltsov, J. Martinez-Garcia, *Stable polarized del Pezzo surfaces*, preprint, arXiv:1606.04370 (2016).

[3] I. Cheltsov, K. Shramov, *Log canonical thresholds of smooth Fano threefolds*, Russian Mathematical Surveys **63** (2008), 859–958.

[4] I. Cheltsov, J. Park, J. Won, *Affine cones over smooth cubic surfaces*, Journal of the European Mathematical Society **18** (2016), 1537–1564.

[5] I. Cheltsov, J. Park, J. Won, *Cylinders in del Pezzo surfaces*, International Mathematics Research Notices **2017** (2017), 1179–1230.

[6] X.-X. Chen, S. Donaldson, S. Sun, *Kähler–Einstein metrics on Fano manifolds. I, II, III*, Journal of the American Mathematical Society **28** (2015), no. 1, 183–197, 199–234, 235–278.

[7] A. Corti, J. Kollár, K. Smith, *Rational and nearly rational varieties*, Cambridge Studies in Advanced Mathematics **92** (2004), Cambridge University Press.

[8] R. Dervan, *Alpha invariants and K-stability for general polarisations of Fano varieties*, International Mathematics Research Notices **16** (2015), 7162–7189.

[9] K. Hong, J. Won, *Alpha invariant for general polarizations of del Pezzo surfaces of degree* 1, Mathematische Nachrichten **290** (2017), 1689–1700.

[10] Y. Odaka, Y. Sano, *Alpha invariant and K-stability of Q-Fano varieties* Advances in Mathematics **229** (2012), 2818–2834.

[11] G. Tian, *On Kähler–Einstein metrics on certain Kähler manifolds with* $\mathrm{c}_1(M) > 0$, Inventiones Mathematicae **89** (1987), 225–246.

[12] G. Tian, *Existence of Einstein metrics on Fano manifolds*, Metric and Differential Geometry, Progress in Mathematics **297** (2012), 119–159.

Ivan Cheltsov
School of Mathematics
University of Edinburgh
Edinburgh EH9 3FD, Scotland

Geometric Methods in Physics. XXXV Workshop 2016
Trends in Mathematics, 175–180

Short-time Behavior of the Exciton-polariton Equations

Cristi D. Guevara and Stephen P. Shipman

Abstract. In the exciton-polariton system, a linear dispersive photon field is coupled to a nonlinear exciton field. Short-time analysis of the lossless system shows that, when the photon field is excited, the time required for that field to exhibit nonlinear effects is longer than the time required for the nonlinear Schrödinger equation, in which the photon field itself is nonlinear

Mathematics Subject Classification (2010). Primary 35Q55; Secondary 35B40, 35A35.

Keywords. Exciton-polariton, nonlinear dispersion, short time evolution, nonlinear Schrödinger.

1. Short-time behavior of the exciton-polariton system

The lossless unforced exciton-polariton system is a quantum-mechanical system involving a linear dispersive photon wave-function $\phi(\mathbf{x}, t)$ and a nonlinear nondispersive exciton wave-function $\psi(\mathbf{x}, t)$ of spatial coordinates $\mathbf{x} \in \mathbb{R}^n$ and time $t \in \mathbb{R}$:

$$\begin{aligned} i\phi_t &= -\Delta\phi + \gamma\psi \\ i\psi_t &= (\omega_0 + g|\psi|^2)\psi + \gamma\phi\,. \end{aligned} \tag{1}$$

For physical discussions of these equations, the reader is referred to [1, 2, 5], among many other references.

The fact that the dispersive term $-\Delta\phi$ and the nonlinear term $g|\psi|^2\psi$ involve different fields results in fundamental differences between the exciton-polariton system (EP) and the nonlinear Schrödinger (NLS) equation

$$i\phi_t = -\Delta\phi + g|\phi|^2\phi\,, \tag{2}$$

in which both terms involve a single field ϕ. The NLS equation is Galilean-invariant, whereas the EP system is not; and NLS admits a frequency-scalable "ground state", whereas the structure of stationary harmonic solutions of EP is

SPS acknowledges the support of NSF research grant DMS-1411393.

complicated [3]. This communication addresses a fundamental difference in the short-time behavior of these two systems.

We take the point of view that the photon field is excited and measured by the observer and that the exciton field is hidden from the observer. Thus we impose initial conditions

$$\begin{aligned} \phi(\mathbf{x},0) &= \phi_0(\mathbf{x}) \\ \psi(\mathbf{x},0) &= 0 \end{aligned} \tag{3}$$

and ask, up to what time can nonlinear effects observed in the photon field through its coupling to the exciton field be considered to be negligible?

At first, the effect of the exciton field on the photon field is altogether negligible and the exciton evolves essentially linearly under the influence of the photon. This is described by the approximate system

$$\begin{aligned} i\phi_t &= -\Delta\phi \\ i\psi_t &= \omega_0\psi + \gamma\phi\,. \end{aligned} \qquad \text{(Approximation A)} \tag{4}$$

After some time, the exciton field grows sufficiently large so that its effect on the photon field becomes non-negligible, but the nonlinear effects remain negligible for a longer period of time. The photon acts as if it were coupled to a linear exciton field:

$$\begin{aligned} i\phi_t &= -\Delta\phi + \gamma\psi \\ i\psi_t &= \omega_0\psi + \gamma\phi\,. \end{aligned} \qquad \text{(Approximation B)} \tag{5}$$

At a later time, the nonlinear effects imparted by the exciton field are observed significantly in the photon field and the linear Approximation B is no longer acceptable.

Theorem 1 makes these assertions precise. The deviation of an approximation $\tilde{\phi}$ to the true photon field ϕ is considered to be negligible if the relative error $\|\tilde{\phi}-\phi\|_{H^s(\mathbb{R}^n)}/\|\phi\|_{H^s(\mathbb{R}^n)}$ is less than a small number ϵ, which is allowed to tend to zero. Our main result is that the deviation of the photon field of the linear polariton system from that of the nonlinear one is negligible up to time $t = C\epsilon^{1/5}$. This result is in contrast to the nonlinear Schrödinger equation, for which nonlinear effects are negligible only up to time $C\epsilon$.

Theorem 1. *Let $(\phi(t),\psi(t))$ be a solution of the polariton system (1) in the interval $0 \le t \le T$, with each field being a continuous function of t with values in $H^s(\mathbb{R}^n)$ with $s > n/2$. Let C_1 and C_2 be real numbers, and for all ϵ such that $C_2\epsilon^{1/5} \le T$, let $(\tilde{\phi}(t),\tilde{\psi}(t))$ be a solution of the equations*

$$(\tilde{\phi}(t),\tilde{\psi}(t)) \text{ satisfies } \begin{cases} \text{approx. A (4) for} & 0 \le t \le C_1\epsilon^{1/2} \\ \text{approx. B (5) for} & C_1\epsilon^{1/2} \le t \le C_2\epsilon^{1/5} \end{cases}$$

with $\tilde{\phi}$ and $\tilde{\psi}$ being continuous function of t with values in $H^s(\mathbb{R}^n)$. Let both systems satisfy the initial conditions

$$(\phi(0),\psi(0)) = (\tilde{\phi}(0),\tilde{\psi}(0)) = (\phi_0,0), \tag{6}$$

with $\|\phi_0\|_{H^s(\mathbb{R}^n)} = \|\phi_0\|_s = M \neq 0$.

The relative error in the photon field is bounded by

$$\frac{\|\tilde{\phi}(t)-\phi(t)\|_s}{\|\phi(t)\|_s} \leq K_1\epsilon + O(\epsilon^2) \quad \textit{for} \qquad 0 \leq t \leq C_1\epsilon^{1/2} \qquad (\epsilon \to 0),$$

$$\frac{\|\tilde{\phi}(t)-\phi(t)\|_s}{\|\phi(t)\|_s} \leq K_2\epsilon + O(\epsilon^{7/5}) \; \textit{for} \;\; C_1\epsilon^{1/2} \leq t \leq C_2\epsilon^{1/5} \qquad (\epsilon \to 0),$$

in which

$$K_1 = \tfrac{1}{2}\gamma^2 C_1^2, \qquad K_2 = \tfrac{1}{2}\gamma^2 C_1^2 + \tfrac{1}{5}|g|K\gamma^3 M^2 C_2^5$$

and K is an absolute constant (defined in the proof below).

The proof of this theorem will be given after existence of solutions and preliminary bounds are established.

2. Existence of solutions to the polariton equations

Theorem 2. *Given $0 < r < 1$, $N > 0$, and $\phi_0 \in H^s(\mathbb{R}^n)$ with $s > n/2$, such that $\|\phi_0\|_s \leq rN$, there exists a unique solution to the polariton equations* (1) *subject to $\|\phi\|_{C(I,H^s(\mathbb{R}^n))} \leq N$ and $\|\psi\|_{C(I,H^s(\mathbb{R}^n))} \leq N$ defined for $t \in [0,T]$, where*

$$T = \frac{1-r}{2\gamma + |g|\tilde{K}N^2}$$

for some constant $\tilde{K}$.

Proof. The proof is a standard contraction argument. Write $u = (\phi, \psi)^t$, and consider the space

$$E_{N,r} = \left\{u \in C(I, H^s(\mathbb{R}^n)) \;:\; \|u\|_{C(I,H^s(\mathbb{R}^n))} \leq N,\; \|u_0\|_s \leq rN\right\},$$

with $I = [0,T]$, equipped with the distance $d\,(u_1 - u_2) = \|u_1 - u_2\|_{C(I,H^s(\mathbb{R}^n))}$. $(E_{N,r}, d)$ is a complete metric space. Define a mapping $\Phi : E_{N,r} \to E_{N,r}$ by

$$\Phi(u)(t) = \begin{pmatrix} e^{it\Delta}\phi_0(x) - i\gamma \displaystyle\int_0^t e^{i(t-\tau)\Delta}\psi(\tau)d\tau \\ -i\displaystyle\int_0^t e^{-i\omega_0(t-\tau)}\left(g|\psi|^2\psi(\tau) + \gamma\phi(\tau)\right)d\tau \end{pmatrix}.$$

Minkowski inequalities and the fact that $e^{it\Delta}$ is an isometry in H^s yields

$$\begin{aligned}\|\Phi(u)(t)\|_s &\leq \|\phi_0\|_s + \gamma T\Big(\sup_{\tau\leq T}\|\psi\|_s + \sup_{\tau\leq T}\|\phi\|_s\Big) + |g|KT\sup_{\tau\leq T}\|\psi\|_s^3 \\ &\leq \|\phi_0\|_s + NT\left(2\gamma + |g|KN^2\right).\end{aligned}$$

The constant K for $s > n/2$ is guaranteed by [4, Theorem 3.4]; it relies on the algebra property of $H^s(\mathbb{R}^n)$. For a different constant K', one obtains

$$\|\Phi(u_1) - \Phi(u_2)\|_s \leq T(2\gamma + |g|K'N^2)\Big(\sup_{\tau\leq T}\|\psi_1 - \psi_2\|_s + \sup_{\tau\leq T}\|\phi_1 - \phi_2\|_s\Big).$$

Set $\tilde{K} = \max\{K, K'\}$. Since $T\left(2\gamma + |g|\tilde{K}N^2\right) = 1 - r < 1$, Φ is a contraction of $(E_{N,r}, d)$ and thus it has a unique fixed point, which, by the definition of Φ, satisfies the exciton-polariton system. Uniqueness of the solution in $C(I, H^s(\mathbb{R}^n))$ follows from Gronwall's Lemma. □

3. Bounds for solutions to the polariton equations

Assume that (ϕ, ψ) is a solution of the polariton equations as in Theorem 2, with initial condition $(\phi(\mathbf{x}, 0), \psi(\mathbf{x}, 0)) = (\phi_0(\mathbf{x}), 0)$ and $\|\phi_0\|_s = rN$.

3.1. Solutions of the polariton equations

The integral form of the system (1), namely $\Phi(u) = u$, Minkowski inequalities, and the fact that $e^{it\Delta}$ is an isometry in $H^s(\mathbb{R}^n)$ yield

$$rN - \gamma \int_0^t \|\psi(\tau)\|_s \, d\tau \;\leq\; \|\phi(t)\|_s \;\leq\; rN + \gamma \int_0^t \|\psi(\tau)\|_s \, d\tau\,, \tag{7}$$

$$\|\psi(t)\|_s \leq \gamma rN\,t \,+\, \gamma^2 \int_0^t \int_0^\tau \|\psi(\sigma)\|_s \, d\sigma\, d\tau \,+\, |g|K \int_0^t \|\psi(\tau)\|_s^3 \, d\tau\,. \tag{8}$$

The constant K is guaranteed by [4, Theorem 3.4]. Hence

$$\sup_{\tau \leq t} \|\psi(t)\|_s \;\leq\; \gamma rN\,t + \tfrac{1}{2}\gamma^2 t^2 \sup_{\tau \leq t} \|\psi(\tau)\|_s + |g|K\,t \sup_{\tau \leq t} \|\psi(\tau)\|_s^3\,.$$

The last estimate can be written as $P(t, y(t)) \geq 0$, where

$$y(t) = \sup_{\tau \leq t} \|\psi(t)\|_s \quad \text{and} \quad P(t, y) := \gamma rN\,t + y\left(|g|K\,ty^2 + \tfrac{1}{2}\gamma^2 t^2 - 1\right).$$

For each t such that $P(t, y)$ has two positive roots as a function of y, denote these roots by $y_1(t) \leq y_2(t)$. One can show that $y_1(t)$ is increasing in t, with $\lim_{t\to 0} y_1(t) = 0$ and $y_2(t)$ is decreasing with $\lim_{t \to 0} y_2(t) = \infty$. Thus $P(t, y(t)) \geq 0$ is equivalent to $\{\, y(t) \leq y_1(t) \text{ or } y(t) \geq y_2(t)\,\}$.

We shall assume from now on that t is small enough so that $y(t) \geq y_2(t)$ is ruled out, so that one has $\sup_{\tau \leq t} \|\psi(\tau)\|_s \leq y_1(t)$, or, equivalently, $\|\psi(t)\|_s \leq y_1(t)$, since $y_1(t)$ is increasing. Hence, (8) yields

$$\|\psi(t)\|_s \;\leq\; \gamma rN\,t \,+\, \gamma^2 \int_0^t \int_0^\tau y_1(\sigma)\, d\sigma\, d\tau \,+\, |g|K \int_0^t y_1(\tau)^3 d\tau\,. \tag{9}$$

The Taylor expansion of $y_1(t)$ around $t = 0$ is

$$y_1(t) \;=\; \gamma rN\,t + \tfrac{1}{2}\gamma^3 rN\,t^3 + |g|K(\gamma rN)^3\,t^4 + \cdots\,. \tag{10}$$

Therefore, from (7), (9) and (10), we have

$$rN - \tfrac{1}{2}\gamma^2 rNt^2 + O(t^4) \;\leq\; \|\phi(t)\|_s \;\leq\; rN + \tfrac{1}{2}\gamma^2 rNt^2 + O(t^4)\,. \tag{11}$$

3.2. Solutions of approximate equation A

Let $(\tilde{\phi}(t), \tilde{\psi}(t))$ be the solution of the approximate system A (4) with initial condition $(\tilde{\phi}(0), \tilde{\psi}(0)) = (\phi_0, 0)$, and $(\phi(t), \psi(t))$ be the solution of the true system

(1). Set $\hat{\phi} := \tilde{\phi} - \phi$ and $\hat{\psi} := \tilde{\psi} - \psi$, so that $(\hat{\phi}(t), \hat{\psi}(t))$ satisfies

$$\begin{cases} i\hat{\phi}_t = -\Delta\hat{\phi} + \gamma\psi(t) \\ i\hat{\psi}_t = \omega_0\hat{\psi} + \gamma\hat{\phi} + g|\psi(t)|^2\psi(t) \end{cases} \qquad \begin{cases} \hat{\phi}(0) = 0 \\ \hat{\psi}(0) = 0\,. \end{cases} \tag{12}$$

One obtains the bounds

$$\|\hat{\phi}(t)\|_s \le \gamma \int_0^t \|\psi(\tau)\| d\tau \le \gamma \int_0^t y_1(\tau)\, d\tau, \tag{13}$$

$$\|\hat{\psi}(t)\|_s \le \gamma^2 \int_0^t \int_0^\tau y_1(\sigma)\, d\sigma\, d\tau \,+\, |g|K \int_0^t y_1(\tau)^3 d\tau\,. \tag{14}$$

3.3. Solutions of approximate equation B

Now let $(\tilde{\phi}(t), \tilde{\psi}(t))$ be the solution of the approximate system B (5) with arbitrary initial conditions, and set again $\hat{\phi} := \tilde{\phi} - \phi$ and $\hat{\psi} := \tilde{\psi} - \psi$; then $(\hat{\phi}(t), \hat{\psi}(t))$ satisfies

$$\begin{cases} i\hat{\phi}_t = -\Delta\hat{\phi} + \gamma\hat{\psi}(t) \\ i\hat{\psi}_t = \omega_0\hat{\psi} + \gamma\hat{\phi} + g|\psi(t)|^2\psi(t) \end{cases} \qquad \begin{cases} \hat{\phi}(t_1) = \hat{\phi}_0 \\ \hat{\psi}(t_1) = \hat{\psi}_0 \end{cases}, \tag{15}$$

and from the integral form of (15), one deduces the bounds

$$\|\hat{\phi}(t)\|_s \le \|\hat{\phi}_0\|_s \,+\, \gamma \int_{t_1}^t \|\hat{\psi}(\tau)\|_s d\tau \tag{16}$$

$$\|\hat{\psi}(t)\|_s \le \|\hat{\psi}_0\|_s \,+\, \gamma t\|\hat{\phi}_0\|_s \,+\, \gamma^2 \int_{t_1}^t \int_{t_1}^\tau \|\hat{\psi}(\sigma)\|_s d\sigma\, d\tau \,+\, |g|K \int_{t_1}^t \|\psi(t)\|_s^3 d\tau.$$

Combining this with (10) yields

$$\|\hat{\psi}(t)\|_s \;\le\; \left(1 - \tfrac{1}{2}\gamma^2 t^2\right)^{-1} \left(\|\hat{\psi}_0\|_s \,+\, \gamma t\|\hat{\phi}_0\|_s \,+\, |g|Kt\, y_1(t)^3\right). \tag{17}$$

3.4. Proof of Theorem 1

For the solutions (ϕ, ψ) and $(\tilde{\phi}, \tilde{\psi})$ in the theorem, define $\hat{\phi} := \tilde{\phi} - \phi$ and $\hat{\psi} := \tilde{\psi} - \psi$, and set $M = rN$.

For $t \in [0, t_1]$ with $t_1 = C_1\epsilon^{1/2}$, (13) yields

$$\|\hat{\phi}(t)\|_s \le \gamma \int_0^t y_1(\tau)\, d\tau \le \tfrac{1}{2}\gamma^2 M t^2 + O(t^4) \le \tfrac{1}{2}\gamma^2 M C_1^2 \epsilon + O(\epsilon^2). \tag{18}$$

Using (11), the relative error is controlled by

$$\frac{\|\hat{\phi}(t)\|_s}{\|\phi(t)\|_s} \;\le\; \tfrac{1}{2}\gamma^2 C_1^2 \epsilon + O(\epsilon^2).$$

For $t \in [t_1, t_2]$ with $t_2 = C_2\epsilon^{1/5}$, (10), (14), and (18) give initial bounds

$$\|\hat{\psi}(t_1)\|_s \;\le\; \tfrac{1}{6}\gamma^3 M t_1^3 + O(t_1^4) \;\le\; \tfrac{1}{6}\gamma^3 M C_1^3\, \epsilon^{3/2} + O(\epsilon^2),$$

$$\|\hat{\phi}(t_1)\|_s \;\le\; \tfrac{1}{2}\gamma M C_1^2\, \epsilon + O(\epsilon^2).$$

Using these in (17) yields

$$\|\hat{\psi}(t)\|_s \leq \left(1+O(t^2)\right) \times \left[\tfrac{1}{6}\gamma^2 M C_1^3 \epsilon^{3/2} + O(\epsilon^2) + \gamma t\left(\tfrac{1}{2}\gamma M C_1^2 \epsilon + O(\epsilon^2)\right) + |g| K (\gamma M)^3 t^4 + O(t^6)\right],$$

and then inserting this into (16) gives

$$\|\hat{\phi}(t)\|_s \leq \tfrac{1}{2}\gamma^2 M C_1^2 \epsilon + O(\epsilon^2) + \left(1+\epsilon^{1/5}\right) \times \left[\tfrac{1}{6}\gamma^2 M C_1^3 \epsilon^{3/2} t + O(\epsilon^2)t + \tfrac{1}{4}\gamma M C_1^2 \epsilon t^2 + O(\epsilon^2)t + \tfrac{1}{5}|g| K (\gamma M)^3 t^5 + O(t^7)\right].$$

In view of $t \leq C_2 \epsilon^{1/5}$, the first four terms in the brackets are $O(\epsilon^{17/10})$, $O(\epsilon^{11/5})$, $O(\epsilon^{7/5})$, and $O(\epsilon^{11/5})$, and the last one is $O(\epsilon^{7/5})$. Therefore

$$\|\hat{\phi}(t)\|_s \leq \left(\tfrac{1}{2}\gamma^2 r N C_1^2 + \tfrac{1}{5}|g| K (\gamma r N)^3 C_2^5\right)\epsilon + O(\epsilon^{7/5}). \tag{19}$$

The relative error is obtained from this and (11),

$$\frac{\|\hat{\phi}(t)\|_s}{\|\phi(t)\|_s} \leq \left(\tfrac{1}{2}\gamma^2 C_1^2 + \tfrac{1}{5}|g| K \gamma^3 (rN)^2 C_2^5\right)\epsilon + O(\epsilon^{7/5}).$$

4. Final remark

The analysis above uses strictly $H^s(\mathbb{R}^n)$ estimates and triangle inequalities and does not address whether the time $t = C\epsilon^{1/5}$ is sharp. A future communication will include a comparison between EP and NLS for initial photon data of order ϵ^α and for nonlinearities not just of order 3 but of any power greater than 1.

References

[1] Iacopo Carusotto and Cristiano Ciuti. Probing microcavity polariton superfluidity through resonant rayleigh scattering. *Phys. Rev. Lett.*, 93(16):166401–1–4, Oct. 2004.

[2] B. Deveaud, editor. *The Physics of Semiconductor Microcavities.* Wiley-VCH, Weinheim, 2007.

[3] Stavros Komineas, Stephen P. Shipman, and Stephanos Venakides, Lossless Polariton Solitons. *Physica D*, 316 (2016) 43–56.

[4] Felipe Linares and Gustavo Ponce. *Introduction to nonlinear dispersive equations.* Springer, 2014.

[5] Francesca Maria Marchetti and Marzena H. Szymanska. Vortices in polariton OPO superfluids. Preprint 2011.

Cristi D. Guevara and Stephen P. Shipman
Louisiana State University
College of Science, Department of Mathematics
Baton Rouge, LA 70803, USA
e-mail: cguevara@lsu.edu, cristi.guevara@asu.edu
shipman@math.lsu.edu

Geometric Methods in Physics. XXXV Workshop 2016
Trends in Mathematics, 181–190

Periods of Mixed Tate Motives over Real Quadratic Number Rings

Ivan Horozov

Abstract. Recently, the author defined multiple Dedekind zeta values [5] associated to a number K field and a cone C. In this paper we construct explicitly non-trivial examples of mixed Tate motives over the ring of integers in K, for a real quadratic number field K and a particular cone C. The period of such a motive is a multiple Dedekind zeta values at $(s_1, s_2) = (1, 2)$, associated to the pair $(K; C)$, times a nonzero element of K.

Mathematics Subject Classification (2010). 11M32, 11R42, 14G10, 14G25.

Keywords. Multiple zeta values, Dedekind zeta values, mixed Tate motives, periods, real quadratic fields.

1. Introduction

The Riemann zeta function

$$\zeta(s) = \sum_{n>0} \frac{1}{n^s}$$

is widely used in number theory, algebraic geometry and quantum field theory. Euler's multiple zeta values

$$\zeta(s_1, \ldots, s_m) = \sum_{0<n_1<\cdots<n_m} \frac{1}{n_1^{s_1} \ldots n_m^{s_m}},$$

where $s_1, \ldots, s_m$ are positive integers and $s_m \geq 2$, appear as values of some Feynman amplitudes, and in algebraic geometry, as periods of mixed Tate motives over $\mathrm{Spec}(\mathbb{Z})$ (see [1, 3, 4, 7]).

Dedekind zeta values

$$\zeta_K(s) = \sum_{\mathfrak{a}\neq(0)} \frac{1}{N(\mathfrak{a})^s},$$

are a generalization of the Riemann zeta function to a number field K. In some Feynman amplitudes one of the summands is $\log(1+\sqrt{2})$ or $\log\left(\frac{1+\sqrt{5}}{2}\right)$. These values are essentially the residues at $s=1$ of Dedekind zeta functions over $\mathbb{Q}(\sqrt{2})$ and over $\mathbb{Q}(\sqrt{5})$, respectively. For $s=2,3,4,\dots$ the values $\zeta_K(s)$ are periods of mixed Tate motives over the ring of algebraic integers in K with ramification only at the discriminant of K (see [2]).

In [5], the author has constructed multiple Dedekind zeta values, which are a generalization of Euler's multiple zeta values to number fields in the same way as Dedekind zeta values generalizes Riemann zeta values. For a quadratic number field K, the key examples of multiple Dedekind zeta values are

$$\begin{aligned}&\zeta_{K;C}(s_1,\dots,s_1;\dots;s_m,\dots,s_m)\\ &\quad=\sum_{\alpha_1,\dots,\alpha_m\in C}\frac{1}{N(\alpha_1)^{s_1}N(\alpha_1+\alpha_2)^{s_2}\cdots N(\alpha_1+\cdots+\alpha_m)^{s_m}},\end{aligned} \tag{1}$$

where $s_1,\dots,s_m$ are positive integers and $s_m\geq 2$ and C is a cone generated by a totally positive unit β in K and 1, defined by

$$C=\mathbb{N}\{1,\beta\}=\{\gamma\in K\,|\,\gamma=a+b\beta,\ \text{for positive integers } a \text{ and } b\}.$$

Similar types of cones were considered by Zagier in [8] and [9].

In [5], the author has proven that multiple Dedekind zeta values can be interpolated to multiple Dedekind zeta functions, which have meromorphic continuation to all complex values of the variables $s_1,\dots,s_m$.

In this paper we prove the following theorem.

Theorem 1. *Let K be a real quadratic field, and let C be a cone generated by a totally positive unit β in K and 1. Then the multiple Dedekind zeta values*

$$(\beta_2-\beta_1)^3\zeta_{K;C}(1,2)$$

is a period of a mixed Tate motive over the ring of integers in K. In particular, it is unramified over the primes dividing the discriminant $\sqrt{D}$.

Remark. The proof of the theorem can easily be generalized to all

$$(\beta_2-\beta_1)^{s_1+\cdots+s_m}\zeta_{K;C}(s_1,\dots,s_m)$$

for the same cone C. The details for the general case will be completed in a sequel to this paper. The choice of considering $\zeta_{K;C}(1,2)$ in this paper is two-fold. First, this is among the simplest non-trivial example of a multiple Dedekind zeta value. Second, for any other (multple) Dedekind zeta value, the proof of the corresponding statement is essentially the same.

2. Background

2.1. Multiple zeta values

The Riemann zeta function at the value $s = 2$ can be expressed in term of an iterated integral in the following way

$$\begin{aligned}\int_0^1 \left(\int_0^y \frac{dx}{1-x}\right)\frac{dy}{y} &= \int_0^1 \left(\int_0^y (1+x+x^2+x^3\dots)dx\right)\frac{dy}{y}\\ &= \int_0^1 \left(y+\frac{y^2}{2}+\frac{y^3}{3}+\frac{y^4}{4}+\cdots\right)\frac{dy}{y} = y+\frac{y^2}{2^2}+\frac{y^3}{3^2}+\frac{y^4}{4^2}\cdots\Bigg|_{y=0}^{y=1}\\ &= 1+\frac{1}{2^2}+\frac{1}{3^2}+\frac{1}{4^2}\cdots = \zeta(2).\end{aligned}$$

Let us examine the domain of integration of the iterated integral. Note that $0 < x < y$ and $0 < y < 1$. We can put both inequalities together. Then we obtain the domain $0 < x < y < 1$, which is a simplex. Thus, we can express the iterated integral as

$$\zeta(2) = \int_0^1 \left(\int_0^y \frac{dx}{1-x}\right)\frac{dy}{y} = \int_{0<x<y<1} \frac{dx}{1-x}\wedge\frac{dy}{y}.$$

Moreover, Goncharov and Manin [4] have expressed all multiple zeta values as periods of motives related to the moduli space of curves of genus zero with $n+3$ marked points, $\mathcal{M}_{0,n+3}$. In particular, $\zeta(2)$ can be expressed as a period of the motive $H^2(\overline{\mathcal{M}}_{0,5} - A, B - A\cap B)$ by pairing of $[\Omega_A] \in Gr_4^W H^2(\overline{\mathcal{M}}_{0,5} - A)$ for $\Omega_A = \frac{dx}{1-x}\wedge\frac{dy}{y}$, with $[\Delta_B] \in \left(Gr_0^W H^2(\overline{\mathcal{M}}_{0,5} - B)\right)^\vee$. The Deligne–Mumford compactification $\overline{\mathcal{M}}_{0,5}$ of the moduli space $\mathcal{M}_{0,5}$ can be obtained by three blow-ups of $\mathbb{P}^1\times\mathbb{P}^1$ at the points $(0,0)$, $(1,1)$ and (∞,∞). Let us name the exceptional divisors at the three points by E_0, E_1 and E_∞, respectively. Then $A = (x = 1)\cup(y=0)\cup(x=\infty)\cup(y=\infty)\cup E_\infty$ and $B = (x=0)\cup(x=y)\cup(y=1)\cup E_0\cup E_1$.

Similarly, one can express $\zeta(3)$ and $\zeta(1,2)$ as iterated integrals

$$\begin{aligned}\zeta(3) &= \int_0^1\left(\int_0^z\left(\int_0^y \frac{dx}{1-x}\right)\frac{dy}{y}\right)\frac{dz}{z} &&= \int_{0<x<y<z<1}\frac{dx}{1-x}\wedge\frac{dy}{y}\wedge\frac{dz}{z},\\ \zeta(1,2) &= \int_0^1\left(\int_0^z\left(\int_0^y \frac{dx}{1-x}\right)\frac{dy}{1-y}\right)\frac{dz}{z} &&= \int_{0<x<y<z<1}\frac{dx}{1-x}\wedge\frac{dy}{1-y}\wedge\frac{dz}{z}.\end{aligned}$$

Again, $\zeta(3)$ and $\zeta(1,2)$ can be expressed as periods of motives related to $\mathcal{M}_{0,6}$. In the same paper, Goncharov and Manin prove that the motives associated to multiple zeta values (MZVs) are mixed Tate motives unramified over Spec($\mathbb{Z}$).

A few years later, Francis Brown [1] proved that periods of mixed Tate motives unramified over Spec($\mathbb{Z}$) can be expressed as a $\mathbb{Q}$-linear combination of MZVs times an integer power of $2\pi i$.

2.2. Multiple Dedekind zeta values (MDZVs)

We recall the construction of MDZVs over a real quadratic field K. (See [5] for definition of MDZVs over any number field.) Let $\mathcal{O}_K$ be the ring of integers in K.

And let β be a totally positive unit in $\mathcal{O}_K$. Let C be the cone defined as $\mathbb{N}$-linear combination of 1 and β, that is,

$$C = \{\gamma \in \mathcal{O}_K \mid \gamma = a + b\beta, \text{ for } a, b \in \mathbb{N}\}.$$

Let $f_0(C; t_1, t_2) = \sum_{\gamma\in C} \exp(-t_1\gamma_1 - t_2\gamma_2)$, where γ_1 and γ_2 are two real embeddings of γ. We express $\zeta_{K;C}(2)$, $\zeta_{K;C}(3)$ and $\zeta_{K;C}(1,2)$ as iterated integrals on a membrane. See [5] and [6], for more examples and properties of iterated integrals on membranes.

$$\begin{aligned}
&\int_0^\infty\int_0^\infty\left(\int_{u_1}^\infty\int_{u_2}^\infty f_0(C;t_1,t_2)dt_1\wedge dt_2\right)du_1\wedge du_2\\
&=\int_0^\infty\int_0^\infty\left(\int_{u_1}^\infty\int_{u_2}^\infty\left(\sum_{\gamma\in C}\exp(-t_1\gamma_1-t_2\gamma_2)\right)dt_1\wedge dt_2\right)du_1\wedge du_2\\
&=\int_0^\infty\int_0^\infty\left(\sum_{\gamma\in C}\frac{\exp(-u_1\gamma_1-u_2\gamma_2)}{\gamma_1\gamma_2}\right)du_1\wedge du_2 \qquad (2)\\
&=\sum_{\gamma\in C}\frac{1}{(\gamma_1\gamma_2)^2}\\
&=\sum_{\gamma\in C}\frac{1}{N(\gamma)^2}=\zeta_{K;C}(2).
\end{aligned}$$

Similarly,

$$\begin{aligned}
\zeta_{K;C}(3) &= \sum_{\gamma\in C}\frac{1}{N(\gamma)^3}\\
&=\int_0^\infty\int_0^\infty\left(\int_{v_1}^\infty\int_{v_2}^\infty\left(\int_{u_1}^\infty\int_{u_2}^\infty f_0(C;t_1,t_2)dt_1\wedge dt_2\right)du_1\wedge du_2\right)dv_1\wedge dv_2,
\end{aligned}$$

and

$$\begin{aligned}
\zeta_{K;C}(1,2) = &\sum_{\gamma,\delta\in C}\frac{1}{N(\gamma)^1N(\gamma+\delta)^2}\\
&=\int_0^\infty\int_0^\infty\left(\int_{v_1}^\infty\int_{v_2}^\infty\left(\int_{u_1}^\infty\int_{u_2}^\infty f_0(C;t_1,t_2)dt_1\wedge dt_2\right)\right.\\
&\qquad\left.\times f_0(C;u_1,u_2)du_1\wedge du_2\right)dv_1\wedge dv_2.
\end{aligned}$$

3. Transition to algebraic geometry

We can write the infinite sum in the definition of f_0 as a product of two geometric series

$$
\begin{aligned}
f_0(C;t_1,t_2) &= \sum_{\gamma\in C}\exp(-\gamma_1t_1-\gamma_2t_2)\\
&= \sum_{a=1}^{\infty}\sum_{b=1}^{\infty}\exp[-(a\alpha_1+b\beta_1)t_1-(a\alpha_2+b\beta_2)t_2]\\
&= \sum_{a=1}^{\infty}\sum_{b=1}^{\infty}\exp[-a(\alpha_1t_1+\alpha_2t_2)]\exp[-b(\beta_1t_1+\beta_2t_2)]\\
&= \frac{\exp[-(\alpha_1t_1+\alpha_2t_2)]}{1-\exp[-(\alpha_1t_1+\alpha_2t_2)]}\times\frac{\exp[-(\beta_1t_1+\beta_2t_2)]}{1-\exp[-(\beta_1t_1+\beta_2t_2)]}.
\end{aligned}
$$

Let $x_1=e^{-t_1}$ and $x_2=e^{-t_2}$. Then

$$
f_0(C;t_1,t_2)=\frac{x_1x_2}{1-x_1x_2}\cdot\frac{x_1^{\beta_1}x_2^{\beta_2}}{1-x_1^{\beta_1}x_2^{\beta_2}}. \tag{3}
$$

Now we are going to express f_0 algebraically. At this point there is a problem of raising the variable x to an integer algebraic power. Note thatβ_1 and β_2 are algebraic integers (in fact totally positive units), which are not rational integers.

How do we raise x to power β_1 and to β_2? We introduce new variables

$$
y_1=x_1^{\beta_1}\text{ and }y_2=x_2^{\beta_2}.
$$

Then $x_1^{a+b\beta_1}=x_1^ay_1^b$, where a and b are integers.

We are going to use the variables x_1,x_2. For each of them we introduce y_1,y_2, so that we write y_1 instead of $x_1^{\beta_1}$ and y_2 instead of $x_2^{\beta_2}$. In terms of x_1, x_2, y_1 and y_2, we can express f_0 as

$$
f_0(C;t_1,t_2)=\frac{x_1x_2}{1-x_1x_2}\cdot\frac{x_1^{\beta_1}x_2^{\beta_2}}{1-x_1^{\beta_1}x_2^{\beta_2}}=\frac{x_1x_2}{1-x_1x_2}\cdot\frac{y_1y_2}{1-y_1y_2}.
$$

Let us also define $\omega_1=\frac{d(x_1x_2)}{1-x_1x_2}\wedge\frac{d(y_1y_2)}{1-y_1y_2}$ and let $\omega_0=\frac{d(x_1x_2)}{x_1x_2}\wedge\frac{d(y_1y_2)}{y_1y_2}$.

Key Remark. The differential forms ω_0 and ω_1 will be used for both algebraic geometry on moduli spaces and for defining multiple Dedekind zeta values.

Lemma 2. *If we substitute $x_1=e^{-t_1}$, $x_2=e^{-t_2}$, $y_1=e^{-\beta_1t_1}$ and $y_2=e^{-\beta_2t_2}$, then*

$$
\omega_0=(\beta_2-\beta_1)dt_1\wedge dt_2.
$$

Proof. Consider x_1,x_2,y_1 and y_2 as functions of t_1 and t_2. Then

$$
y_1y_2=x_1^{\beta_1}x_2^{\beta_2}
$$

and

$$
\frac{d(y_1y_2)}{y_1y_2}=\frac{d(x_1^{\beta_1}x_2^{\beta_2})}{x_1^{\beta_1}x_2^{\beta_2}}=\beta_1\frac{dx_1}{x_1}+\beta_2\frac{dx_2}{x_2}=-\beta_1dt_1-\beta_2dt_2.
$$

Similarly,

$$\frac{d(x_1x_2)}{x_1x_2} = -dt_1 - dt_2.$$

Again, as functions of t_1 and t_2, we have

$$\begin{aligned}\omega_0 = \frac{d(x_1x_2)}{x_1x_2} \wedge \frac{d(y_1y_2)}{y_1y_2} &= (dt_1 + dt_2) \wedge (\beta_1 dt_1 + \beta_2 dt_2) \\ &= (\beta_2 - \beta_1) dt_1 \wedge dt_2.\end{aligned}$$

Now let us write $\omega_0(x_1, x_2)$ and $\omega_1(x_1, x_2)$, when we want to specify the dependence on the variables. In fact, both forms depend also on y_1 and y_2; however, we will take care of that by choosing a region of integration together with tangential base points. □

4. Tangential base points

Let $x_1 = e^{-t_1}$ and let $y_1 = e^{-\beta_1 t_1}$ We would like to find an algebraic relation among the variables x_1 and y_1 when they approach $(0,0)$ or when they approach $(1,1)$. That occurs when t_1 approaches ∞ or when t_1 approaches 0, respectively. If $\beta_1 > 1$ then

$$\lim_{t_1 \to \infty} \frac{dy_1}{dx_1} = \lim_{t_1 \to \infty} \frac{de^{-\beta_1 t_1}}{de^{-t_1}} = \lim_{t_1 \to \infty} \beta_1 \frac{e^{t_1}}{(e^{t_1})^{\beta_1}} = 0.$$

Also

$$\lim_{t_1 \to 0} \frac{dy_1}{dx_1} = \lim_{t_1 \to 0} \beta_1 \frac{e^{-\beta_1 t_1}}{e^{-t_1}} = \beta_1.$$

Let

$$\gamma_1 : (0, \infty) \to \mathcal{M}_{0,5},$$
$$\gamma_1(t_1) = (e^{-t_1}, e^{-\beta_1 t_1}) = (x_1, y_1).$$

For a vector $v = (a, b)$, consider $[v] = [a : b]$ as an element of $\mathbb{P}^1$.

We have proven the following lemma.

Lemma 3.

(a) $\lim_{t_1 \to \infty} \left[\frac{d\gamma_1}{dt_1}\right] = [1 : 0]$,

(b) $\lim_{t_1 \to 0} \left[\frac{d\gamma_1}{dt_1}\right] = [1 : \beta_1]$.

Similarly, we have $x_2 = e^{-t_2}$ and $y_2 = e^{-\beta_2 t_2}$ with $0 < \beta_2 < 1$. Let

$$\gamma_2 : (0, \infty) \to \mathcal{M}_{0,5}, \qquad \gamma_2(t_2) = (e^{-t_2}, e^{-\beta_2 t_2}) = (x_2, y_2).$$

The following lemma could be proven in the same way.

Lemma 4.

(a) $\lim_{t_2 \to \infty} \left[\frac{d\gamma_2}{dt_2}\right] = [0 : 1]$,

(b) $\lim_{t_2 \to 0} \left[\frac{d\gamma_1}{dt_2}\right] = [1 : \beta_2]$.

Remark. The paths γ_1 and γ_2 can be used to define a membrane $m = \gamma_1 \times \gamma_2$ by taking a Cartesian products of both the domains and the targets

$$m = \gamma_1 \times \gamma_2 : (0,1)^2 \to (\mathcal{M}_{0,5})^2 .$$

The definition of multiple Dedekind zeta values via iterated integrals on a membrane use exactly the membrane m in the case of quadratic fields (see [5]).

Proposition 5. *With the above choice of tangential base points, we have*

$$\int_{0<x_1<x_3<1;\ 0<x_2<x_4<1} \omega_1(x_1,x_2) \wedge \omega_0(x_3,x_4) = (\beta_2 - \beta_1)^2 \zeta_{K;C}(2).$$

Proof. The differential forms ω_0 and ω_1 are closed. Thus we can vary the paths γ_1 and γ_2 without changing the value of the integral as long as the tangential base points remain the same. Thus, we can choose the parametrization $x_i = e^{-t_i}$ and $y_i = e^{-\beta_i t_i}$, keeping the tangential points fixed. Using Formulas (2) and (3), we obtain

$$\frac{d(x_3x_4)}{x_3x_4} \wedge \frac{d(y_3y_4)}{y_3y_4} = (\beta_2 - \beta_1) dt_3 \wedge dt_4$$

Similarly, we have that

$$\frac{x_1x_2}{1-x_1x_2} \cdot \frac{y_1y_2}{1-y_1y_2} \cdot \left(\frac{d(x_3x_4)}{x_3x_4} \wedge \frac{d(y_3y_4)}{y_3y_4} \right) = f_0(C;t_1,t_2)(\beta_2 - \beta_1) dt_1 \wedge dt_2.$$

Thus, with the above choice of tangential base points, we have

$$\begin{aligned} &\int_{0<x_1<x_3<1;\ 0<x_2<x_4<1} \omega_1(x_1,x_2) \wedge \omega_0(x_3,x_4) \\ &= (\beta_2 - \beta_1)^2 \int_{t_1>t_3>0;\ t_2>t_4>0} f_0(C;t_1,t_2) dt_1 \wedge dt_2 \wedge t_3 \wedge dt_4 \\ &= (\beta_2 - \beta_1)^2 \zeta_{K;C}(2). \end{aligned}$$

□

Corollary 6. *With the above choice of tangential base points, we have*

$$\begin{aligned} &(\beta_2 - \beta_1)^3 \zeta_{K;C}(1,2) \\ &= \int_{0<x_1<x_3<x_5<1;\ 0<x_2<x_4<x_6<1} \omega_1(x_1,x_2) \wedge \omega_1(x_3,x_4) \wedge \omega_0(x_5,x_6). \end{aligned}$$

Theorem 7. *In Corollary 6, the integral on the right-hand side is a period of a mixed Tate motive unramified over a real quadratic number ring.*

Proof. In this proof we are going to follow closely the paper by Goncharov and Manin [4]. The period will be a pairing between $[\Omega_A] \in Gr^W_{12} H^6(\overline{\mathcal{M}}_{0,15} - A)$ and $[\Delta_B] \in \left(Gr^W_0 H^6(\overline{\mathcal{M}}_{0,15} - B)\right)^\vee$ associated to a mixed Tate motive $H^6(\overline{\mathcal{M}}_{0,15} - A; B - A \cap B)$.

Let the $(4n)$-coordinates $x_{2i-1}, y_{2i-1}, z_{2i-1}, w_{2i-1}$ for indices $i = 1, 2, \ldots, n$, be a coordinate of a point on $\mathcal{M}_{0,4n+3}$. One can think of $\mathcal{M}_{0,4n+3}$ as $(\mathbb{P}^1)^{4n} - D$

where the divisor D is obtained by setting any of the coordinates to be 0, 1, ∞ or setting any two of the coordinates to be equal. Let us define

$$x_{2i} = \frac{1}{z_{2i-1}} \quad \text{and} \quad y_{2i} = \frac{1}{w_{2i-1}}.$$

Now the coordinates of any point on $\mathcal{M}_{0,4n+3}$ can be written as $(x_1, y_1, x_2, y_2, \ldots, x_{2n}, y_{2n})$. In terms of the new coordinates, we have the following components of D:

$$\begin{aligned} x_i &= 0, \quad x_i = 1, \;\; x_i = \infty, \\ y_i &= 0, \quad y_i = 1, \;\; y_i = \infty, \\ x_1 &= x_3, \; x_3 = x_5, \\ y_1 &= y_3, \; y_3 = y_5, \\ x_1x_2 &= 1, \\ x_3x_4 &= 1, \\ y_1y_2 &= 1, \\ y_3y_4 &= 1. \end{aligned}$$

The last four components can be realized in terms of the previous coordinates as $x_1 = z_1$, $x_3 = z_3$, $y_1 = w_1$ and $y_3 = w_3$.

Let $n = 3$. Let $\overline{\mathcal{M}}_{0,4n+3} = \overline{\mathcal{M}}_{0,15}$ be the Deligne–Mumford compactification of the moduli space of curves of genus 0 with 15 marked points. The ambient space will be $\overline{\mathcal{M}}_{0,15}$. From it we will remove a divisor A whose components occur as poles of the differential forms under the integral. Explicitly, the differential forms are

$$\omega_1(x_1, x_2) = \frac{d(x_1x_2)}{1 - x_1x_2} \wedge \frac{d(y_1y_2)}{1 - y_1y_2}, \tag{4}$$

$$\omega_1(x_3, x_4) = \frac{d(x_3x_4)}{1 - x_3x_4} \wedge \frac{d(y_3y_4)}{1 - y_3y_4}, \tag{5}$$

$$\omega_0(x_5, x_6) = \frac{d(x_5x_6)}{x_5x_6} \wedge \frac{d(y_5y_6)}{y_5y_6}. \tag{6}$$

The components of the divisor A consists of the union of

$$\begin{aligned} &(x_1x_2 = 1), \; (y_1y_2 = 1), \; (x_3x_4 = 1), \; (y_3y_4 = 1), \\ &(x_5 = 0), \; (x_6 = 0), \; (y_5 = 0), \; (y_6 = 0), \\ &(x_i = \infty), \; (y_i = \infty), \; \text{for } i = 1, 2, \ldots, 6, \end{aligned}$$

together with the exceptional divisors obtained via blow-up at the intersections of two components that both contain the same variable or the same constant 0, 1 or ∞ on the right-hand side of the equalities.

Thus, the differential form

$$\Omega_A = \omega_1(x_1, x_2) \wedge \omega_1(x_3, x_4) \wedge \omega_0(x_5, x_6)$$

is well defined on $\overline{\mathcal{M}}_{0,15} - A$.

Now we proceed to defining B. The key part will be to include the tangential base points in the definition of B.

The components of B consist of a union of codimension 1 subvarieties and codimension 2 subvarieties. The latter ones correspond to the tangential base points.

The codimension 1 components are the following:

$$(x_1 = 0),\ (x_1 = x_3),\ (x_3 = x_5),\ (x_5 = 1),$$
$$(x_2 = 0),\ (x_2 = x_4),\ (x_4 = x_6),\ (x_6 = 1),$$
$$(y_1 = 0),\ (y_1 = y_3),\ (y_3 = y_5),\ (y_5 = 1),$$
$$(y_2 = 0),\ (y_2 = y_4),\ (y_4 = y_6),\ (y_6 = 1),$$

together with the exceptional divisors of the blow-up at an intersection of two subvarieties such that the two polynomials contain the same variable or the same constant 0 or 1 on the right-hand side of the equalities, except the following 4 double intersections of components

$$(x_1 = 0) \quad \text{and} \quad (y_1 = 0),$$
$$(x_2 = 0) \quad \text{and} \quad (y_2 = 0),$$
$$(x_5 = 1) \quad \text{and} \quad (y_5 = 1),$$
$$(x_6 = 1) \quad \text{and} \quad (y_6 = 1),$$

to which we associate a codimension 2 subvarieties of $\overline{\mathcal{M}}_{0,15}$, using the tangential base points.

- For the blow-up at the intersection $(x_1 = 0)$ and $(y_1 = 0)$ we choose a divisor B_1 on the exceptional divisor defined by $[x_1 : y_1] = [1 : 0]$. Note that B_1 is of codimension 2 in $\overline{\mathcal{M}}_{0,15}$.
- For the blow-up at the intersection $(x_2 = 0)$ and $(= 0y_2)$ we choose a divisor B_2 on the exceptional divisor defined by $[x_2 : y_2] = [0 : 1]$.
- For the blow-up at the intersection $(x_5 = 1)$ and $(y_5 = 1)$ we choose a divisor B_5 on the exceptional divisor defined by $[x_5 : y_5] = [1 : \beta_1]$.
- For the blow-up at the intersection $(x_6 = 1)$ and $(y_6 = 1)$ we choose a divisor B_6 on the exceptional divisor defined by $[x_6 : y_6] = [1 : \beta_2]$.

The tangential base points define the components B_1, B_2, B_5, B_6. Thus, $(\beta_2 - \beta_1)^3\zeta_{K,C}(1,2)$ occurs as a period of $H^6(\overline{\mathcal{M}}_{0,15} - A; B - A \cap B)$ when $[\Omega_A] \in Gr^W_{12} H^6(\overline{\mathcal{M}}_{0,15} - A)$ is paired with $[\Delta_B] \in \left(Gr^W_0 H^6(\overline{\mathcal{M}}_{0,15} - B)\right)^\vee$

Note that B_1 and B_2 are defined over $\mathbb{Z}$, and B_5 and B_6 are defined over the ring of integers $\mathcal{O}_K$ of the field K. Each of them is naturally isomorphic to $\overline{\mathcal{M}}_{0,13}$ as a variety over $\mathcal{O}_K$. Similarly, any intersection of the components of B is isomorphic over $\mathcal{O}_K$ to $\overline{\mathcal{M}}_{0,n}$ for some integer n. Using that $H^i(\overline{\mathcal{M}}_{0,n})$ is a mixed Tate motive over $\mathrm{Spec}(\mathcal{O}_K)$, we obtain that the motivic cohomology of the components of B are mixed Tate motives. Using Proposition 1.7 from Deligne and Goncharov, [3], we conclude that for $l \neq char(\nu)$ the l-adic cohomology of the reduction of B_j modulo ν of the motive $H^i(B_j)$ is unramified for any component B_j of B, since B_j

is isomorphic to $\overline{\mathcal{M}}_{0,n}$ over $\mathrm{Spec}(\mathcal{O}_K)$ for some n. We conclude that for $l \neq char(\nu)$ the l-adic cohomology of the reduction modulo any $\nu \in \mathrm{Spec}(\mathcal{O}_K)$ of the motive $H^6(\overline{\mathcal{M}}_{0,15} - A; B - A \cap B)$ is unramified. Thus, $H^6(\overline{\mathcal{M}}_{0,15} - A; B - A \cap B)$ is a mixed Tate motive unramified over $\mathrm{Spec}(\mathcal{O}_K)$. □

Acknowledgement

I would like to thank the organizers of the XXXV Workshop on Geometry and Mathematical Physics in Białowieża, Poland, especially, Emma Previato, for the opportunity to present this result, and for the encouragement and interest in my work. I would also like to thank the referees of the paper for their questions, suggestions and corrections. Due to their comments, the mathematical content of the paper has been improved immensely.

References

[1] Brown, F.: *Mixed Tate motives over* $\mathbb{Z}$. Ann. of Math. (2) 175 (2012), no. 2, 949–976.

[2] Brown, F.: *Dedekind zeta motives for totally real number fields.* Invent. Math. 194 (2013), no. 2, 257–311.

[3] Deligne, P. and Goncharov, A: *Groupes fondamentaux motiviques de Tate mixte.* (French) [Mixed Tate motivic fundamental groups], Ann. Sci. École Norm. Sup. (4) 38 (2005), no. 1, 1–56.

[4] Goncharov, A. and Manin, Yu.: *Multiple ζ-motives and moduli spaces $\overline{\mathcal{M}}_{0,n}$*. Compos. Math. 140 (2004), no. 1, 1–14.

[5] Horozov, I.: *Multiple Dedekind Zeta Functions*, J. Reine Angew. Math. 722 (2017), 65–104. (Crelle's Journal), DOI: 10.1515/crelle-2014-0055

[6] Horozov, I.: *Non-commutative Hilbert modular symbols*, Algebra and Number Theory Vol. 9 (2015), No. 2, 317–370 DOI: 10.2140/ant.2015.9.317

[7] Kontsevich, M. and Zagier, D.: *Periods*, Mathematics unlimited – 2001 and beyond, 771–808, Springer, Berlin, 2001.

[8] Zagier, D.: *A Kronecker limit formula for real quadratic fields.* Math. Ann. 213 (1975), 153–184.

[9] Zagier, D.: *On the values at negative integers of the zeta-function of a real quadratic field*, L'Enseignement Mathématique, 22 (1976) 55–95.

Ivan Horozov
City University of New York
Bronx Community College
Dept. of Mathematics and
Computer Science, CP 315,
2155 University Avenue
Bronx, New York 10453, USA
e-mail: ivan.horozov@bcc.cuny.edu

Geometric Methods in Physics. XXXV Workshop 2016
Trends in Mathematics, 191–200

Soliton Hierarchies from Matrix Loop Algebras

Wen-Xiu Ma and Xing Lü

Abstract. Matrix loop algebras, both semisimple and non-semisimple, are used to generate soliton hierarchies. Hamiltonian structures to guarantee the Liouville integrability are determined by using the trace identity or the variational identity. An application example is presented from a perturbed Kaup–Newell matrix spectral problem associated with the three-dimensional real special linear algebra.

Mathematics Subject Classification (2010). Primary 37K10; Secondary 35Q53.

Keywords. Zero curvature equation, soliton hierarchy, Hamiltonian structure.

1. Introduction

Soliton hierarchies possessing Hamiltonian structures or bi-Hamiltonian structures provide examples of integrable systems. Within given matrix loop algebras, zero curvature equations associated with matrix spectral problems (or equivalently, Lax pairs) are essential objects in generating soliton hierarchies and their Hamiltonian structures (see, e.g., [1–7]).

Among celebrated examples are the Korteweg–de Vries hierarchy [8], the Ablowitz–Kaup–Newell–Segur hierarchy [9], the Dirac hierarchy [10], the Kaup–Newell hierarchy [11], the Wadati–Konno–Ichikawa hierarchy [12] and the Heisenberg hierarchy [13]. All those soliton hierarchies are generated from the three-dimensional real special linear algebra sl(2,$\mathbb{R}$). This Lie algebra is simple and has the basis

$$e_1 = \begin{bmatrix} 1 & 0 \\ 0 & -1 \end{bmatrix}, \; e_2 = \begin{bmatrix} 0 & 1 \\ 0 & 0 \end{bmatrix}, \; e_3 = \begin{bmatrix} 0 & 0 \\ 1 & 0 \end{bmatrix}, \tag{1}$$

with the standard commutation relations:

$$[e_1, e_2] = 2e_2, \; [e_2, e_3] = e_1, \; [e_3, e_1] = 2e_3. \tag{2}$$

Its derived algebra is itself, and so, it is 3-dimensional, too. The only other three-dimensional real Lie algebras with a three-dimensional derived algebra is the special orthogonal algebra so(3,$\mathbb{R}$), whose basis $\{e_1, e_2, e_3\}$ satisfying the circular commutation relations: $[e_1, e_2] = e_3,\ [e_2, e_3] = e_1,\ [e_3, e_1] = e_2$, is called a standard basis. Those two Lie algebras have been widely used in generating soliton hierarchies in integrable systems (see, e.g., [8–16] using sl(2,$\mathbb{R}$) and [17–21] using so(3,$\mathbb{R}$)).

For a given matrix Lie algebra g, its loop algebra $\tilde{g}$ adopted in this paper is defined as

$$\tilde{g} = \Big\{ \sum_{i\geq 0} M_i \lambda^{n-i} \,|\, M_i \in g,\ i \geq 0,\ n \in \mathbb{Z} \Big\}, \tag{3}$$

that is, the space of all Laurent series in λ with coefficients in g and a finite regular part. Particular examples of a matrix loop algebra contain the linear combinations: $\lambda^m d_1 f_1 + \lambda^n d_2 f_2 + \lambda^l d_3 f_3$ with arbitrary integers m, n, l, real constants d_1, d_2, d_3 and elements f_1, f_2, f_3 in g. Matrix loop algebras provide a structural basis for our study of soliton hierarchies.

Let us also recall the Liouville integrability of PDEs (see, e.g., [14, 15, 21]). Let $x = (x^1, \ldots, x^p)$ be the vector of spatial variables and $u = (u^1, \ldots, u^q)^T$ the vector of dependent variables. A Hamiltonian system of evolutionary PDEs is

$$u_t = J\frac{\delta \mathcal{H}}{\delta u},\ \ u = u(x,t), \tag{4}$$

where $J = J(x, t, u)$ is a Hamiltonian operator and $\frac{\delta}{\delta u}$ stands for the variational derivative [22]. A conserved functional of a Hamiltonian system (4) is a functional $\mathcal{T} = \int T\,dx$ which determines a conservation law of (4): $D_t T + \mathrm{Div}\, X = 0$, in which Div denotes spatial divergence. For a given differential function F, its corresponding one-form is given by

$$dF := \sum_{i=1}^{p} \frac{\partial F}{\partial x^i} dx^i + \frac{\partial F}{\partial t} dt + \sum_{\alpha=1}^{q} \sum_{\#L\geq 0} \frac{\partial F}{\partial u_L^\alpha} du_L^\alpha,$$

where if $\#L = 0$, then $u_L^\alpha = u^\alpha$, and if $\#L = k \geq 1$, then $u_L^\alpha = \frac{\partial^k u^\alpha}{\partial x^{l_1} \cdots \partial x^{l_k}}$, for $L = (l_1, \ldots, l_k),\ 1 \leq l_i \leq p,\ 1 \leq i \leq k$, with $\#L = l_1 + \cdots + l_k$.

Definition 1. Let I be a set of integers and $r \geq 1$ a natural number. We say that a set of r-tuples of differential functions $\{S_n = (S_n^1, \ldots, S_n^r)^T \,|\, n \in I\}$ is independent, if all r-tuples of one-forms, $dS_n = (dS_n^1, \ldots, dS_n^r)^T,\ n \in I$, are linearly independent at every point in the infinite jet space. A set of conserved functionals $\{\mathcal{H}_n \,|\, n \in I\}$ of a Hamiltonian system (4) is said to be independent, if all characteristics $\{J\frac{\delta \mathcal{H}_n}{\delta u} \,|\, n \in I\}$ of the associated Hamiltonian vector fields are independent.

By the differential order of an r-tuple S of differential functions, we mean the order of the highest-order derivative of u with respect to x in S. It is obvious to see that if a set of r-tuples of differential functions has distinct differential orders, then it is independent.

Definition 2. A Hamiltonian system of evolutionary PDEs, (4), is called to be Liouville integrable, if there exists infinitely many conserved functionals $\{\mathcal{H}_n\}_{n=0}^{\infty}$, which are in involution with respect to the Poisson bracket:

$$\{\mathcal{H}_m, \mathcal{H}_n\}_J := \int \left(\frac{\delta \mathcal{H}_m}{\delta u}\right)^T J \frac{\delta \mathcal{H}_n}{\delta u}\, dx = 0,\ m, n \geq 0, \tag{5}$$

and the characteristics of whose associated Hamiltonian vector fields

$$K_n := J\frac{\delta \mathcal{H}_n}{\delta u},\ n \geq 0, \tag{6}$$

are independent.

In this paper, we would like to focus on an application of the matrix loop algebra $\widetilde{\mathrm{sl}}(2,\mathbb{R})$ within the zero curvature formulation. We will introduce a perturbed Kaup–Newell matrix spectral problem, based on $\widetilde{\mathrm{sl}}(2,\mathbb{R})$, and construct its associated integrable Hamiltonian hierarchy through zero curvature equations. The corresponding Hamiltonian structures will be furnished by using the trace identity, and all systems in the resulting perturbed Kaup–Newell hierarchy will be shown to be Liouville integrable. A few concluding remarks will be given in the last section.

2. Zero curvature formulation

Lax proposed an operator pair approach for studying the Korteweg–de Vries equation [8], and such an involved pair is nowadays called a Lax pair. It is realized (see, e.g., [23, 24]) that a Lax pair presentation is generally equivalent to a zero curvature presentation. We say that an integrable system of PDEs possesses a zero curvature representation, if it can be generated from a zero curvature equation

$$U_t - V_x + [U, V] = 0, \tag{7}$$

where $x, t \in \mathbb{R}$, and the two matrices U and V, called a spectral matrix and a Lax matrix (or operator), are taken from a given matrix loop algebra [3, 25].

As soon as a spectral matrix U is well selected, in order to present a soliton hierarchy, we start to solve a stationary zero curvature equation

$$W_x = [U, W] \tag{8}$$

in $\tilde{g}$. Then, introduce a series of Lax matrices

$$V^{[m]} = (\lambda^m W)_+ + \Delta_m,\ \Delta_m \in \tilde{g},\ m \geq 0, \tag{9}$$

where P_+ denotes the polynomial part of P in λ, such that the corresponding zero curvature equations

$$U_{t_m} - V_x^{[m]} + [U, V^{[m]}] = 0,\ m \geq 0, \tag{10}$$

yield a hierarchy of soliton equations

$$u_{t_m} = K_m,\ m \geq 0. \tag{11}$$

The structure of W often tells how to determine the modification terms Δ_m, $m \geq 0$. The associated Lax pairs are starting points to find soliton solutions by the inverse scattering transform [1, 2].

One of our tasks in the study of integrable systems is to construct Hamiltonian structures or bi-Hamiltonian structures [26],

$$u_{t_m} = K_m = J\frac{\delta \mathcal{H}_m}{\delta u} = M\frac{\delta \mathcal{H}_{m-1}}{\delta u}, \ m \geq 1, \tag{12}$$

which naturally generate a hereditary recursion operator $\Phi = MJ^{-1}$, and thus, infinitely many commuting conserved functionals and symmetries [27, 28]. The basic tool for constructing Hamiltonian functionals is the trace identity in the semisimple case [14]:

$$\frac{\delta}{\delta u}\int \mathrm{tr}\left(\frac{\partial U}{\partial \lambda}W\right)dx = \lambda^{-\gamma}\frac{\partial}{\partial \lambda}\lambda^{\gamma}\mathrm{tr}\left(\frac{\partial U}{\partial u}W\right), \ \gamma = -\frac{\lambda}{2}\frac{d}{d\lambda}\ln|\mathrm{tr}(W^2)|, \tag{13}$$

or generally, the variational identity in the non-semisimple case [29]:

$$\frac{\delta}{\delta u}\int \left\langle\frac{\partial U}{\partial \lambda}, W\right\rangle dx = \lambda^{-\gamma}\frac{\partial}{\partial \lambda}\lambda^{\gamma}\left\langle\frac{\partial U}{\partial u}, W\right\rangle, \ \gamma = -\frac{\lambda}{2}\frac{d}{d\lambda}\ln|\langle W, W\rangle|, \tag{14}$$

where $\langle \cdot, \cdot \rangle$ is a symmetric, non-degenerate and ad-invariant bilinear form over the matrix loop algebra $\tilde{g}$.

3. An example: a perturbed integrable Kaup–Newell hierarchy

3.1. A perturbed Kaup–Newell hierarchy

We apply the zero curvature formulation to present a perturbed integrable Kaup–Newell hierarchy. We start with a new 2×2 matrix spectral problem:

$$\phi_x = U\phi = U(u,\lambda)\phi, \ U = \begin{bmatrix} \lambda + \alpha p & \lambda p \\ q & -\lambda - \alpha p \end{bmatrix}, \ u = \begin{bmatrix} p \\ q \end{bmatrix}, \tag{15}$$

where λ is the spectral parameter and α is a fixed constant. If $\alpha = 0$, then (15) reduces to the standard Kaup–Newell spectral problem [11], and thus, (15) is called a perturbed Kaup–Newell spectral problem and the corresponding soliton hierarchy is called a perturbed Kaup–Newell hierarchy.

Once a matrix spectral problem is chosen, it is inherently feasible to calculate the corresponding soliton hierarchy. First, we solve the stationary zero curvature equation (8) for $W \in \widetilde{\mathrm{sl}}(2,\mathbb{R})$. When W is assumed to be

$$W = \begin{bmatrix} a & b \\ \lambda^{-1}c & -a \end{bmatrix}, \tag{16}$$

the stationary zero curvature equation (8) becomes

$$a_x = pc - qb, \ b_x = 2\lambda b - 2\lambda pa + 2\alpha pb, \ c_x = -2\lambda c + 2\lambda qa - 2\alpha pc. \tag{17}$$

This leads to

$$pc_x + qb_x = -2\lambda(pc - qb) - 2\alpha p^2 c + 2\alpha pqb. \tag{18}$$

Further, expand a, b and c as the Laurent series in λ:

$$a = \sum_{i\geq 0} a_i\lambda^{-i},\ b = \sum_{i\geq 0} b_i\lambda^{-i},\ c = \sum_{i\geq 0} c_i\lambda^{-i}, \tag{19}$$

and take the initial data

$$a_0 = 1,\ b_0 = p,\ c_0 = q, \tag{20}$$

to fix a solution to the equations from the highest powers of λ in (17):

$$a_{0,x} = pc_0 - qb_0,\ b_0 = pa_0,\ c_0 = qa_0.$$

Then, based on (18), we see that the system (17) gives rise to

$$\begin{cases} a_{i+1,x} = -\dfrac{1}{2}(pc_{i,x} + qb_{i,x}) - \alpha p^2 c_i + \alpha pqb_i, \\ b_{i+1} = \dfrac{1}{2}b_{i,x} + pa_{i+1} - \alpha pb_i, \qquad\qquad i \geq 0. \\ c_{i+1} = -\dfrac{1}{2}c_{i,x} + qa_{i+1} - \alpha pc_i, \end{cases} \tag{21}$$

While using the above recursion relations, we impose the condition that the constants of integration take the value of zero:

$$a_i|_{u=0} = b_i|_{u=0} = c_i|_{u=0} = 0,\ i \geq 1, \tag{22}$$

to uniquely determine the sequence of $\{a_i, b_i, c_i |\, i \geq 1\}$. This way, the first two sets can be worked out:

$$\begin{aligned} a_1 &= -\frac{pq}{2},\ b_1 = \frac{1}{2}(p_x - 2\alpha p^2 - p^2 q),\ c_1 = -\frac{1}{2}(q_x + 2\alpha pq + pq^2); \\ a_2 &= -\frac{1}{4}(qp_x - pq_x) + \alpha p^2 q + \frac{3}{8}p^2q^2, \\ b_2 &= \frac{1}{4}p_{xx} - \frac{3}{4}qpp_x - \frac{3}{2}\alpha pp_x + p^3\left(\alpha^2 + \frac{3}{2}\alpha q + \frac{3}{8}q^2\right), \\ c_2 &= \frac{1}{4}q_{xx} + \frac{3}{4}qpq_x + \frac{1}{2}\alpha qp_x + \alpha pq_x + p^2q\left(\alpha^2 + \frac{3}{2}\alpha q + \frac{3}{8}q^2\right). \end{aligned}$$

We saw above the localness of the first three sets of $\{a_i, b_i, c_i |\, i \geq 1\}$. This is not an accident, and the functions $a_i, b_i, c_i,\ i \geq 1$, are all local, indeed. We can verify this fact as follows. First from $W_x = [U, W]$, we get

$$\frac{d}{dx}\mathrm{tr}(W^2) = 2\mathrm{tr}(WW_x) = 2\mathrm{tr}(W[U, W]) = 0,$$

and so, due to $\mathrm{tr}(W^2) = 2(a^2 + \lambda^{-1}bc)$, we can compute that

$$a^2 + \lambda^{-1}bc = (a^2 + \lambda^{-1}bc)|_{u=0} = 1, \tag{23}$$

the second step of which follows from the initial data in (20) and the recursion relations in (21). Then, by using the Laurent expansions in (19) and noting the initial data in (20) again, balancing the coefficients of λ^i in (23) for each $i \geq 1$ yields

$$a_i = -\frac{1}{2}\left(\sum_{k+l=i,\, k,l\geq 1} a_k a_l + \sum_{k+l=i-1,\, k,l\geq 0} b_k c_l\right),\ i \geq 1. \tag{24}$$

Based on this recursion relation (24) and the last two recursion relations in (21), applying the mathematical induction finally tells that all functions $a_i, b_i, c_i,\ i \geq 1$, are differential polynomials in p and q, i.e., they are all local; and that for each $i \geq 2$, the differential orders of the differential functions a_i, b_i and c_i are $i-2, i-1$ and $i-1$, respectively.

Now as usual, we introduce

$$V^{[m]} = \lambda(\lambda^m W)_+ + \delta_m e_1,\ m \geq 0, \tag{25}$$

where δ_m are differential functions to be determined later. A direct computation shows that $V_x^{[m]} - [U, V^{[m]}]$ is equal to

$$\begin{bmatrix} \delta_{m,x} & 2\lambda(b_{m+1} - pa_{m+1} + p\delta_m) \\ 2(-c_{m+1} + qa_{m+1} - q\delta_m) & -\delta_{m,x} \end{bmatrix}. \tag{26}$$

Therefore, the corresponding zero curvature equations (10) precisely present

$$\begin{cases} \alpha p_{t_m} = \delta_{m,x},\ p_{t_m} = 2(b_{m+1} - pa_{m+1} + p\delta_m), \\ q_{t_m} = 2(-c_{m+1} + qa_{m+1} - q\delta_m), \end{cases} m \geq 0. \tag{27}$$

To satisfy the above third equation, we choose, based on (21), that

$$\delta_m = \alpha b_m,\ m \geq 0, \tag{28}$$

and then, all the systems in (27) determine a soliton hierarchy

$$u_{t_m} = K_m = \begin{bmatrix} p \\ q \end{bmatrix}_{t_m} = \begin{bmatrix} b_{m,x} \\ c_{m,x} + 2\alpha p c_m - 2\alpha q b_m \end{bmatrix},\ m \geq 0, \tag{29}$$

which is the required perturbed Kaup–Newell hierarchy. The first nonlinear system in this perturbed hierarchy is given by

$$u_{t_1} = \begin{bmatrix} p \\ q \end{bmatrix}_{t_1} = \begin{bmatrix} \frac{1}{2}(p_{xx} - 2pp_x q - p^2 q_x - 4\alpha p p_x) \\ -\frac{1}{2}q_{xx} - \frac{1}{2}p_x q^2 - pqq_x - 2\alpha(p_x q + pq_x) \end{bmatrix}. \tag{30}$$

3.2. Hamiltonian structures and Liouville integrability

We shall show that all systems in the perturbed Kaup–Newell hierarchy (29) are Liouville integrable. Towards this end, let us first establish Hamiltonian structures for the perturbed hierarchy (29) by using the trace identity (13).

In the perturbed Kaup–Newell case discussed above, the trace identity (13) reads

$$\frac{\delta}{\delta u}\int (2a + \lambda^{-1}pc)\, dx = \lambda^{-\gamma}\frac{\partial}{\partial\lambda}\lambda^{\gamma}\begin{bmatrix} 2\alpha a + c \\ b \end{bmatrix}. \tag{31}$$

Balancing the coefficients of λ^{-m-1} for each $m \geq 0$ in this equality tells that $\gamma = 0$ and that

$$\frac{\delta}{\delta u}\mathcal{H}_m = \begin{bmatrix} 2\alpha a_m + c_m \\ b_m \end{bmatrix}, \quad m \geq 0, \tag{32}$$

with the Hamiltonian functionals being defined by

$$\mathcal{H}_0 = \int (2\alpha p + pq)\, dx, \quad \mathcal{H}_m = \int \left(-\frac{2a_{m+1} + pc_m}{m} \right) dx, \quad m \geq 1. \tag{33}$$

It follows now that the hierarchy (29) has the Hamiltonian structures:

$$u_{t_m} = K_m = J\frac{\delta \mathcal{H}_m}{\delta u}, \quad J = \begin{bmatrix} 0 & \partial \\ \partial & 0 \end{bmatrix}, \quad m \geq 0. \tag{34}$$

From the recursion relations in (21), we can obtain

$$K_{m+1} = \Phi K_m, \quad m \geq 0, \tag{35}$$

where Φ is the recursion operator

$$\Phi = \begin{bmatrix} \frac{1}{2}\partial - \frac{1}{2}\partial p \partial^{-1} q - \alpha \partial p \partial^{-1} & -\frac{1}{2}\partial p \partial^{-1} p \\ -\frac{1}{2}\partial q \partial^{-1} q - \alpha \partial q \partial^{-1} - \alpha q & -\frac{1}{2}\partial - \frac{1}{2}\partial q \partial^{-1} p - \alpha p \end{bmatrix}. \tag{36}$$

We readily check that $J\Psi = \Phi J$, where Ψ is the adjoint operator of Φ, and thus, all systems, except the first one, in the perturbed Kaup–Newell hierarchy (29) are bi-Hamiltonian:

$$u_{t_m} = K_m = J\frac{\delta H_m}{\delta u} = M\frac{\delta H_{m-1}}{\delta u}, \quad m \geq 1, \tag{37}$$

where the second Hamiltonian operator is defined by

$$M = \Phi J = \begin{bmatrix} -\frac{1}{2}\partial p \partial^{-1} p \partial & \frac{1}{2}\partial^2 - \frac{1}{2}\partial p \partial^{-1} q \partial - \alpha \partial p \\ -\frac{1}{2}\partial^2 - \frac{1}{2}\partial q \partial^{-1} p \partial - \alpha p \partial & -\frac{1}{2}\partial q \partial^{-1} q \partial - \alpha \partial q - \alpha q \partial \end{bmatrix}.$$

Now from an observation of the Hamiltonian structures presented in (34) and the differential orders of the sequence $\{a_i, b_i, c_i | \, i \geq 1\}$ shown in the last subsection, it follows that the perturbed Kaup–Newell hierarchy (29) is Liouville integrable. Namely, every system in the perturbed hierarchy (29) possesses infinitely many independent commuting conserved functionals:

$$\{\mathcal{H}_k, \mathcal{H}_l\}_J := \int \left(\frac{\delta \mathcal{H}_k}{\delta u}\right)^T J \frac{\delta \mathcal{H}_l}{\delta u}\, dx = 0, \quad k, l \geq 0, \tag{38}$$

and infinitely many independent commuting symmetries:

$$[K_k, K_l] := K_k'(u)[K_l] - K_l'(u)[K_k] = J\frac{\delta}{\delta u}\{\mathcal{H}_k, \mathcal{H}_l\}_J = 0, \quad k, l \geq 0, \tag{39}$$

where K' is the Gateaux derivative. These commuting relations are also consequences of the Virasoro algebra of Lax matrices (see, e.g., [30] for details).

4. Concluding remarks

Matrix loop algebras were used to search for integrable Hamiltonian equations, which come in hierarchies. Within the matrix loop algebra $\widetilde{\mathrm{sl}}(2,\mathbb{R})$, the Kaup–Newell spectral problem was perturbed by linear perturbation, and a hierarchy of associated integrable bi-Hamiltonian equations was successfully generated. Their Hamiltonian structures and Liouville integrability were established by the trace identity.

The spectral problem (15) is a special reduction of general matrix Lax pairs associated with semisimple Lie algebras (see, e.g., [3]–[7]). However, determination of all integrable reductions within the category of semisimple Lie algebras is one of the most important problems in the theory of integrable system [2], and it is still very interesting to see concrete examples of soliton hierarchies of integrable Hamiltonian equations. Among typical discussed spectral matrices associated with $\widetilde{\mathrm{sl}}(2,\mathbb{R})$ and $\widetilde{\mathrm{so}}(3,\mathbb{R})$ are the following three cases:

$$U(u,\lambda) = \lambda e_1 + pe_2 + qe_3,\ \lambda^2 e_1 + \lambda pe_2 + \lambda qe_3,\ \lambda e_1 + \lambda pe_2 + \lambda qe_3,$$

where $u = (p,q)^T$ and e_1, e_2, e_3 are three matrices in a standard basis. These correspond to the Ablowitz–Kaup–Newell–Segur type hierarchy [17], the Kaup–Newell type hierarchy [18] and the Wadati–Konno–Ichikawa type hierarchy [19], when the underlying matrix loop algebra is $\widetilde{\mathrm{so}}(3,\mathbb{R})$. In those three examples, the vector u consists of only two dependent variables, p and q. There are various examples of soliton hierarchies with three or more dependent variables (see, e.g., [22, 25, 31]).

We also point out that given initial matrix loop algebras, it still requires a considerable amount of time to compute soliton hierarchies within the zero curvature formulation, and it is much more complicated in the case of higher spatial dimensions. The study of integrable couplings [22], associated with non-semisimple matrix loop algebras, provides specific examples of soliton hierarchies generated from higher-order matrix spectral problems. The resulting soliton hierarchies can be solved by applying Darboux transformations associated with the underlying matrix spectral problems (see, e.g., [32, 33]), possibly yielding lump solutions [34, 35]. It is, however, known that the variational identity [25, 29] does not present Hamiltonian structures for the bi-integrable couplings:

$$u_t = K(u),\ v_t = K'(u)[v],\ w_t = K'(u)[w],$$

where K' stands for the Gateaux derivative. It remains open how to generalize the variational identity such that we can furnish Hamiltonian structures for such integrable couplings.

Acknowledgment. The work was supported in part by the Open Fund of IPOC (BUPT) under grant no. IPOC2016B008, Natural Science Fund for Colleges and Universities of Jiangsu Province of China under the grant 17KJB110020, the National Natural Science Foundation of China under the grants 11371326, 11271008, 11301454 and 11271168, NSF under the grant DMS-1664561, and the distinguished professorships of the Shanghai University of Electric Power and the Shanghai Second Polytechnic University.

References

[1] M.J. Ablowitz and P.A. Clarkson, *Solitons, Nonlinear Evolution Equations and Inverse Scattering.* Cambridge University Press, Cambridge, 1991.

[2] S. Novikov, S.V. Manakov, L.P. Pitaevskii and V.E. Zakharov, *Theory of Solitons – The Inverse Scattering Method.* Consultants Bureau/A Division of Plenum Publishing Corporation, New York, 1984.

[3] V.G. Drinfel'd and V.V. Sokolov, *Equations of Korteweg–de Vries type and simple Lie algebras.* Soviet Math. Dokl. **23** (1981), 457–462.

[4] G. Wilson, *The modified Lax and two-dimensional Toda lattice equations associated with simple Lie algebras.* Erg. Theory Dynam. Sys. **1** (1981), 361–380.

[5] L.A. Dickey, *Soliton Equations and Hamiltonian Systems.* World Scientific, Singapore, 2003.

[6] M.F. de Groot, T.J. Hollowood and J.L. Miramontes, *Generalized Drinfel'd–Sokolov hierarchies.* Comm. Math. Phys. **145** (1992), 57–84.

[7] N.J. Burroughs, M.F. de Groot, T.J. Hollowood and J.L. Miramontes, *Generalized Drinfel'd–Sokolov hierarchies II – The Hamiltonian structures.* Comm. Math. Phys. **153** (1993), 187–215.

[8] P.D. Lax, *Integrals of nonlinear equations of evolution and solitary waves.* Comm. Pure Appl. Math. **21** (1968), 467–490.

[9] M.J. Ablowitz, D.J. Kaup, A.C. Newell and H. Segur, *The inverse scattering transform-Fourier analysis for nonlinear problems.* Stud. Appl. Math. **53** (1974), 249–315.

[10] H. Grosse, *New solitons connected to the Dirac equation.* Phys. Rep. **134** (1986), 297–304.

[11] D.J. Kaup and A.C. Newell, *An exact solution for a derivative nonlinear Schrödinger equation.* J. Math. Phys. **19** (1978), 798–801.

[12] M. Wadati, K. Konno and Y.H. Ichikawa, *New integrable nonlinear evolution equations.* J. Phys. Soc. Jpn. **47** (1979), 1698–1700.

[13] D.Y. Chen and Y.S. Li, *The transformation operator V of nonlinear evolution equation.* Acta Math. Sin. **2** (1986), 343–356.

[14] G.Z. Tu, *On Liouville integrability of zero-curvature equations and the Yang hierarchy.* J. Phys. A: Math. Gen. **22** (1989), 2375–2392.

[15] W.X. Ma, *A new hierarchy of Liouville integrable generalized Hamiltonian equations and its reduction.* Chin. Ann. Math. A **13** (1992), 115–123; Chin. J. Contemp. Math. **13** (1992), 79–89.

[16] L.Y. Jiang, Y.Y. Jin, W.X. Ma, S.F. Shen and H.Y. Zhu, *Novel soliton hierarchies of Levi type and their bi-Hamiltonian structures.* Commun. Nonlinear Sci. Numer. Simul. **23** (2015), 388–396.

[17] W.X. Ma, *A soliton hierarchy associated with* $so(3,\mathbb{R})$. Appl. Math. Comput. **220** (2013), 117–122.

[18] W.X. Ma, *A spectral problem based on* $so(3,\mathbb{R})$ *and its associated commuting soliton equation.* J. Math. Phys. **54** (2013), 103509, 8 pp.

[19] W.X. Ma, S. Manukure and H.C. Zheng, *A counterpart of the Wadati–Konno–Ichikawa soliton hierarchy associated with* $so(3,\mathbb{R})$. Z. Naturforsch. A **69** (2014), 411–419.

[20] W.X. Ma, S.F. Shen, S.M. Yu, H.Q. Zhang and W.Y. Zhang, *An integrable $SO(3, R)$-counterpart of the Heisenberg soliton hierarchy.* Rep. Math. Phys. **74** (2014), 283–299.

[21] W.X. Ma, *An integrable counterpart of the D-AKNS soliton hierarchy from $so(3, \mathbb{R})$.* Phys. Lett. A **378** (2014), 1717–1720.

[22] W.X. Ma and B. Fuchssteiner, *Integrable theory of the perturbation equations.* Chaos, Solitons Fractals **7** (1996), 1227–1250.

[23] B.A. Kupershmidt and G. Wilson, *Conservation laws and symmetries of generalized sine-Gordon equations.* Comm. Math. Phys. **81** (1981), 189–202.

[24] W.X. Ma, *Lax representations of hierarchies of evolution equations with Lienard recursive structures.* Acta Math. Sci. **14** (1994), 409–418.

[25] W.X. Ma, *Integrable couplings and matrix loop algebras.* In: W.X. Ma and D. Kaup (eds.), *Proceedings of the 2nd Intenational Workshop on Nonlinear and Modern Mathematical Physics*, AIP Conference Proceedings **1562**, American Institute of Physics, Melville, NY, 2013, pp. 105–122.

[26] F. Magri, *A simple model of the integrable Hamiltonian equation.* J. Math. Phys. **19** (1978), 1156–1162.

[27] P.J. Olver, *Evolution equations possessing infinitely many symmetries.* J. Math. Phys. **18** (1977), 1212–1215.

[28] B. Fuchssteiner and A.S. Fokas, *Symplectic structures, their Bäcklund transformations and hereditary symmetries.* Phys. D. **4** (1981), 47–66.

[29] W.X. Ma and M. Chen, *Hamiltonian and quasi-Hamiltonian structures associated with semi-direct sums of Lie algebras.* J. Phys. A: Math. Gen. **39** (2006), 10787–10801.

[30] W.X. Ma, *The algebraic structures of isospectral Lax operators and applications to integrable equations.* J. Phys. A: Math. Gen. **25** (1992), 5329–5343.

[31] Y.F. Zhang and H.W. Tam, *A few new higher-dimensional Lie algebras and two types of coupling integrable couplings of the AKNS hierarchy and the KN hierarchy.* Commun. Nonlinear Sci. Numer. Simul. **16** (2011), 76–85.

[32] V.B. Matveev and M.A. Salle, *Darboux Transformations and Solitons.* Springer-Verlag, Berlin, 1991.

[33] W.X. Ma, *Darboux transformations for a Lax integrable system in $2n$-dimensions.* Lett. Math. Phys. **39** (1997), 33–49.

[34] K. Imai and K. Nozaki, *Lump solutions of the Ishimori-II equation.* Progr. Theoret. Phys. **96** (1996), 521–526.

[35] X. Lü and W.X. Ma, *Study of lump dynamics based on a dimensionally reduced Hirota bilinear equation.* Nonlinear Dynam. **85** (2016), 1217–1222.

Wen-Xiu Ma
Department of Mathematics and Statistics
University of South Florida
Tampa, Florida 33620-5700, USA
e-mail: mawx@cas.usf.edu

Xing Lü
School of Computer and
Communication Engineering
Univ. of Science and Technology Beijing
Beijing 100083, China
e-mail: xinglv655@aliyun.com

Geometric Methods in Physics. XXXV Workshop 2016
Trends in Mathematics, 201–212

On the Ground State Energy of the Delta-function Fermi Gas II: Further Asymptotics

Craig A. Tracy and Harold Widom

Abstract. Building on previous work of the authors, we here derive the weak coupling asymptotics to order γ^2 of the ground state energy of the delta-function Fermi gas. We use a method that can be applied to a large class of finite convolution operators.

Mathematics Subject Classification (2010). 47B35, 81V80.

Keywords. Bethe Ansatz, delta-function Fermi gas, Gaudin integral equation, ground state energy, convolution operators.

1. Introduction

One of the most widely studied Bethe Ansatz solvable models is the quantum, many-body system in one-dimension with delta-function two-body interaction [8] with Hamiltonian

$$H_N = -\sum_{j=1}^{N} \frac{\partial^2}{\partial x_j^2} + 2c \sum_{i<j} \delta(x_i - x_j).$$

Here N is the number of particles and $2c$ is the coupling constant. A basic quantity is the ground state energy per particle in the thermodynamic limit: If $E_0(N, L)$ is the ground state energy for the finite system of N particles on a circle of length L, then in the limit $N \to \infty$, $L \to \infty$, such that $\rho := N/L$ is fixed, the ground state energy per particle is

$$\varepsilon_0 := \lim \frac{E_0(N, L)}{N}.$$

This work was supported by the National Science Foundation through grants DMS–1207995 (first author) and DMS–1400248 (second author).

Lieb and Liniger [8] showed, for particles with Bose statistics and repulsive interaction ($c > 0$), that $\mathfrak{e}_B := \varepsilon_0/\rho^2$ is a function only of $\gamma := c/\rho$. To state their result, we define the *Lieb–Liniger operator*

$$\mathcal{L}_\kappa f(x) := \frac{\kappa}{\pi} \int_{-1}^{1} \frac{f(y)}{(x-y)^2+\kappa^2}\, dy, \quad -1 < x < 1. \tag{1}$$

If $f_B(x;\kappa)$ solves the Lieb–Liniger integral equation

$$f(x) - \mathcal{L}_\kappa f(x) = 1, \tag{2}$$

then $\mathfrak{e}_B(\gamma)$ is determined, by elimination of κ, from the relations

$$\frac{\kappa}{\gamma} = \frac{1}{2\pi} \int_{-1}^{1} f_B(x;\kappa)\, dx, \quad \mathfrak{e}_B(\gamma) = \frac{1}{2\pi} \left(\frac{\gamma}{\kappa}\right)^3 \int_{-1}^{1} x^2 f_B(x;\kappa)\, dx.$$

The asymptotics of $\mathfrak{e}_B(\gamma)$ as $\gamma \to 0$ have been derived in the literature. See [9] and references therein.

A natural question to ask is how the problem changes when the particles obey Fermi statistics. The generalization of Bethe Ansatz to this case was solved by Gaudin [2, 3] and Yang [12]. For spin-1/2 particles with attractive interaction ($c < 0$) with total spin zero, the ground state energy per particle in the thermodynamic limit is given by [3]

$$\frac{\varepsilon_0}{\rho^2} = -\frac{\gamma^2}{4} + \mathfrak{e}_F(\gamma)$$

where $\gamma = |c|/\rho$ and the equation is now the *Gaudin integral equation*

$$f(x) + \mathcal{L}_\kappa f(x) = 1. \tag{3}$$

If $f_F(x;\kappa)$ solves this equation, then $\mathfrak{e}_F(\gamma)$ is determined by elimination of κ from the equations

$$\frac{\kappa}{\gamma} = \frac{2}{\pi} \int_{-1}^{1} f_F(x;\kappa)\, dx, \quad \mathfrak{e}_F(\gamma) = \frac{2}{\pi} \left(\frac{\gamma}{\kappa}\right)^3 \int_{-1}^{1} x^2 f_F(x;\kappa)\, dx. \tag{4}$$

Equation (3) also arises in the computation of the charge Q on each of two coaxial conducting discs of radius one separated by a distance κ and each maintained at the same unit potential. For the Lieb–Liniger equation (2), the discs are maintained at equal but opposite potentials. In both cases the charge Q is given by a constant times the zeroth moment of f. For the case of equal potentials (the Fermi case), the charge is given by

$$Q = \frac{1}{\pi} \int_{-1}^{1} f_F(x;\kappa)\, dx, \tag{5}$$

and Leppington and Levine [7] proved rigorously that as $\kappa \to 0$,

$$Q = \frac{1}{\pi} + \frac{\kappa}{2\pi^2} \left(\log \kappa^{-1} + \log \pi + 1\right) + o(\kappa). \tag{6}$$

The authors derived this by finding an approximate solution of the related boundary value problem. In later work, Atkinson and Leppington [1] analyzed the integral equation directly and reproduced this result.

As for the ground state energy, Gaudin [3] used an approximate solution to (3) to obtain

$$\mathfrak{e}_F(\gamma) = \frac{\pi^2}{12} - \frac{\gamma}{2} + \mathrm{o}(\gamma). \tag{7}$$

Guan and Ma [4] derived (7) with the error bound $O(\gamma^2)$, although Krivnov and Ovchinnikov [6] had predicted earlier that the term $\gamma^2 \log^2 \gamma^{-1}$ appears. Using different methods to analyze (3), Iida and Wadati [5] found

$$\mathfrak{e}_F(\gamma) = \frac{\pi^2}{12} - \frac{\gamma}{2} + \frac{\gamma^2}{6} + o(\gamma^2). \tag{8}$$

The methods used to derive the above-mentioned results for $\mathfrak{e}_F(\gamma)$ were heuristic. In [10] we applied a rigorous analysis to the integral equation (3) to derive the first-order results (6) and (7). It was indicated there that one could in principle derive further asymptotics, and this is what we do here.

We use the notation $O(\kappa^{n+})$ to denote a bound $O(\kappa^n \log^m \kappa^{-1})$ for some $m \geq 0$, and similarly for $O(\gamma^{n+})$. What we have found is that as $\kappa \to 0$,

$$Q = \frac{1}{\pi} + \frac{\kappa}{2\pi^2}(\log \kappa^{-1} + \log \pi + 1) + \frac{\kappa^2}{4\pi^3}(\log \kappa^{-1} + \log \pi + 1/2) + O(\kappa^{3+}), \tag{9}$$

and as $\gamma \to 0$,

$$\mathfrak{e}_F(\gamma) = \frac{\pi^2}{12} - \frac{\gamma}{2} + \frac{\gamma^2}{6} + O(\gamma^{3+}), \tag{10}$$

thus confirming the Iida–Wadati result (8).

The derivation of these asymptotics involved some straightforward but tedious computations that were done by Maple, and so we cannot claim complete rigor. Until the end we shall present the results of only a few of the preliminary computations; but we shall get to the points where it is clear that those computations were routine. The reader, if he or she so chooses, can check the outcomes that we exhibit.

In the next section we summarize the results of [10]. In the following sections we show how to go further.

2. Asymptotic solution of the Gaudin equation

The method used in this section to analyze the Gaudin operator will be seen to be quite general and applicable to a large class of finite convolution operators. It was used earlier by one of the authors [11] to derive asymptotics for Toeplitz matrices, which are the discrete analogue of convolution operators.

We first replace the operator $\mathcal{L}_\kappa$ with kernel

$$\frac{\kappa}{\pi} \frac{1}{(x-y)^2 + \kappa^2}$$

on the fixed interval $(-1, 1)$ by the operator with kernel

$$\frac{1}{\pi}\frac{1}{(x-y)^2+1}$$

on the variable interval $(-1/\kappa,\, 1/\kappa)$. For convenience we set $r = 2/\kappa$ and consider the convolution equation

$$\frac{f(x)}{2} + \frac{1}{2\pi}\int_{-r/2}^{r/2} \frac{f(y)}{(x-y)^2+1}\, dy = 1, \quad -r/2 < x < r/2.$$

(The factors 1/2 here avoid factors $\sqrt{2}$ later.)

The solution $f_F(x;\kappa)$ of (3) and our $f(x)$ are related by $f(rx/2) = 2\, f_F(x;\kappa)$. From (5) we get

$$Q = \frac{1}{r\pi}\int_{-r/2}^{r/2} f(x)\, dx = \frac{\kappa}{2\pi}\int_{-r/2}^{r/2} f(x)\, dx. \tag{11}$$

From the first part of (4) we find that

$$\gamma = \left(\frac{1}{\pi}\int_{-r/2}^{r/2} f(x)\, dx\right)^{-1} = \frac{1}{2}\kappa\, Q^{-1}. \tag{12}$$

From (4) and a little computation we find that

$$\mathfrak{e}_F(\gamma) = \pi^2 \frac{\displaystyle\int_{-r/2}^{r/2} x^2\, f(x)\, dx}{\left(\displaystyle\int_{-r/2}^{r/2} f(x)\, dx\right)^3}. \tag{13}$$

Now we go to our integral equation. If we extend the function $f(x)$ to be zero outside the interval $(-r/2, r/2)$ then the equation may be written

$$\int_{-\infty}^{\infty} k(x-y)\, f(y)\, dy = g(x), \quad x \in (-r/2, r/2),$$

where

$$k(x) = \frac{1}{2}\,\delta(x) + \frac{1}{2\pi}\frac{1}{x^2+1}, \quad g(x) = \chi_{(-r/2,r/2)}(x).$$

The Fourier transforms[1] $\sigma(\xi)$ of k and $\hat{g}(\xi)$ of g are given by

$$\sigma(\xi) = (1 + e^{-|\xi|})/2, \quad \hat{g}(\xi) = 2\, \sin(r\xi/2)/\xi. \tag{14}$$

If $\hat{f}$ is the Fourier transform of f then $\sigma\hat{f} - \hat{g}$ is the Fourier transform of an L^1 function supported outside the interval $(-r/2, r/2)$. Such a function may be written as $e^{ir\xi/2}\, h^+(\xi) + e^{-ir\xi/2}\, h^-(\xi)$, where $h^\pm$ is the Fourier transform of an

---

[1] In our notation, the $x \to \xi$ Fourier transform has $e^{ix\xi}$ in the integrand; the $\xi \to x$ inverse Fourier transform has $e^{-ix\xi}$ in the integrand and the factor $1/2\pi$

L^1 function supported on $\mathbb{R}^{\pm}$. Thus, by taking Fourier transforms we may rewrite the equation as

$$\sigma \hat{f} = \hat{g} + e^{ir\xi/2}\, h^+ + e^{-ir\xi/2}\, h^-.$$

We consider $h^{\pm}$ the unknown functions; once they are determined, so is $\hat{f}$.

We denote by $\psi \to \psi_{\pm}$ the conjugate by the Fourier transform of multiplication by $\chi_{\mathbb{R}^{\pm}}$, the characteristic functions of $\mathbb{R}^{\pm}$. These are given by

$$\psi_{\pm}(\xi) = \frac{1}{2}\psi(\xi) \pm \frac{1}{2\pi i} \int_{-\infty}^{\infty} \frac{\psi(\eta)}{\eta - \xi}\, d\eta,$$

where the integral is a principal value. The functions extend analytically to the upper and lower half-planes by the formulas

$$\psi_{\pm}(\xi) = \pm\frac{1}{2\pi i} \int_{-\infty}^{\infty} \frac{\psi(\eta)}{\eta - \xi}\, d\eta, \tag{15}$$

where ξ is in the upper half-plane for ψ_+ and the lower half-plane for ψ_-.

The *Wiener–Hopf factors* of σ, which confusingly we denote by $\sigma_{\pm}$, are given by

$$\sigma_{\pm} = e^{(\log \sigma)_{\pm}},$$

where the $\pm$ on the right are the projection operators defined above. The function $\log \sigma$ is a constant plus the Fourier transform of an L^1 function. It follows that $\sigma_{\pm}$ and their reciprocals are constants plus Fourier transforms of L^1 functions supported on $\mathbb{R}^{\pm}$. It follows that by changing notation we may replace our equation by

$$\sigma_- \sigma_+ \hat{f} = \hat{g} + e^{ir\xi/2}\, \sigma_+\, h^+ + e^{-ir\xi/2}\, \sigma_- h^-. \tag{16}$$

The factors are given explicitly by [1]

$$\sigma_+(\xi) = \pi^{1/2} \exp\left\{ \frac{\xi}{2\pi i} \Big[\log(-i\xi) - \log 2\pi - 1 \Big] \right\} \Gamma\left(\frac{1}{2} + \frac{\xi}{2\pi i} \right)^{-1},$$

$$\sigma_-(\xi) = \pi^{1/2} \exp\left\{ -\frac{\xi}{2\pi i} \Big[\log(i\xi) - \log 2\pi - 1 \Big] \right\} \Gamma\left(\frac{1}{2} - \frac{\xi}{2\pi i} \right)^{-1}.$$

For ξ in the upper resp. lower half-plane, $-i\xi$ resp. $i\xi$ lies in the right half-plane and the principal values of the logarithms are taken.

Since $\sigma_{\pm}(0) = 1$ and $\hat{g}(0) = r$, we have

$$\int_{-r/2}^{r/2} f(x)\, dx = \hat{f}(0) = r + h^+(0) + h^-(0), \tag{17}$$

which by (11) determines Q from $h^{\pm}(0)$. Observe also that

$$\begin{aligned} &\int_{-r/2}^{r/2} x^2\, f(x)\, dx = -\hat{f}''(0) \\ &= \hat{f}(0)/2 - 2 \times \text{the coefficient of } \xi^2 \text{ in the expansion of } \sigma(\xi)\hat{f}(\xi). \end{aligned} \tag{18}$$

So the goal is to find the coefficients in the expansions of $h^{\pm}(\xi)$ as $\xi \to 0$.

Here is how we do this. The inverse Fourier transform of $\hat{f}$ is supported on $(-r/2,\infty)$, so the inverse Fourier transform of $e^{ir\xi/2}\sigma_+\hat{f}$ is supported on $\mathbb{R}^+$. Therefore if we multiply (16) by $e^{ir\xi/2}/\sigma_-$ and apply the minus operator we get

$$0=(e^{ir\xi/2}\,\sigma_+\,\hat{f})_-=\left(\frac{e^{ir\xi/2}\,\hat{g}}{\sigma_-}\right)_-+\left(e^{ir\xi}\,\frac{\sigma_+}{\sigma_-}\,h^+\right)_-+h^-.$$

Similarly

$$0=\left(\frac{e^{-ir\xi/2}\,\hat{g}}{\sigma_+}\right)_++h^++\left(e^{-ir\xi}\,\frac{\sigma_-}{\sigma_+}\,h^-\right)_+.$$

Define the operators U and V by

$$Uu^-=\left(e^{-ir\xi}\,\frac{\sigma_-}{\sigma_+}\,u^-\right)_+,\quad Vv^+=\left(e^{ir\xi}\,\frac{\sigma_+}{\sigma_-}\,v^+\right)_-. \tag{19}$$

The operator U takes Fourier transforms of functions in $L^1(\mathbb{R}^-)$ to Fourier transforms of functions in $L^1(\mathbb{R}^+)$, and V does the opposite. If we define

$$G^-=-\left(\frac{e^{ir\xi/2}\,\hat{g}}{\sigma_-}\right)_-,\quad G^+=-\left(\frac{e^{-ir\xi/2}\,\hat{g}}{\sigma_+}\right)_+, \tag{20}$$

our two relations may be written

$$h^-+Vh^+=G^-,\quad h^++Uh^-=G^+.$$

The solution is given, formally, by

$$\begin{aligned}\begin{pmatrix}h^-\\ h^+\end{pmatrix}&=\left(I+\begin{pmatrix}0&V\\ U&0\end{pmatrix}\right)^{-1}\begin{pmatrix}G^-\\ G^+\end{pmatrix}\\ &=\sum_{j=0}^{\infty}(-1)^j\begin{pmatrix}0&V\\ U&0\end{pmatrix}^j\begin{pmatrix}G^-\\ G^+\end{pmatrix}.\end{aligned} \tag{21}$$

This is quite general. For the Gaudin equation we have a precise statement. Using a different notation than in [10], we define $\mathcal{F}^+$ to be those families of Fourier transforms of functions in $L^1(\mathbb{R}^+)$, depending on the parameter r, for which there is an asymptotic expansion as $\xi\to 0$,

$$\varphi(\xi)\sim\sum_{0\le m\le n}c_{n,m,r}\,\xi^n\,\log^m(-i\xi),$$

where each $c_{n,m,r}=O(r^{n+})$ as $r\to\infty$. Similarly we define $\mathcal{F}^-$. It was shown in [10] that $G^\pm$ and $h^\pm$ belong to $\mathcal{F}^\pm$, and that truncating the series in (21) at $j=k-1$ leads to an error in $h^\pm$ belonging to $r^{-k}\mathcal{F}^\pm$. (In other words the error equals r^{-k} times a function in $\mathcal{F}^\pm$.)

Taking $k=1$ leads to the conclusion that if in (16) we replace $h^\pm$ on the right side by $G^\pm$ the error in the constant term is $O(r^{-1+})$ and the error in the coefficient of ξ^2 is $O(r^{1+})$. This implies[2] that the error in $f(0)$ is $O(r^{-1+})$ and the error in $f''(0)$ is $O(r^{1+})$. Having computed the coefficients in the expansions of $G^\pm$ to the right order, this led us to the first-order asymptotics (6) and (7).

Here we truncate the series at $j = 1$, in other words we make the replacements

$$h^- \to G^- - VG^+, \quad h^+ \to G^+ - UG^-,$$

knowing that this will lead to an error $O(r^{-2+})$ in $f(0)$ and an error $O(r^{0+})$ in $f''(0)$,[3] which will give the next-order asymptotics of Q and of $\mathfrak{e}_F(\gamma)$.

3. Expansion of $G^+(\xi)$ near $\xi = 0$

We do this differently than in [10]. From (14), (15), and (20), we see that the expression for $G^+(\xi)$ for ξ in the upper half-plane is

$$G^+(\xi) = \frac{1}{2\pi}\int_{-\infty}^{\infty} \frac{1-e^{-ir\eta}}{\eta\,\sigma_+(\eta)}\,\frac{d\eta}{\eta-\xi}.$$

Using $\sigma_+(0) = 1$ we write this as the difference

$$\frac{1}{2\pi}\int_{-\infty}^{\infty} \frac{1}{\eta}\left[\frac{1}{\sigma_+(\eta)} - 1\right]\frac{d\eta}{\eta-\xi} - \frac{1}{2\pi}\int_{-\infty}^{\infty} \frac{1}{\eta}\left[\frac{e^{-ir\eta}}{\sigma_+(\eta)} - 1\right]\frac{d\eta}{\eta-\xi}.$$

For the first integral we push the contour up. We get contributions from the pole at $\eta = \xi$, with the result

$$\frac{i}{\xi}\left(\frac{1}{\sigma_+(\xi)} - 1\right).$$

For the second integral we use a trick from [10]. We swing the $\mathbb{R}^+$ part of the contour down to the right side of the negative imaginary axis and the $\mathbb{R}^-$ part of the contour down to the left side of the negative imaginary axis, making there the substitution $\eta = -ix$. We use $1/\sigma_+ = \sigma_-/\sigma$ and that the analytic continuation of $1/\sigma(\eta)$ to the right side of the imaginary axis minus its analytic continuation to the left side of the imaginary axis equals

$$\frac{2}{1+e^{ix}} - \frac{2}{1+e^{-ix}} = -2i\tan(x/2).$$

Thus the second integral with its factor becomes

$$\int_0^{\infty} \frac{e^{-rx}}{x}\,\psi(x)\,\frac{dx}{x-i\xi}, \qquad \psi(x) = \frac{1}{\pi}\,\sigma_-(-ix)\,\tan(x/2).$$

(This is a principal value integral at each odd multiple of π. The contributions of the integrals over the little semicircles on either side of the imaginary axis cancel each other.) Observe that with ξ in the upper half-plane $-i\xi$ is in the right half-plane.

---

[2] We use that $\mathcal{F}^\pm$ is closed under multiplication by $e^{\pm ir\xi/2}$ or $\sigma_\pm(\xi)$.

[3] We refer to these as "acceptable errors".

We have obtained the representation

$$G^{+}(\xi)=\frac{i}{\xi}\left(\frac{1}{\sigma_{+}(\xi)}-1\right)-\int_{0}^{\infty}\frac{e^{-rx}}{x}\,\psi(x)\,\frac{dx}{x-i\xi}. \tag{22}$$

We are interested first in the first few terms in the expansion of this as $\xi\to 0$, and for $k\le 2$ we allow an error $O(r^{k-2+})$ in the coefficient of terms involving ξ^{k}.[4] The expansion of the first summand can be found and has terms independent of r. For the integral, if we replace $\psi(x)$ by the terms up to powers less than N of its expansion near $x=0$ the error will be an integral in which $\psi(x)$ is replaced by $O(x^{N+})$.[5] In the ξ-expansion of the resulting integral the coefficients involving ξ^{k} would be $O(r^{-N+k+1+})$. This shows that with acceptable errors we allow in the coefficients we may replace $\psi(x)$ by the terms in its expansion up to powers less than three, and these are

$$\frac{x}{2\pi}-\frac{x^{2}}{4\pi^{2}}\,(\log x^{-1}+\log(\pi/2)-\gamma_{E}+1).$$

(Here and below γ_E denotes the Euler gamma.) From the general formula

$$\int_{0}^{\infty}e^{-x}\,x^{a-1}\,\frac{dx}{x+z}=\Gamma(a)\,e^{z}\,E_{a}(z), \tag{23}$$

we see that the integrals that arise can be expressed in terms of generalized exponential integrals (and a derivative of one of them) evaluated at $-ir\xi$, and their expansions are known.

4. Expansion of $VG^{+}(\xi)$ near $\xi=0$

From (15) and the definition of the operator V in (19) we have

$$VG^{+}(\xi)=-\frac{1}{2\pi i}\int_{-\infty}^{\infty}e^{ir\eta}\,\frac{\sigma_{+}(\eta)}{\sigma_{-}(\eta)}\,G^{+}(\eta)\,\frac{d\eta}{\eta-\xi},$$

where ξ is now in the lower half-plane. We use the same trick as in the last section. We rewrite σ_{+}/σ_{-} as σ_{+}^{2}/σ and swing the half-lines up to the imaginary axis in the upper half-plane, where we make the substitution $\eta=iy$. The result is

$$VG^{+}(\xi)=\int_{0}^{\infty}e^{-ry}\,\varphi(y)\,G^{+}(iy)\,\frac{dy}{y+i\xi},\qquad \varphi(y)=-\frac{1}{\pi}\sigma_{+}(iy)^{2}\,\tan(y/2).$$

Now $i\xi$ is in the right half-plane.

---

[4]Recall that the expansion of $G^{+}(\xi)$ involves powers of ξ times powers of logarithms. Powers of logarithms also occur in the expansion of $\psi(x)$,

[5]The reader may be concerned about the factor $\tan x/2$ in $\psi(x)$. A representation using a somewhat different contour resolves this issue. We deform the original contour to the left and right parts of the negative imaginary axis only down to $-i\pi/2$, say, and then slightly off-vertical rays from $-i\pi/2$ downward. There are no singularities on the modified contour. The integrals over the rays are analytic in ξ near zero with the coefficient of each power of ξ exponentially small in r. The upper limit on the integral in (22) becomes $\pi/2$, which does not affect the argument that follows.

In formula (22) $G^+(\xi)$ is given as the sum of two terms, the first of which is independent of r. Its contribution to the integral for $VG^+(\xi)$ is like the integral in (22) and we can treat it analogously. In this case for the error we accept in the coefficients we may replace $\varphi(y)$ times the first term in (22) (with ξ replaced by iy) by the terms in its expansion up to powers less than two. Thus we may replace this product by the single term

$$-\frac{1}{4\pi^2}(\log y^{-1} + \log(\pi/2) - \gamma_E + 1)\, y.$$

Again the corresponding integral over y is expressible in terms of exponential integrals.

There remains the double integral

$$\int_0^\infty e^{-ry}\, \varphi(y)\, \frac{dy}{y+i\xi} \int_0^\infty \frac{e^{-rx}}{x}\, \psi(x)\, \frac{dx}{x+y}.$$

As in footnote 5 we may replace the upper limits of integration by $\pi/2$, so the integrands have no singularities. Now, if replace $\psi(x)$ in the inner integral by the terms up to powers less than N of its expansion near $x = 0$ the error in the inner integral can be seen to be $O(\min(y^{N-1+}, r^{-N+1+}))$. The resulting double integral would be a function of ξ for whose expansion the coefficient of ξ^k (for fixed $k < N$) would be $O(r^{-N+k+})$. So given our acceptable error we may replace $\psi(x)$ by finitely many terms in its expansion. Then we may replace $\varphi(y)$ by finitely many terms in its expansion. (And we may replace the upper limits by ∞.)

To see what the individual summands will be we consider the integral

$$\int_0^\infty e^{-ry}\, y^p\, \frac{dy}{y+i\xi} \int_0^\infty e^{-rx}\, x^q\, \frac{dx}{x+y}$$

with $p \geq 1,\ q \geq 0$. (When logarithms appear we differentiate some number of times with respect to p or q.) We make the variable changes $y \to y/r,\ x \to x/r$ and set $X = ir\xi$. We obtain, using (23),

$$\begin{aligned} r^{-p-q} & \int_0^\infty e^{-y}\, y^p\, \frac{dy}{y+X} \int_0^\infty e^{-x}\, x^q\, \frac{dx}{x+y} \\ &= r^{-p-q}\, \Gamma(q+1) \int_0^\infty y^p\, E_{q+1}(y) \frac{dy}{y+X}. \end{aligned}$$

From the expansion of $E_{q+1}(y)$ as $y \to 0$ we see that there is an expansion as $X \to 0$ with summands that are nonnegative powers times logarithms. In terms of ξ (recall that $X = ir\xi$), the summands involving ξ^k have coefficients $O(r^{k+})$. Recalling that r^{-p-q} multiplies this integral, and the errors in the coefficients that are acceptable, we see that we need consider only the term with $p = 1,\ q = 0$. The integral, which eventually gets the factor $-1/(4\pi^2 r)$, becomes

$$\int_0^\infty y\, E_1(y)\, \frac{dy}{y+X} = X \int_0^\infty E_1(Xy)\, \frac{y\, dy}{y+1}.$$

Integrating by parts using $E_1'(y) = -y^{-1} e^{-y}$ gives

$$1 - X \int_0^\infty e^{-Xy}\, y^{-1} \log(y+1)\, dy. \tag{24}$$

The following may be a needlessly complicated way of finding the asymptotics of this as $X \to 0$. If we call the last integral $I(X)$ then

$$\begin{aligned} I'(X) &= -\int_0^\infty e^{-X\,y} \log(1+y)\, dy \\ &= \frac{d}{ds} \int_0^\infty e^{-X\,y} (1+y)^{-s}\, dy \Big|_{s=0} = e^X \frac{d}{ds} E_s(X) \Big|_{s=0}. \end{aligned}$$

From the known expansion of $E_s(X)$ we find that as $X \to 0$,

$$I'(X) = X^{-1}(\log X + \gamma_E) + \log X - 1 + \gamma_E + O(X^{1+}).$$

Integrating from 1 to X gives

$$I(X) = \frac{1}{2} \log^2 X + \gamma_E \log X + C + X \log X - (2 - \gamma_E) X + O(X^{2+}),$$

for some constant C. To evaluate C we use

$$-X^{-1}(\log X + \gamma_E) = \int_0^\infty e^{-Xy} \log y\, dy,$$

and integrate from X to 1 to obtain

$$\frac{1}{2} \log^2 X + \gamma_E \log X = \int_0^\infty \frac{e^{-Xy} - e^{-y}}{y} \log y\, dy.$$

Subtracting this from $I(X)$ and taking the $X \to 0$ limit gives

$$C = \int_0^\infty [\log(1+y) - (1 - e^{-y}) \log y] \frac{dy}{y}.$$

We leave as an exercise for the reader that $C = \pi^2/4 + \gamma_E^2/2$. Thus (24), which gets the factor $-1/(4\pi^2 r)$, has the asymptotics as $X \to 0$,

$$1 - [(\log^2 X)/2 + \gamma_E \log X + \pi^2/4 + \gamma_E^2/2]\, X - [\log X - 2 + \gamma_E]\, X^2 + O(X^{3+}).$$

After setting $X = ir\xi$ this gives the asymptotics as $\xi \to 0$, with terms up to those involving ξ^2, and with acceptable error for the coefficients.

5. The final results

We know that with acceptable error we may replace $h^-(\xi)$ in the right side of (16) by $G^-(\xi) - VG^+(\xi)$. In Section 3 we showed how to compute the series for $G^+(\xi)$ up to terms involving ξ^2, and $G^-(\xi)$ is the complex conjugate of $G^+(\xi)$ for real ξ. In section 4 we showed how to compute the series for $VG^+(\xi)$ up to terms involving ξ^2. All with acceptable errors in the coefficients. Then we multiply by $e^{-ir\xi/2}\, \sigma_-(\xi)$ to obtain the last term on the right in (16). The next-to-last term is the complex conjugate of the last, so we know that, also. Then we add the series for $\hat{g}(\xi)$. From these computations and (17) we find that

$$\int_{-r/2}^{r/2} f(x)\,dx = r + \frac{1}{\pi}\,[\log(\pi r/2) + 1] + \frac{1}{\pi^2 r}\,[\log(\pi r/2) + 1/2] + O(r^{-2+}).$$

(Observe that all terms involving γ_E have canceled.) Setting $r = 2/\kappa$ and using (11), we obtain (9).

From the computations and (18) we find that

$$\int_{-r/2}^{r/2} x^2\, f(x)\,dx = \frac{r^3}{12} + \frac{r^2}{4\pi}\,[\log(\pi r/2) - 1]$$

$$+\frac{r}{4\pi^2}\,[\log^2(\pi r/2) - \log(\pi r/2) - 5/2 + 2\pi^2/3] + O(r^{0+}).$$

Then from these and (13) we obtain

$$\mathfrak{e}_F(\gamma) = \frac{\pi^2}{12} - \frac{\pi}{2r} + \frac{1}{2r^2}\,[\log(\pi r/2) + 1 + \pi^2/3] + O(r^{-3+}).$$

From (12) we find that

$$\gamma = \frac{\pi}{r} - \frac{1}{r^2}\,[\log(\pi r/2) + 1] + O(r^{-3+}),$$

and (10) follows.

References

[1] C. Atkinson and F.G. Leppington, The asymptotic solution of some integral equations, IMA J. Appl. Math. **31** (1983) 169–182.

[2] M. Gaudin, Un système à une dimension de fermions en interaction, Phys. Lett. A, **24** (1967), 55–56.

[3] M. Gaudin, *The Bethe Wavefunction*, Cambridge Univ. Press, 2014. (English translation by J.-S. Caux.)

[4] X-W. Guan and Z-Q. Ma, One-dimensional multicomponent fermions with δ-function interaction in strong- and weak-coupling limits: Two-component Fermi gas, Phys. Rev. A **85** (2012), 033632.

[5] T. Iida and M. Wadati, Exact analysis of a δ-function spin-1/2 attractive Fermi gas with arbitrary polarization, J. Stat. Mech. (2007) P06011.

[6] V.Ya. Krivnov and A.A. Ovchinnikov, One-dimensional Fermi gas with attraction between the electrons, Sov. Phys. JETP **40** (1975), 781–786.

[7] F. Leppington and H. Levine, On the problem of closely separated circular discs at equal potential, Quarterly J. Mech. Appl. Math. **25** (1972), 225–245.

[8] E.H. Lieb and W. Liniger, Exact analysis of an interacting Bose gas. I. The general solution and ground state energy, Phys. Rev. **130**(4) (1963), 1605–1616.

[9] C.A. Tracy and H. Widom, On the ground state energy of the delta-function Bose gas, J. Phys. A: Math. Theor. **49** (2016), 29400.

[10] C.A. Tracy and H. Widom, On the ground state energy of the delta-function Fermi Gas, to appear in J. Math. Phys., arXiv:1604.02440.

[11] H. Widom, Toeplitz determinants with singular generating functions, Amer. J. Math. **95** (1973) 333–383.

[12] C.N. Yang, Some exact results for the many-body problem in one dimension with repulsive delta-function interaction, Phys. Rev. Lett. **19** (1967), 1312–1315.

Craig A. Tracy
Department of Mathematics
University of California
Davis, CA 95616, USA
e-mail: `tracy@math.ucdavis.edu`

Harold Widom
Department of Mathematics
University of California
Santa Cruz, CA 95064, USA
e-mail: `widom@ucsc.edu`

Geometric Methods in Physics. XXXV Workshop 2016
Trends in Mathematics, 213–225

Non-periodic One-gap Potentials in Quantum Mechanics

Dmitry Zakharov and Vladimir Zakharov

Abstract. We construct a broad class of bounded potentials of the one-dimensional Schrödinger operator that have the same spectral structure as periodic finite-gap potentials, but that are neither periodic nor quasi-periodic. Such potentials, which we call primitive, are non-uniquely parametrized by a pair of positive Hölder continuous functions defined on the allowed bands. Primitive potentials are constructed as solutions of a system of singular integral equations, which can be efficiently solved numerically. Simulations show that these potentials can have a disordered structure. Primitive potentials generate a broad class of bounded non-vanishing solutions of the KdV hierarchy, and we interpret them as an example of integrable turbulence in the framework of the KdV equation.

Mathematics Subject Classification (2010). 37K15, 70H06.

Keywords. Schrödinger operator, KdV equation, Riemann–Hilbert problem, integrable turbulence.

1. Introduction

We consider the Schrödinger equation on the real axis

$$-\psi'' + u(x)\psi = E\psi, \quad -\infty < x < \infty, \tag{1}$$

with a bounded potential $u(x)$.

A value of E belongs to the spectrum of $u(x)$ if there exist one or two independent bounded wave functions $\psi(x, E)$:

$$|\psi(x, E)| < 1, \quad -\infty < x < \infty. \tag{2}$$

The spectrum is a subset of the axis $-\infty < E < \infty$, and can have a quite complicated structure. We only consider the case when the spectrum is purely continuous, i.e., consists of a finite or infinite collection of segments (allowed bands), whose length is bounded from below. We pose the question of determining all potentials having a purely continuous spectrum.

The simplest examples of such potentials are those that are periodic in x. A dense subset of the periodic potentials are the finite-gap potentials, expressible in terms of Riemann θ-functions of hyperelliptic curves. An N-gap potential has N finite and one infinite allowed bands, interspersed with N forbidden gaps. However, there also exist quasi-periodic N-gap potentials. We pose the question of describing a wider class of N-gap potentials, which have the same spectrum as the algebro-geometric potentials, but which are neither periodic nor quasi-periodic. In this paper we limit ourselves to studying one-gap potentials, with spectrum consisting of the positive semiaxis $E > 0$ and a segment $-k_2^2 < E < -k_1^2$ on the negative semiaxis.

A periodic one-gap potential is determined up to translation by the formula

$$u(x) = 2\wp(x + i\omega' - x_0) + e_3. \tag{3}$$

Here $\wp(x)$ is the elliptic Weierstrass function with periods 2ω and $2i\omega'$. The spectrum is equal to $[-k_2^2, -k_1^2] \cup [0, \infty)$, where

$$e_1 - e_3 = k_2^2, \quad e_2 - e_3 = k_1^2, \quad e_1 > e_2 > e_3, \quad e_1 + e_2 + e_3 = 0, \tag{4}$$

and the e_i are the values of $\wp$ at the half-periods. The spectrum is doubly degenerate and reflectionless, and within the allowed bands a quantum particle moves freely in both directions. In this paper we construct a family of potentials that have the same spectrum and that are reflectionless, but that are not periodic. Such a potential is determined by two positive Hölder-continuous functions R_1 and R_2 defined on $[k_1, k_2]$. The spectrum of the corresponding Schrödinger operator is doubly degenerate inside the allowed gap $[-k_2^2, -k_1^2]$.

To construct these one-gap potentials, we consider the closure of the set of reflectionless Bargmann potentials, also known as N-soliton potentials, as $N \to \infty$. This problem was posed and formally solved in the works of Marchenko and his students [2–4], but the obtained results are not effective. In this paper we consider a new technique for constructing the closure of the Bargmann potentials, using an associated $\overline{\partial}$-problem. This technique proves to be quite effective. In particular, we construct the periodic potential (3) as a limit of N-soliton solutions.

2. Bargmann potentials via the dressing method

Bargmann potentials were first constructed in 1948 as a class of potentials of the one-dimensional Schrödinger operator (1) having N bound states with negative energy and zero reflection coefficient for all positive energies. From the point of view of the KdV equation, Bargmann potentials correspond to N-soliton solutions at fixed moments of time, and hence can be explicitly constructed using the inverse spectral transform for the operator (1). In this section, we give an alternate construction of the Bargmann potentials using the so-called dressing method, following Zakharov and Manakov [5]. Compared with the IST, this method gives us additional flexibility that will later prove crucial when we generalize the associated Riemann–Hilbert problem.

We consider a $\overline{\partial}$-problem on the complex k-plane of the following kind:

$$\frac{\partial\chi}{\partial\overline{k}} = ie^{2ikx}T(k)\chi(x,-k). \tag{5}$$

Here $T(k)$ is a compactly supported distribution called the *dressing function* of the $\overline{\partial}$-problem. A solution of (5) is defined up to multiplication by a function of x, hence if a solution exists we can normalize it by the condition $\chi \to 1$ as $|k| \to \infty$. Such a solution satisfies the integral equation

$$\chi(x,k) = 1 + \frac{i}{\pi}\iint \frac{e^{-2iqx}T(-q)\chi(x,q)}{k+q}\,dq d\overline{q}, \tag{6}$$

where we normalize the integral in the following way:

$$\frac{1}{k} = \lim_{\varepsilon\to 0}\frac{\overline{k}}{|k|^2+\varepsilon^2}, \quad \frac{\partial}{\partial\overline{k}}\left(\frac{1}{k}\right) = \pi\delta(k). \tag{7}$$

Here $\delta(k)$ is the two-dimensional δ-function.

We now show that a solution of the $\overline{\partial}$-problem (5) gives rise to a solution of the Schrödinger equation (1).

Theorem 1. *Suppose that the dressing function $T(k)$ has the property that the $\overline{\partial}$-problem (5) has a unique solution χ normalized by the condition*

$$\chi(x,k) = 1 + o(1) \text{ as } |k| \to \infty \tag{8}$$

on the set $\mathfrak{U}\times\mathbb{C}$, where $\mathfrak{U}\subset\mathbb{R}$ is an open subset. Denote

$$\chi(x,k) = 1 + \frac{i\chi_0(x)}{k} + O(k^{-2}), \quad u(x) = 2\frac{d}{dx}\chi_0(x). \tag{9}$$

Then the function $\chi(x,k)$ is a solution of the differential equation

$$\chi_{xx} - 2ik\chi_x - u(x)\chi = 0, \tag{10}$$

and the function $\psi(x,k) = \chi(x,k)e^{-ikx}$ is a solution of the Schrödinger equation (1) with $E = k^2$.

Proof. Let χ be the unique solution of (5) satisfying the normalization condition (8). Define the function

$$\widetilde{\chi}(x,k) = \chi_{xx} - 2ik\chi_x - u(x).$$

It is straightforward to check that $\widetilde{\chi}$ also satisfies the $\overline{\partial}$-problem (5), and the choice of $u(x)$ guarantees that $\widetilde{\chi}\to 0$ as $|k|\to\infty$. By the uniqueness assumption, it follows that $\widetilde{\chi}$ is identically equal to zero, which completes the proof. □

We obtain the class of reflectionless Bargmann potentials by considering a $\overline{\partial}$-problem whose solution χ is a rational function of k with simple poles along the imaginary axis.

Theorem 2. *Let $\kappa_1, \ldots, \kappa_N$ and $c_1, \ldots, c_N$ be a collection of real numbers satisfying the following properties:*

1. $\kappa_m \neq \pm\kappa_n$ *for all* m *and* n.
2. $c_n/\kappa_n > 0$ *for all* n.

Consider the dressing function

$$T(k) = \pi \sum_{n=1}^{N} c_n \delta(k - i\kappa_n). \tag{11}$$

Then the $\overline{\partial}$-problem (5) has a unique solution χ satisfying the normalization condition $\chi \to 1$ as $|k| \to \infty$. This solution is a rational function of k having simple poles at the points $k = i\kappa_n$ for $n = 1, \ldots, N$, and has the following form:

$$\chi(x, k) = 1 + i \sum_{n=1}^{N} \frac{\chi_n(x)}{k - i\kappa_n}, \tag{12}$$

where the $\chi_n(x)$ are real-valued functions. The corresponding potential

$$u(x) = 2\frac{d}{dx} \sum_{n=1}^{N} \chi_n(x)$$

is a reflectionless Bargmann potential having the finite discrete spectrum $-\kappa_1^2, \ldots, -\kappa_N^2$, and $\psi_n(x) = \chi_n(x) e^{\kappa_n x}$ are the corresponding eigenfunctions. Furthermore, for each $m = 1, \ldots, N$, if we define $c_n^{(m)}$ and $\kappa_n^{(m)}$ by

$$\kappa_n^{(m)} = \begin{cases} \kappa_n, & n \neq m \\ -\kappa_n, & n = m \end{cases}, \quad c_n^{(m)} = \begin{cases} \left(\dfrac{\kappa_n - \kappa_m}{\kappa_n + \kappa_m}\right)^2 c_n, & n \neq m \\ -4\kappa_n^2/c_n, & n = m \end{cases} \tag{13}$$

the potential $u(x)$ corresponding to the data $\{\kappa_n^{(m)}, c_n^{(m)}\}$ is the same as for $\{\kappa_n, c_n\}$.

Remark 3. Given a reflectionless Bargmann potential $u(x)$ with a finite negative discrete spectrum $-\kappa_1^2, \ldots, -\kappa_N^2$, the direct spectral transform proceeds by constructing a solution $\psi(x, k)$ of the Schrödinger equation (1) that is analytic in the k-upper half-plane. In the k-lower half-plane, the function $\psi(x, k)$, and hence the function $\chi(x, k) = \psi(x, k)e^{ikx}$, then has poles on the negative imaginary axis at the points $-i|\kappa_1|, \ldots, -i|\kappa_N|$ corresponding to the discrete spectrum. To construct $u(x)$ using the dressing method, we can place the poles of χ on both the positive and negative imaginary axes, so long as the poles have distinct absolute values, and every N-soliton Bargmann potential can be constructed in 2^N different ways by arbitrarily choosing the signs of the κ_n.

Remark 4. It is possible to relax the condition that c_n and κ_n have the same sign for each n, but the corresponding potentials $u(x)$ will be singular functions of x.

Proof. Given the dressing function (11), the identity (7) implies that a solution χ of (5) has simple poles at the points $k = i\kappa_n$ and no other singularities. The condition $\chi \to 1$ as $|k| \to \infty$ then implies that χ has the form (12). Substituting this into the integral equation (6), we obtain a system of linear equations on the residues $\chi_n(x)$:

$$\chi_n(x) = e^{-2\kappa_n x} c_n \chi(x, -i\kappa_n). \tag{14}$$

Writing this system out explicitly, and replacing $\chi_n(x) = \psi_n(x)e^{-\kappa_n x}$, we obtain the following system:

$$\psi_n(x) + c_n \sum_{m=1}^{N} \frac{e^{-(\kappa_n+\kappa_m)x}}{\kappa_n + \kappa_m} \psi_m(x) = c_n e^{-\kappa_n x} \tag{15}$$

The matrix of this system

$$A_{nm} = \delta_{nm} + \frac{c_n e^{-(\kappa_n+\kappa_m)x}}{\kappa_n + \kappa_m}$$

is the sum of an identity matrix and a Cauchy-like matrix, therefore its determinant is the sum of the principal minors of the Cauchy-like matrix. This sum is indexed by subsets $I = \{i_1, \ldots, i_n\}$ of the index set $\{1, \ldots, N\}$ and can be explicitly evaluated as follows:

$$A = \det[A_{nm}] = \sum_{I \subset \{1,\ldots,N\}} \left[\prod_{\{i,j\}\subset I, i<j} \left(\frac{\kappa_i - \kappa_j}{\kappa_i + \kappa_j} \right)^2 \prod_{i\in I} \frac{c_i}{2\kappa_i} e^{-2\kappa_i x} \right].$$

By assumption, the quantities c_i/κ_i and $(\kappa_i - \kappa_j)^2$ are all positive, therefore each summand and hence all of A is positive, so the system (15) has a unique solution. By Theorem 1, χ satisfies equation (10), and the corresponding potential $u(x)$ is

$$u(x) = 2\frac{d\chi_0}{dx} = 2\frac{d}{dx} \sum_{n=1}^{N} \chi_n(x). \tag{16}$$

To evaluate $u(x)$, we note that the derivative of the nth column of the matrix $[A_{nm}]$ is equal to the right-hand side of equation (15) multiplied by $-e^{-\kappa_n x}$. Therefore, by Cramer's rule we have

$$\begin{aligned} u(x) = 2\frac{d}{dx} \sum_{n=1}^{N} \chi_n(x) &= 2\frac{d}{dx} \sum_{n=1}^{N} \psi_n(x) e^{\kappa_n x} \\ = 2\frac{d}{dx} \left[-\frac{1}{A} \frac{d}{dx} A \right] &= -2\frac{d^2}{dx^2} \ln A. \end{aligned} \tag{17}$$

When all the κ_n are positive, this is the familiar formula for the N-soliton reflectionless potentials (see formula (1.5) in [6]).

To finish the proof, we consider what happens to formula (17) when we replace $\{c_n, \kappa_n\}$ with $\{c_n^{(m)}, \kappa_n^{(m)}\}$ according to (13). A direct calculation shows that

$$A = \frac{c_m}{2\kappa_m} e^{-2\kappa_m x} A^{(m)},$$

where $A^{(m)}$ is the determinant of the matrix (15) corresponding to the data $\{c_n^{(m)}, \kappa_n^{(m)}\}$. By formula (17), the data $\{\kappa_n, c_n\}$ and $\{\kappa_n^{(m)}, c_n^{(m)}\}$ determine the same potential $u(x)$. Hence, starting with an arbitrary $\{\kappa_n, c_n\}$, we can replace all κ_n with $|\kappa_n|$ and make the corresponding changes to the c_n while preserving $u(x)$, so in fact all of the potentials that we obtain in this way are reflectionless Bargmann potentials.

Finally, considering the leading term in equation (10) near the poles, we see that ψ_n are eigenfunctions of the Schrödinger operator with potential $u(x)$ corresponding to the eigenvalues $-\kappa_n^2$. We also note that all principal minors of A are positive, hence A is a positive definite matrix. □

3. The symmetric Riemann–Hilbert problem

In this section, we consider a Riemann–Hilbert problem that is a continuous analogue of the finite $\overline{\partial}$-problem of Theorem 2 that generates the Bargmann potentials.

Theorem 5. *Let $0 < k_1 < k_2$ be real numbers, and let R_1 and R_2 be two positive Hölder-continuous functions on the interval $[k_1, k_2]$. Consider the dressing function*

$$T(k) = \pi \int_{k_1}^{k_2} R_1(p)\delta(k - ip)dp - \pi \int_{k_1}^{k_2} R_2(p)\delta(k + ip)dp. \tag{18}$$

Then the corresponding $\overline{\partial}$-problem (5) has a unique solution χ satisfying the normalization condition $\chi \to 1$ as $|k| \to \infty$. This function is analytic on the k-plane away from two cuts $[ik_1, ik_2]$ and $[-ik_2, -ik_1]$ on the imaginary axis. Denoting by χ^+ and χ^- the right and left boundary values of χ along the cuts

$$\chi^{\pm}(x, k) = \lim_{\varepsilon \to 0^+} \chi(x, k \pm \varepsilon), \quad k \in [-ik_2, -ik_1] \cup [ik_1, ik_2],$$

the function χ satisfies a symmetric Riemann–Hilbert problem on the cuts:

$$\chi^+(x, ip) - \chi^-(x, ip) = \pi i R_1(p) e^{-2px}[\chi^+(x, -ip) + \chi^-(x, -ip)], \tag{19}$$

$$\chi^+(x, -ip) - \chi^-(x, -ip) = -\pi i R_2(p) e^{2px}[\chi^+(x, ip) + \chi^-(x, ip)]. \tag{20}$$

The function χ can be explicitly given as

$$\chi(x, k) = 1 + i \int_{k_1}^{k_2} \frac{f(x, p)}{k - ip} dp + i \int_{k_1}^{k_2} \frac{g(x, p)}{k + ip} dp, \tag{21}$$

where $f(x, p)$ and $g(x, p)$ are real-valued functions defined for real x and for $p \in [k_1, k_2]$. The corresponding potential of the Schrödinger operator (1) is

$$u(x) = 2\frac{d}{dx} \int_{k_1}^{k_2} [f(x, p) + g(x, p)]dp.$$

Proof. Given R_1 and R_2, we look for a solution of the $\overline{\partial}$-problem (5) in the form (21), where f and g are unknown functions of x and $p \in [k_1, k_2]$. The jumps of χ

along the cuts are then equal to

$$\chi^+(x, ip) - \chi^-(x, ip) = 2\pi i f(x, p),$$
$$\chi^+(x, -ip) - \chi^-(x, -ip) = 2\pi i g(x, p).$$

Plugging (21) into (5), we see that χ satisfies the Riemann–Hilbert problem (19)–(20) if f and g satisfy the following system of singular integral equations:

$$f(x,p) + R_1(p)e^{-2px}\left[\int_{k_1}^{k_2} \frac{f(x,q)}{p+q}dq + \fint_{k_1}^{k_2} \frac{g(x,q)}{p-q}dq\right] = R_1(p)e^{-2px} \tag{22}$$

$$g(x,p) + R_2(p)e^{2px}\left[\fint_{k_1}^{k_2} \frac{f(x,q)}{p-q}dq + \int_{k_1}^{k_2} \frac{g(x,q)}{p+q}dq\right] = -R_2(p)e^{2px}. \tag{23}$$

We note that the Riemann–Hilbert problem (19)–(20) is a continuous generalization of equation (14).

We need to show that the system (22)–(23) has a unique solution on the entire real axis. To do this, we approximate these equations by Riemann sums. Fix an integer N, and let $\Delta = (k_2 - k_1)/2N$. We subdivide the segment $[k_1, k_2]$ into $2N$ equal parts and denote

$$\lambda_n = k_1 + (2n-2)\Delta, \quad \alpha_n = R_1(\lambda_n) \quad n = 1, \ldots, N+1,$$
$$\mu_n = k_1 + (2n-1)\Delta, \quad \beta_n = R_2(\mu_n). \quad n = 1, \ldots, N.$$

We note that all these quantities are positive. Approximating the Riemann–Hilbert problem (19)–(20) by replacing the integrals containing f with their Riemann sums at the λ_n and the integrals containing g with their Riemann sums at the μ_n, we obtain the system

$$f_n(x) + \alpha_n e^{-2\lambda_n x}\left(\sum_{m=1}^{N+1} \frac{f_m(x)}{\lambda_n + \lambda_m} + \sum_{m=1}^{N} \frac{g_m(x)}{\lambda_n - \mu_m}\right) = \alpha_n e^{-2\lambda_n x}, \tag{24}$$

$$g_n(x) - \beta_n e^{2\mu_n x}\left(\sum_{m=1}^{N+1} \frac{f_m(x)}{-\mu_n + \lambda_m} + \sum_{m=1}^{N} \frac{g_m(x)}{-\mu_n - \mu_m}\right) = -\beta_n e^{2\mu_n x}, \tag{25}$$

where $f_n(x) = f(x, \lambda_n)$ and $g_n(x) = g(x, \mu_n)$. We see that this system is equivalent to the system (15) on the eigenfunctions of a Bargmann potential having $2N+1$ solitons corresponding to the poles $(\lambda_1, \ldots, \lambda_{N+1}, -\mu_1, \ldots, -\mu_N)$ and the constants $(\alpha_1, \ldots, \alpha_{N+1}, -\beta_1, \ldots, -\beta_N)$. According to Theorem 2, this system has a unique solution for all x and gives a Bargmann potential with $2N+1$ solitons.

We claim that, given $L > 0$ and $\varepsilon > 0$, there exists N large enough so that the sums in Eqs. (24)–(25) are Riemann sums approximating the integrals (22)–(23) to within ε for all $x \in (-L, L)$. To show this, we solve our equations by iteration. Define

$$f(x,p) = R_1(p)e^{-2px}\widetilde{f}(x,p), \quad g(x,p) = R_2(p)e^{2px}\widetilde{g}(x,p),$$
$$f_n(x) = \alpha_n e^{-2\lambda_n x}\widetilde{f}_n(x), \qquad g_n(x) = \beta_n e^{2\mu_n x}\widetilde{g}_n(x).$$

We now solve equations (24)–(25) iteratively:

$$\begin{aligned} \widetilde{f}(x,p) &= \widetilde{f}^{(1)}(x,p) + \cdots, & \widetilde{g}(x,p) &= \widetilde{g}^{(1)}(x,p) + \cdots, \\ \widetilde{f}_n(x) &= \widetilde{f}_n^{(1)}(x) + \cdots, & \widetilde{g}_n(x) &= \widetilde{g}_n^{(1)}(x) + \cdots \end{aligned} \tag{26}$$

We see that

$$\begin{aligned} \widetilde{f}^{(1)}(x,p) &= 1 - \int_{k_1}^{k_2} \frac{R_1(q)e^{2qx}}{p+q} dq - \fint_{k_1}^{k_2} \frac{R_2(q)e^{-2qx}}{p-q} dq, \\ \widetilde{f}_n^{(1)}(x) &= 1 - \sum_{m=1}^{N+1} \frac{\alpha_m e^{-2\alpha_m x}}{\lambda_n + \lambda_m} - \sum_{m=1}^{N} \frac{\beta_m e^{2\beta_m x}}{\lambda_n - \mu_m}. \end{aligned} \tag{27}$$

Recalling that $\alpha_m = R_1(\lambda_m)$ and $\beta_m = R_2(\mu_m)$, we see that the sums approximate the integrals when

$$\alpha_m e^{2\lambda_m x} \Delta \ll 1, \quad \beta_m e^{-2\mu_m x} \Delta \ll 1.$$

We see that this condition holds for all $x \in (-L, L)$ if

$$\Delta \ll \frac{1}{R} e^{-2k_2 L}$$

where $R = \max(R_1, R_2)$. We note that, in order to maintain the same degree of accuracy when increasing the length L of our interval, we need to exponentially increase the number N of approximation points.

We do not have a strict proof of the above statements, but they are confirmed by numerical experiments [8–10]. We hope to soon publish a complete proof that equations (22)–(23) are uniquely solvable.

If R_1 and R_2 are positive on (k_1, k_2), then the spectrum of the corresponding Schrödinger operator is doubly degenerate, and there are two orthogonal eigenfunctions

$$\varphi(x,k) = \frac{e^{ikx}}{\sqrt{R_1(k)}} f(x,k), \quad \psi(x,p) = \frac{e^{-ikx}}{\sqrt{R_2(k)}} g(x,k). \tag{28}$$

Consider the nonstationary Schrödinger equation

$$i\frac{\partial \Psi}{\partial t} = -\Psi_{xx} + u\Psi \tag{29}$$

and the corresponding continuity equation

$$\frac{\partial}{\partial t}|\Psi|^2 + \frac{\partial}{\partial x}\rho = 0, \tag{30}$$

where $\rho = i(\Psi\overline{\Psi}_x - \overline{\Psi}\Psi_x)$ is the particle density. Setting

$$\Psi = e^{ik^2 t}(\varphi(x,k) + i\psi(x,k)), \tag{31}$$

we observe that the particle density ρ is the Wronskian of the solutions φ, ψ, hence is independent of x. Therefore a particle with wave function $\varphi + i\psi$ undergoes ballistic transport with no resistance, and the potential $u(x)$ describes an ideal conductor.

We also observe that the nonlocal symmetric Riemann–Hilbert problem (19)–(20) can be converted to a local vector Riemann–Hilbert problem. Define

$$\Sigma(x,k) = \begin{bmatrix} \chi_1(x,k) \\ \chi_1(x,-k) \end{bmatrix},$$

$$M(x,k) = \frac{1}{1+\pi^2 R_1(k)R_2(k)} \begin{bmatrix} 1-\pi^2 R_1(k)R_2(k) & 2\pi i R_1(k)e^{-2kx} \\ 2\pi i R_2(k)e^{2kx} & 1-\pi^2 R_1(k)R_2(k) \end{bmatrix}.$$

Then (19)–(20) is equivalent to the system

$$\Sigma^+(k) = M(k)\Sigma^-(k), \tag{32}$$

where $\Sigma^+(k)$ and $\Sigma^-(k)$ are respectively the right and left limit values of Σ on the cuts $[ik_1, ik_2]$ and $[-ik_2, -ik_1]$. □

4. Periodic one-gap potentials

In this section, we show that periodic one-gap potentials of the Schrödinger operator can be constructed from the symmetric Riemann–Hilbert problem.

Let ω and ω' be positive real numbers, and consider the elliptic curve $E = \mathbb{C}/\Lambda$, where Λ is the period lattice generated by 2ω and $2i\omega'$. Denote by $\wp(z)$ the Weierstrass elliptic function associated to the lattice Λ. It satisfies the differential equation

$$[\wp'(z)]^2 = 4\wp(z)^3 - g_2\wp(z) - g_3 = 4(\wp(z)-e_1)(\wp(z)-e_2)(\wp(z)-e_3),$$

where the zeroes e_1, e_2, e_3 are real-valued, satisfy $e_1+e_2+e_3=0$, and we assume that $e_3 < e_2 < e_1$.

The function

$$u(x) = 2\wp(x-\omega-i\omega') + e_3 \tag{33}$$

is a real-valued potential of the Schrödinger operator (1) with period 2ω. Our goal is to construct a solution of (1) that gives a solution of the symmetric Riemann–Hilbert problem.

We consider the following function $\varphi(x,z)$, where x is real and z is defined on the curve E:

$$\varphi(x,z) = \frac{\sigma(x-\omega-i\omega'+z)\sigma(\omega+i\omega')}{\sigma(x-\omega-i\omega')\sigma(\omega+i\omega'-z)}\exp[-\zeta(z)x]. \tag{34}$$

A direct calculation shows that φ satisfies the Lamé equation

$$\varphi'' - [2\wp(x-\omega-i\omega')+\wp(z)]\varphi = 0.$$

Hence we see that φ is a solution of the Schrödinger equation (1) with potential (33) if the parameter z satisfies the relation

$$k^2 = e_3 - \wp(z), \quad k = \frac{1}{\operatorname{sn} z}. \tag{35}$$

The Weierstrass function $\wp$ has degree two, hence for a generic complex value of k there are two values of z on E that satisfy (35). In order to make the function

(34) a single-valued function of k, we need to choose a branch of z. We choose the solution $z(k)$ of (35) that satisfies

$$z(k) = \frac{i}{k} + O\left(\frac{1}{k^2}\right) \text{ as } |k| \to \infty. \tag{36}$$

This branch defines a single-sheeted map from the complex k-plane with two cuts on the imaginary axis to a period rectangle of the lattice Λ centered at 0. The cuts on the imaginary axis are $[-ik_2, -ik_1]$ and $[ik_1, ik_2]$, where

$$k_1 = \sqrt{e_2 - e_3}, \quad k_2 = \sqrt{e_1 - e_3}.$$

The right and left sides of the top cut $[ik_1, ik_2]$ are mapped to the line segments joining ω to $\omega + i\omega'$ and $\omega - i\omega'$, respectively, and the right and left sides of the bottom cut $[-ik_2, -ik_1]$ are respectively mapped to the segments joining $-\omega$ to $-\omega + i\omega'$ and $-\omega - i\omega'$.

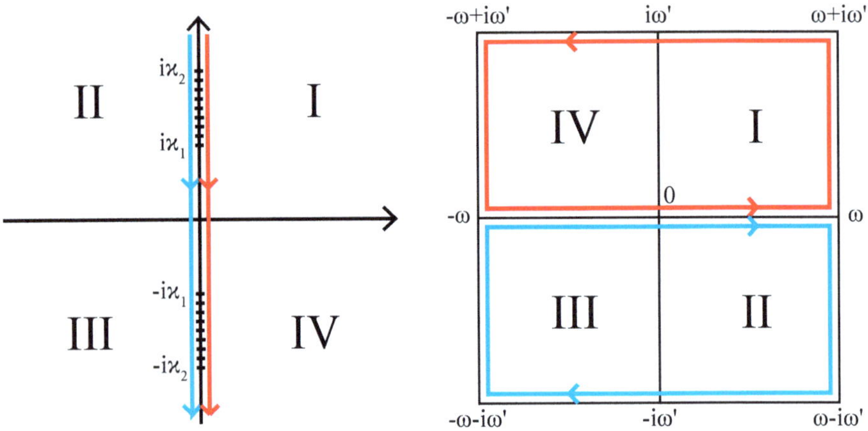

The k-plane The z-plane

The function φ satisfies the following properties:

$$\varphi(x, z + 2\omega) = \varphi(x, z), \quad \varphi(x, z + 2i\omega') = \varphi(x, z),$$
$$\overline{\varphi}(x, z) = \varphi(x, z) \text{ when } \overline{z} = z, \overline{x} = x.$$

In addition,

$$\overline{\varphi}(x, z) = \varphi(x, \overline{z})$$

for all z having real part ω.

Theorem 6. *Let $z(k)$ be the branch of the solution of* (35) *satisfying* (36). *Let $f(k)$ be the branch of the function*

$$f(k) = \sqrt{\frac{k + ik_1}{k + ik_2}}$$

satisfying $f(k) \to 1$ *as* $|k| \to \infty$. *On the complex* k*-plane with two cuts* $[ik_1, ik_2]$ *and* $[-ik_2, -ik_1]$ *along the imaginary axis, define the function*

$$\xi(x,k) = f(k)\varphi(x, z(k))e^{-ikx}.$$

Then the function $\xi(x,k)$ *satisfies the equation*

$$\xi'' + 2ik\xi' - u(x)\xi = 0, \quad \xi \to 1 \text{ as } |k| \to \infty$$

with potential $u(x)$ *given by* (33). *On the cuts, the function* ξ *satisfies the Riemann–Hilbert problem*

$$\xi^+(x, ip) - \xi^-(x, ip) = i\pi R_1(p)e^{2px}\left[\xi^+(x, -ip) + \xi^-(x, -ip)\right],$$
$$\xi^+(x, -ip) - \xi^-(x, -ip) = -i\pi R_2(p)e^{-2px}\left[\xi^+(x, ip) + \xi^-(x, ip)\right].$$

Here $p \in [k_1, k_2]$, *and* $\xi^\pm(x, \pm ip)$ *are the right- and left-hand values of the upper and lower cuts. The functions* R_1 *and* R_2 *are equal to*

$$R_1(p) = \frac{1}{\pi}h(p), \quad R_2(p) = \frac{1}{\pi h(p)}, \quad h(p) = \sqrt{\frac{(p-k_1)(p+k_2)}{(k_2-p)(p+k_1)}}$$

We note that in this case we have

$$R_1(p)R_2(p) = \frac{1}{\pi^2}, \quad \xi^+(x, ip) = ih(p)e^{-2px}\xi^-(x, -ip).$$

5. Unitary equivalent potentials

We have already noted that the representation of a potential of the Schrödinger operator using the dressing method is not unique. For example, a Bargmann potential with N solitons can be represented by the dressing method in 2^N different ways. It is therefore natural to pose the question of describing all dressings that define the same potential. However, it is more productive to consider the following more general question.

Let L be the Schrödinger operator with potential $u(x)$, and let U be a unitary operator. If

$$\widetilde{L} = U^+ L U \tag{37}$$

is a Schrödinger operator with potential $\widetilde{u}(x)$, we say that $u(x)$ and $\widetilde{u}(x)$ are unitary equivalent. It is clear that $u(x)$ and $\widetilde{u}(x)$ have the same spectra.

We construct unitary equivalent potentials following Lax' seminal paper [11]. We assume that the potential depends on time according to one of the equations of the KdV hierarchy

$$\frac{\partial u}{\partial t} = \frac{\partial}{\partial x}\frac{\delta H}{\delta u}, \quad H = \sum_{n=0}^{\infty} c_n I_n, \tag{38}$$

where c_n are real numbers, only finitely many of which are nonzero, and I_n are the integrals of the KdV equation, normalized as follows:

$$I_n = \frac{1}{2}\int_{-\infty}^{\infty} \left[u^{(n)}(x)^2 + \cdots\right] dx. \tag{39}$$

Introducing time evolution has the following effect on the dressing kernel:

$$T(k,t) = T(k,0)\exp\left(it\sum_{n=0}^{\infty} c_n(-1)^n(2k)^{n+1}\right) \tag{40}$$

and the functions R_1 and R_2 (18) are modified as follows:

$$R_1(p,t) = R_1(p)e^{-S(p)t}, \quad R_2(p,t) = R_2(p)e^{S(p)t}, \quad S(p) = \sum_{n=0}^{\infty} c_n(2p)^{n+1}. \tag{41}$$

Applying time evolution (41) to (18) produces a unitary-equivalent potential. We note that the function $W(p) = R_1(p)R_2(p)$ is unitary invariant.

Consider the one-gap case, and assume that R_1 and R_2 are positive for $k_1 < p < k_2$. Suppose that at $t = 0$ we have

$$\ln R_1(p,0) - \ln R_2(p,0) = 2s_0(p). \tag{42}$$

Introducing time evolution, we obtain

$$\ln R_1(p,t) - \ln R_2(p,t) = 2(s_0(p) - ts(p)). \tag{43}$$

Any real-valued function on the interval $[k_1, k_2]$ can be approximated using a polynomial of odd degree, so choosing c_n and t appropriately, we can assume that $R_1(p) = R_2(p)$. In this case, the only invariant of unitary transformations are k_1, k_2 and the function $W(p) = R_1(p)R_2(p)$.

Suppose that $W(p) = 1/\pi^2$. We have seen that the cnoidal wave can be constructed using the dressing method with $R_1(p)R_2(p) = 1/\pi^2$. It is well known that the time evolution of a cnoidal wave according to a higher KdV equation is a cnoidal wave with the same velocity. Hence the dressing $R_1(p) = R_2(p) = 1/\pi$ also results in a cnoidal wave. This is confirmed by numerical experiments [8–10].

Finally, if one of the two functions R_1 or R_2 vanishes at a point $p \in (k_1, k_2)$, then so does $W(p)$, and the corresponding primitive potential has a simple spectrum at p.

Acknowledgment

The authors would like to thank Sergey Dyachenko for the numerical experiments. The first author is thankful for the support of NSF grant DMS-1716822. The second author is grateful for the support of NSF grant DMS-1715323 and RSF grant No. 14-22-00174.

References

[1] V. Bargmann, On the connection between phase shifts and scattering potential, *Rev. Mod. Phys.* 21 (1949), 488–493.

[2] V. Marchenko, Nonlinear equations and operator algebras, *Mathematics and its Applications* (*Soviet Series*), 17, D. Reidel Publishing Co., Dordrecht, 1988.

[3] D. Lundina, Compactness of sets of reflectionless potentials, *J. Soviet Math.* 48 (1990), no. 3, 290–297.

[4] V. Marchenko, The Cauchy problem for the KdV equation with non-decreasing initial data, in *What is integrability?*, edited by V. E. Zakharov. Springer Series in Nonlinear Dynamics. Springer-Verlag, Berlin, 1991.

[5] V. Zakharov, S. Manakov, Construction of multidimensional nonlinear integrable systems and their solutions, *Funktsional. Anal. i Prilozhen.* (in Russian) 19 (1985), no. 2, 11–25, 96.

[6] S. Novikov, S. Manakov, L. Pitaevskii, V. Zakharov, Theory of solitons. The inverse scattering method. Translated from the Russian. Contemporary Soviet Mathematics. Consultants Bureau [Plenum], New York, 1984.

[7] V. Zakharov, Turbulence in integrable systems, *Stud. Appl. Math.* 122 (2009), no. 3, 219–234.

[8] S. Dyachenko, D. Zakharov, V. Zakharov, Primitive potentials and bounded solutions of the KdV equation, *Phys. D* 333 (2016), 148–156.

[9] D. Zakharov, S. Dyachenko, V. Zakharov, Non-periodic one-dimensional ideal conductors and integrable turbulence, *Phys. Lett. A* 380 (2016), no. 46, 3881–3885.

[10] D. Zakharov, S. Dyachenko, V. Zakharov, Bounded solutions of KdV and non-periodic one-gap potentials in quantum mechanics, *Lett. Math. Phys.* 106 (2016) no. 6, 731–740.

[11] P. Lax, Integrals of nonlinear equations of evolution and solitary waves, *Comm. Pure Appl. Math.* 21 (1968) 467–490.

Dmitry Zakharov
Department of Mathematics
Central Michigan University
Mount Pleasant, MI, 48858, USA
e-mail: `dvzakharov@gmail.com`

Vladimir Zakharov
Department of Mathematics
University of Arizona
Tucson, AZ 85791, USA

Lebedev Physical Institute RAS
Leninsky 53
Moscow 119991, Russia

Novosibirsk State University
Novosibirsk 630090, Russia
e-mail: `zakharov@math.arizona.edu`

Part III

Abstracts of the Lectures at "School on Geometry and Physics"

Geometric Methods in Physics. XXXV Workshop 2016
Trends in Mathematics, 229–235

Integrable Geodesic Flows

Alexey Bolsinov

Mathematics Subject Classification (2010). 53D25, 53C22, 37J3.

Keywords. Geodesic flow, integrability, Hamiltonian representation.

Lecture 1. Integrability of geodesic flows

What are geodesics?

Let M be a manifold with a Riemannian metric g. In local coordinates, $g = \sum g_{ij}(x) dx^i dx^j$.

Consider a point moving on M and let $\gamma(t)$ be its trajectory. According to the second Newton law, its motion is defined by the equation $ma = F$. What is the meaning of the acceleration a in this case? The answer is: a is the derivative of the velocity $v = \frac{d\gamma}{dt}$, but one should consider the covariant derivative. If the point is moving by inertia, then $F = 0$ and we come to the following equation:

$$\nabla_{\frac{d\gamma}{dt}} \frac{d\gamma}{dt} = 0$$

or, in local coordinates,

$$\frac{d^2 x^i}{dt^2} + \Gamma^i_{jk}(x) \frac{dx^j}{dt} \frac{dx^k}{dt} = 0, \tag{1}$$

where $\gamma(t) = (x^1(t), \dots, x^n(t))$ and $\Gamma^i_{jk}(x)$ are the Christoffel symbols of the Levi-Civita connection of the metric g.

This is the equation of geodesics on a Riemannian manifold (M, g).

Properties of geodesics:

- if the metric is Euclidean, then $\Gamma^i_{jk}(x) = 0$ and the geodesics are straight lines, i.e., $x_i(t) = a_i t + b_i$;
- Eq. (1) is a non-linear second-order system of ODEs;
- Existence and Uniqueness Theorem;
- geodesic completeness;
- Hopf–Rinow theorem;
- arc-length parametrisation and admissible reparametrisations $s' = as + b$;

- kinetic energy $H = \sum g_{ij}(x)\dot{x}^i\dot{x}^j$ as a first integral (in particular, the unit (co-)tangent bundle is $2n-1$ invariant submanifold).

Hamiltonian form of the equation of geodesics

Consider T^*M as a symplectic manifold with canonical coordinates (x,p) and the symplectic form $\omega = \sum dp_i \wedge dx^i$.

In coordinates (x,p), the equation of geodesics (1) can be rewritten in Hamiltonian form:

$$\frac{dp_i}{dt} = \frac{\partial H}{\partial x^i}, \quad \frac{dx^i}{dt} = -\frac{\partial H}{\partial p_i}, \tag{2}$$

where $H = \frac{1}{2}\sum g^{ij}(x)p_ip_j$.

If we accept this Hamiltonian representation as a fundamental property of geodesics, then we may simply define a geodesic flow to be a Hamiltonian system on the cotangent bundle T^*M whose Hamiltonian is a positive definite quadratic form in momenta p_i. We can naturally generalize this viewpoint:

- indefinite non-degenerate quadratic form $\mapsto$ pseudo-Riemannian geodesic flows
- non-negative quadratic forms $H = \sum g^{ij}p_ip_j \geq 0$ with the additional property that $H = 0$ iff p belongs a totally non-integrable co-distribution $\mapsto$ sub-Riemannian geodesic flows
- convex homogeneous Hamiltonians $\mapsto$ Finsler geodesic flows

Definition of Integrability [1]

Definition 1. We say that a geodesic flow on (M,g) is Liouville integrable, if the Hamiltonian system (2) admits n commuting first integrals $f_1,\dots,f_n$ that are independent almost everywhere on T^*M.

Some issues to discuss: almost everywhere, what kind of integrals, what about the zero-section.

Local and global aspects

Proposition 1. *Every geodesic flow is locally completely integrable.*

Basically, this follows from the Darboux theorem and homogeneity (locally means "on $T^*U(x_0)$", i.e., "locality" is meant in the sense of coordinates only (in the sense of p, the neighborhood is not local but global).

The best example explaining why integrability should be understood as a global phenomenon is the geodesic flow on the surface M^2_m of genus m with a constant negative curvature metric. This geodesic flows is known to be chaotic/ergodic, so definitely non-integrable. However, locally this geodesic flows admit a polynomial integral that commutes with H (there are even 3 independent linear integrals so that this geodesic flow is locally super-integrable). In particular, in any reasonable local coordinates the geodesic flow can be integrated explicitly. The problem is that when trying to extend local polynomial integrals up to global ones we obtain multivalued functions (but multivalued integrals make no sense).

Proposition 2. *Assume that F is a real analytic (commuting with and independent of H) integral of the geodesic flow on (M, g). Then this geodesic flow admits a non-trivial polynomial integral*[1].

Proof. It is easy to see that for a series of the form

$$F = \sum a_{n_1 n_2 \dots n_k}(x) p_1^{n_1} p_2^{n_2} \dots p_k^{n_k} = F_0 + F_1 + F_k + \cdots ,$$

where F_k is the homogeneous component of this power series of degree k, the condition $\{H, F\} = 0$ immediately implies $\{H, F_k\} = 0$ for each $k \in \mathbb{N}$. □

Also, one needs to distinguish between "abstract integrability" and "practical integrability". Existence of commuting integrals and possibility to describe them explicitly, these are two different stories.

Zoll metrics is an example of metrics whose geodesic flows are integrable in abstract but not practical sense.

The most natural and strongest type of integrability is polynomial integrability. Let us discuss this property in dim = 2.

- A generic Riemannian metric does not admit any non-trivial polynomial integrals[2] (except for the powers of the Hamiltonian).
- For any $n \in \mathbb{N}$, there exist (local) examples of geodesic flows that admit polynomial integrals of degree n and do not admit any non-trivial integrals of smaller degree (V. Ten, V. Kozlov, K. Kiyohara).
- All such metrics are defined by an integrable system of PDE's and, in principle, can be described "explicitly" (M. Bialy, A. Mironov, S. Tsarev, M. Pavlov).

Lecture 2: Integrable geodesic flows on two-dimensional surfaces

Examples: Sphere, flat torus, surface of revolution, ellipsoid [3]

Proposition 3.

1. *Assume that the geodesic flow of g admits a linear integral $F = a_1(x)p_1 + a_2(x)p_2$. If F is not zero at a point $x_0 \in M$, then there exists a local coordinate system (u, v) such that*

$$g = du^2 + G(u)dv^2 \quad (\textit{equivalently, } H = \tfrac{1}{2}(p_u^2 + G^{-1}(u)p_v^2)) \quad \textit{and} \quad F = p_v.$$

2. *Assume that the geodesic flow of g admits a quadratic integral $F = A(x)p_1^2 + 2B(x)p_1p_2 + C(x)p_2^2$. If F (as a quadratic form) is not proportional to the*

---

[1]In this theory, by saying "polynomial integrals" we always mean polynomiality in momenta p, i.e., functions of the form $\sum a_{n_1 n_2 \dots n_k}(x) p_1^{n_1} p_2^{n_2} \dots p_k^{n_k}$ with coefficients being some functions on M. This functions will be automatically smooth or real analytic depending on the properties of g.

[2]The statement is more or less obvious as the existence of a polynomial integral for a given g is defined by an overdetermined system of PDEs and solutions exist only under some special assumptions on g. However, rigorous proof requires serious efforts, see [2].

Hamiltonian at a point $x_0 \in M$, then there exists a local coordinate system (u, v) such that

$$g = (f(u) + g(v))(du^2 + dv^2)$$

$$\left(\textit{equivalently, } H = \frac{p_u^2 + p_v^2}{2(f(u) + g(v))} \textit{ and } F = \frac{g(v)p_u^2 - f(u)p_v^2}{f(u) + g(v)} \right).$$

Similar statements can be formulated for singular points.

From now on, we assume that M is a closed 2-dimensional surface. Consider the unit (co)-tangent bundle $Q^3 = \{H(x,p) = \frac{1}{2}\} \subset T^*M^2$.

Theorem 1. *There are only 4 closed surfaces which admit integrable geodesic flows with polynomial (real-analytic on Q^3, geometrically simple, tame) integrals, namely, S^2, $\mathbb{R}P^2$, T^2 and the Klein bottle.*

Lecture 3: Integrability of geodesic flows, homogeneous spaces and bi-quotients of Lie groups

The theory of integrable geodesic flows on Lie groups and their homogeneous spaces is based on some quite natural and simple constructions from Poisson Geometry. I will follow the classical approach developed by Sophus Lie: *Poisson algebras* and *"groups of functions"*.

We consider a symplectic manifold (M, ω) and $C^\infty(M)$ as a Poisson algebra.

By saying "Poisson (sub)algebra" $\mathcal{F}$, I mean basically three things:

- elements of $\mathcal{F}$ are functions of some kind;
- $\mathcal{F}$ is closed under a Poisson bracket, i.e., $f, g \in \mathcal{F}$ implies $\{f, g\} \in \mathcal{F}$;
- if $f_1, \dots, f_k \in \mathcal{F}$, then $h(f_1, \dots, f_n) \in \mathcal{F}$.

Let $\mathcal{F} \in C^\infty(M)$ be a Poisson algebra. Consider a generic point $x \in M$. This means that in a neighborhood of x there are generators $f_1, \dots, f_k$ such that they are independent and any function $h \in \mathcal{F}$ can be written as $h = h(f_1, \dots, f_k)$ (the number k can be understood as the (differential) dimension of $\mathcal{F}$). Also, we assume that the matrix of brackets $(P_{ij}) = (\{f_i, f_j\})$ is of (locally) constant rank.

Theorem 2 (Classification of groups of functions (S. Lie)). *There exist functions $p_1, \dots, p_r, q_1, \dots, q_r, z_1, \dots, z_s \in \mathcal{F}$, $2r + s = \dim \mathcal{F}$, such that the matrix of brackets is of the form*

$$\begin{pmatrix} 0 & \mathrm{id} & 0 \\ -\mathrm{id} & 0 & 0 \\ 0 & 0 & 0 \end{pmatrix}$$

Moreover, these functions can be completed up to a canonical coordinate system $p_1, \dots, p_n, q_1, \dots, q_n$ on M^{2n} so that $z_i = p_{s+i}$.

If the rank is not locally constant then we get the famous Weinstein splitting theorem:

$$\begin{pmatrix} 0 & \mathrm{id} & 0 \\ -\mathrm{id} & 0 & 0 \\ 0 & 0 & P(z) \end{pmatrix}$$

The next theorem is even more interesting. For each Poisson algebra $\mathcal{F}$ we can define a dual (polar, reciprocal) Poisson algebra $\widetilde{\mathcal{F}} = \{f \in C^\infty(M),\ \{f,h\} = 0 \text{ for all } h \in \mathcal{F}\}$. It is obvious that $\widetilde{\mathcal{F}}$ is indeed a Poisson algebra.

Can we classify such pairs? In local setting there is no problem at all.

Corollary 1. *There is a canonical coordinate system on M such that, if coordinates $P_I, P_{II}, P_{III}, Q_I, Q_{II}, Q_{III}$ have the property that $\mathcal{F}$ is generated by P_I, Q_I, P_{II}, then its dual $\widetilde{\mathcal{F}}$ is generated by P_{II}, P_{III}, Q_{III} and the Poisson matrix in coordinates $Q_I, P_I, P_{II}, P_{III}, Q_{III}, Q_{II}$ (order is slightly changed!) takes the form:*

$$\begin{pmatrix} 0 & -\mathrm{id}_r & 0 & 0 & 0 & 0 \\ \mathrm{id}_r & 0 & 0 & 0 & 0 & 0 \\ 0 & 0 & 0 & 0 & 0 & \mathrm{id}_s \\ 0 & 0 & 0 & 0 & \mathrm{id}_m & 0 \\ 0 & 0 & 0 & -\mathrm{id}_m & 0 & 0 \\ 0 & 0 & -\mathrm{id}_s & 0 & 0 & 0 \end{pmatrix}$$

Different types of Poisson (sub)algebras

We will need several natural definitions. For each Poisson algebra $\mathcal{F}$ and each point $x \in M$ we define $d\mathcal{F}(x) = \mathrm{span}\{df(x),\ f \in \mathcal{F}\} \subset T_x^*M$.

Important: the idea of a generic point

Definition 2. A commutative subalgebra $\mathcal{A} \subset C^\infty(M)$ is called complete if at a generic point $x \in M$, the subspace $d\mathcal{A}(x) \subset T_x^*M$ is maximal isotropic.

Similarly, we say that a commutative subalgebra $\mathcal{A} \subset \mathcal{F}$ is complete in $\mathcal{F}$ if $d\mathcal{A}(x)$ is a maximal isotropic subspace of $d\mathcal{F}(x)$.

- commutative, i.e., $d\mathcal{F}(x)$ is isotropic (equivalently, $\mathcal{F} \subset \widetilde{\mathcal{F}}$);
- coisotropic (complete in the non-commutative sense) $\widetilde{\mathcal{F}} \subset \mathcal{F}$;
- complete commutative $\mathcal{F} = \widetilde{\mathcal{F}}$.

Casimirs of $\mathcal{F}$ = centre of $\mathcal{F}$ (Lie: ausgezeichnet).

Liouville integrability: the algebra of integrals is complete commutative.

Non-commutative integrability: the algebra of integrals is co-isotropic.

Corollary 2. *Let $\mathcal{F}$ be a Poisson subalgebra of $C^\infty(M)$ and $\widetilde{\mathcal{F}}$ be its dual. Then*

- *$\mathcal{F} + \widetilde{\mathcal{F}}$ is a Poisson subalgebra;*
- *$\mathcal{F} + \widetilde{\mathcal{F}}$ is co-isotropic;*
- *the Casimirs of $\mathcal{F}$ coincides with Casimirs of $\tilde{\mathcal{F}}$ and also with Casimirs of $\mathcal{F} + \widetilde{\mathcal{F}}$;*
- *if $\mathcal{A}$ is complete commutative subalgebra of $\mathcal{F}$, then $\mathcal{A} + \widetilde{\mathcal{F}}$ is coisotropic, the Casimirs of $\mathcal{A} + \widetilde{\mathcal{F}}$ is $\mathcal{A}$;*
- *if $\mathcal{B}$ is complete commutative subalgebra of $\widetilde{\mathcal{F}}$, then $\mathcal{B} + \mathcal{F}$ is coisotropic, the Casimirs of $\mathcal{B} + \mathcal{F}$ is $\mathcal{B}$;*

- $\widetilde{\widetilde{\mathcal{F}}} = \mathcal{F}$;
- $\mathcal{A} + \mathcal{B}$ *is complete commutative subalgebra of* $C^\infty(M)$;
- *for any* $f \in \mathcal{F}$, *the dual algebra* $\widetilde{\mathcal{F}}$ *consists of the first integrals of* X_f;
- *for any* $h \in \widetilde{\mathcal{F}}$, *the algebra* $\mathcal{F}$ *consists of the first integrals of* X_h.

Theorem 3. *Let* $\mathcal{F}$ *be a co-isotropic algebra of integrals of a certain Hamiltonian system. Consider a common level of the integrals passing through a point* $x_0 \in M$, *i.e.,* $X = \{f_i = f_i(x_0)\}$. *Assume that this level is generic in the sense that we can choose generators* $f_1, \dots, f_m$ *of* $\mathcal{F}$, $m = \dim\mathcal{F}$, *in such a way that they are independent and the Poisson matrix is of constant rank nearby* X *(notice that this condition is local but not semi-local).*

Then X *(and all neighbouring fibers) are tori with quasi periodic dynamics and there exists a natural action-angle coordinate system* $p_1, \dots, p_r, q_1, \dots, q_r, I_1, \dots, I_s, \phi_1, \dots, \phi_s$ *(here* $p_1, \dots, p_r$, $q_1, \dots, q_r$, $I_1, \dots, I_s \in \mathcal{F}$ *and are local generators of it).*

This means that such a system can be understood as the direct product of an integrable system with s degrees of freedom and $2r$*-dimensional symplectic manifold which does not really play any essential role.*

Example. Assume that G acts on a symplectic manifold in a Hamiltonian way, which means that for every $\xi \in \mathfrak{g}$ the corresponding generator $\hat{\xi}$ is a Hamiltonian vector field with a Hamiltonian H_ξ and we have $\{H_\xi, H_\eta\} = H_{[\xi,\eta]}$ (symplectic action with a good momentum mapping).

This action defines two natural subalgebras $\mathcal{F}$, the algebra of Noether integrals generated by H_ξ, $\xi \in \mathfrak{g}$ and the algebra of G-invariant functions which is nothing else than $\widetilde{\mathcal{F}}$.

We have a natural map $\mathfrak{g} \to C^\infty(M)$ and more generally $C^\infty(g^*) \to C^\infty(M)$. The dual map $M \to \mathfrak{g}^*$ is known as the momentum mapping associated with the action of G on M. In fact for our purposes the map between the Poisson algebras is more important than the momentum mapping between the manifolds.

If G-invariant functions distinguish generic orbits[3], then (at a generic point) we are in the situation described by the above Lie theorem.

One more remark: the algebra $\mathcal{F}$ is "isomorphic" to the algebra of polynomials on $\mathfrak{g}^*$ with the standard Lie–Poisson bracket. We can simply identify H_ξ as a function on M with ξ as a function on $\mathfrak{g}^*$.

Immediately, we get the following result.

Theorem 4. *Let* G *act on* M *as above and* G*-invariant functions distinguish generic orbits. Assume that* H *is a Casimir of* $\mathfrak{g}$. *Then the Hamiltonian system with the Hamiltonian* H *is completely integrable in the non-commutative sense. The algebra of integral is* $\mathcal{F} + \widetilde{\mathcal{F}}$. *More generally, let* $H \in C^\infty(g^*)$ *be a function that defines a completely integrable Hamiltonian system on* $\mathfrak{g}^*$ *(more precisely on those orbits*

[3]This always happens if G is compact.

which belong to the image of the momentum mapping). Then H (considered as a (lifted) function on T^*M) defines a completely integrable system on T^*M.

Corollary 3. *Let $M = G/H$ be a homogeneous space of a compact Lie group H and g be the normal metric on M. Then the geodesic flow on (M, g) is always completely integrable in non-commutative sense. The same is true for any bi-quotient $K\backslash G/H$.*

References

[1] Bolsinov, Alexey V.; Jovanović, Božidar, Integrable geodesic flows on Riemannian manifolds: construction and obstructions. Contemporary geometry and related topics (eds. Bokan N., Djoric M., Rakic Z., Fomenko A.T., Wess J.), 57–103, World Sci. Publ., River Edge, NJ, 2004.

[2] Kruglikov, Boris; Matveev, Vladimir S. *The geodesic flow of a generic metric does not admit nontrivial integrals polynomial in momenta.* Nonlinearity **29** (2016), no. 6, 1755–1768.

[3] Bolsinov, A.V.; Fomenko, A.T. Integrable geodesic flows on two-dimensional surfaces. Monographs in Contemporary Mathematics. Consultants Bureau, New York, 2000.

Alexey Bolsinov
Department of Mathematical Sciences
Loughborough University
Loughborough
Leicestershire LE11 3TU, UK
e-mail: A.Bolsinov@lboro.ac.uk

Geometric Methods in Physics. XXXV Workshop 2016
Trends in Mathematics, 237–247

Anyonic Fock Spaces, q-CCR Relations for $|q| = 1$ and Relations with Yang–Baxter Operators

Marek Bozejko

Mathematics Subject Classification (2010). 47L10, 47L55, 47L90, 81R10, 16T25.

Keywords. Fock space, anyonic space, Yang–Baxter operator, Hecke operator, deformation.

1. Fock spaces of Yang–Baxter type

1.1. Yang–Baxter symmetrizator

Let $T : \mathcal{H} \otimes \mathcal{H} \to \mathcal{H} \otimes \mathcal{H}$ be a Yang–Baxter operator, i.e.,

$$T_1 T_2 T_1 = T_2 T_1 T_2, \quad T - T^*, \quad T \geq -\mathrm{I}$$

on $\mathcal{H} \otimes \mathcal{H} \otimes \mathcal{H}$, where $T_1 = T \otimes \mathrm{I}$, $T_2 = \mathrm{I} \otimes \mathrm{T}$. We define the T-symmetrizator operator

$$P_T^{(n)}(T_1, T_2, \ldots, T_{n-1}) = P_T^{(n)} : \mathcal{H}^{\otimes n} \to \mathcal{H}^{\otimes n}$$

as follows:

$$P_T^{(n)} = (1 + T_1 + T_2 T_1 + T_3 T_2 T_1 + \cdots + T_{n-1} \ldots T_1) P_T^{(n-1)}(T_2, T_3, \ldots, T_{n-1}),$$

1.2. Positivity of T-symmetrizators

Here $P_T^{(1)} = 1$, $P_T^{(2)} = 1 + T_1$ and

$$T_i = \underbrace{1 \otimes \cdots \otimes 1}_{i-1 \text{ times}} \otimes T \otimes \underbrace{1 \otimes \cdots \otimes 1}_{n-i-1 \text{ times}} : \mathcal{H}^{\otimes n} \to \mathcal{H}^{\otimes n}.$$

Under the assumption that $\|T\| \leq 1$ we proved (see [1, 2]), that $P_T^{(n)} \geq 0$ for each n and hence we can form a new pre-scalar product on $\mathcal{H}^{\otimes n}$ as follows: for $\xi, \eta \in \mathcal{H}^{\otimes n}$

$$\langle \xi | \eta \rangle_T := \langle P_T^{(n)} \xi | \eta \rangle,$$

This paper was partially supported by NCN grant 2016/21/B/ST1/00628.

where $\langle\cdot|\cdot\rangle$ is the natural scalar product on $\mathcal{H}^{\otimes n}$. Then we can form the creation operator

$$a^+(f)\xi = f\otimes\xi$$

and the annihilation operator

$$a(f)\xi = l(f)(1+T_1+T_1T_2+\cdots+T_1T_2\dots T_{n-1})\xi \qquad \text{for } \xi\in\mathcal{H}^{\otimes n},$$

where $l(f)$ is the free annihilation operator defined as

$$l(f)(x_1\otimes\cdots\otimes x_n) = \langle f|x_1\rangle\, x_2\otimes\cdots\otimes x_n.$$

The main object of this abstract is the structure of the von Neumann algebra

$$\Gamma_T(\mathcal{H}) = \{G_T(f) : f\in\mathcal{H}_{\mathbb{R}}\}''$$

generated by the T-Gaussian field $G_T(f) = a^+(f)+a(f)$, where $\mathcal{H}_{\mathbb{R}}$ denotes the real part of $\mathcal{H}$.

As was shown earlier – (Voiculescu, Bożejko-Speicher, Ricard, Nou) this von Neumann algebra is a non-injective II_1-factor.

The linear span of T-Gaussian random variables is completely isomorphic to the operator space called row and column, as we will show later. This is an extension of the results of Haagerup and Pisier, A. Buchholz and of our results with R. Speicher.

2. Hecke operators and positivity of T-symmetrizators

Now we answer the question posed by L. Accardi, namely, when are the T-symmetrizator operators $P_T^{(n)}$ similar to self-adjoint projections, i.e.,

$$\left(P_T^{(n)}\right)^2 = \alpha(n)\; P_T^{(n)} \qquad \text{for some } \alpha(n)>0. \tag{1}$$

First, let us see that if $P_T^{(2)} = 1+T$ satisfies (1) then

$$(1+T)^2 = \alpha(1+T) \qquad \text{for } \alpha=\alpha(2) \tag{2}$$

which implies that

$$T^2 = (q-1)T + q\,1, \tag{3}$$

where $q=\alpha-1$. Such an operator satisfying (3) is called *Hecke operator* with parameter q.

2.1. Examples of Hecke operators

(H_1) The flip $T=\sigma:\mathcal{H}\otimes\mathcal{H}\to\mathcal{H}\otimes\mathcal{H}$ given by an exchange of the factors $\sigma(x\otimes y) = y\otimes x$ is a Hecke operator with $q=1$ and the corresponding "projection"

$$P_T^{(n)} = \sum_{\pi\in S_n}\pi$$

is the classical symmetrizator operator on $\mathcal{H}^{\otimes n}$.

(H_2) For $T = -\sigma$ we obtain the anti-symmetrizator

$$P_T^{(n)} = \sum_{\pi \in S_n} \operatorname{sgn}(\pi)\, \pi,$$

where $\operatorname{sgn}(\pi)$ is the classical sign of a permutation $\pi \in S_n$.

(H_3) If we take $\epsilon = \pm 1$ and we define the operator

$$T = T_\epsilon = \frac{q-1}{2} + \epsilon \frac{q+1}{2} \sigma$$

then we get the *Hecke operator* with parameter q, i.e.,

$$T^2 = (q-1)T + q\, 1.$$

This operator is a Yang–Baxter operator if and only if $q = 1$, which means that T_ϵ is the symmetrizator ($\epsilon = 1$) or the anti-symmetrizator ($\epsilon = -1$).

(H_4) We get a very interesting example of a Yang–Baxter–Hecke operator for a Hilbert space $\mathcal{H}$ of finite dimension $\dim \mathcal{H} = m$ with an orthonormal basis $(e_1, e_2, \ldots, e_m)$. We consider the operator $\tilde{P} : \mathcal{H} \otimes \mathcal{H} \to \mathcal{H} \otimes \mathcal{H}$ given by

$$\tilde{P}(e_i \otimes e_j) = -\frac{1}{m} \delta_{ij} \sum_{k=1}^{m} e_k \otimes e_k.$$

One can see that $P = (-\tilde{P})$ is the projector operator of the following form:

$$P(x \otimes y) = \frac{1}{m} \langle x|y \rangle\, \theta, \qquad \text{where } \theta = \sum_{k=1}^{m} e_k \otimes e_k, \quad x, y \in \mathcal{H}$$

(see [15, page 449]).

(H_5) The main example of that theory is Pusz–Woronowicz twisted CCR (CAR) operators: $T_\mu^{\mathrm{CCR}}, T_\mu^{\mathrm{CAR}}$ defined as:

$$T_\mu^{\mathrm{CCR}}(e_i \otimes e_j) = \begin{cases} \mu(e_j \otimes e_i) & \text{if } i < j, \\ \mu^2(e_i \otimes e_i) & \text{if } i = j, \\ -(1-\mu^2)(e_i \otimes e_j) + \mu(e_j \otimes e_i) & \text{if } i > j, \end{cases}$$

$$T_\mu^{\mathrm{CAR}}(e_i \otimes e_j) = \begin{cases} -\mu(e_j \otimes e_i) & \text{if } i < j, \\ -(e_i \otimes e_i) & \text{if } i = j, \\ -(1-\mu^2)(e_i \otimes e_j) - \mu(e_j \otimes e_i) & \text{if } i > j. \end{cases}$$

Both the twisted CCR and twisted CAR are Yang–Baxter–Hecke operators with the parameter $q = \mu^2$, which means that $T^2 = (\mu^2 - 1)T + \mu^2\, 1$.

(H_6) As a special case we consider $T_0^{\mathrm{CAR}} = T^M$, where T^M is of the following form:

$$T^M(e_i \otimes e_j) = \begin{cases} 0 & \text{if } i < j, \\ -(e_i \otimes e_j) & \text{if } i \geq j. \end{cases}$$

It is connected with Muraki–Lu monotone Fock space, as we will see later.

(H_7) Also it will be interesting to see the corresponding T-Fock space in the case when the twisted CCR operator has parameter $\mu = 0$ and then we get the following operator:

$$T_0^{\mathrm{CCR}}(e_i \otimes e_j) = \begin{cases} 0 & \text{if } i \leq j, \\ -(e_i \otimes e_j) & \text{if } i > j. \end{cases}$$

Later we will use this operator to construct the Bose monotone Fock space.

(H_8) In the paper [11] we introduced another type (called anyonic) of the Yang–Baxter–Hecke operator T_z on $L^2(\mathbb{R}, \sigma)$, where σ is a non-atomic Radon measure on $\mathbb{R}$ defined for $f \in L^2(\mathbb{R}^2, \sigma \otimes \sigma)$ as follows:

$$T_z f(x,y) = Q(x,y) f(y,x),$$

where $|z| = 1$ and

$$Q(x,y) = \begin{cases} z & \text{if } x < y, \\ \bar{z} & \text{if } x > y. \end{cases}$$

Then T_z is a Yang–Baxter–Hecke operator with parameter $q = 1$.

2.2. Positivity of $P_T^{(n)}$ for Yang–Baxter–Hecke operators

Proposition 1. *Let $T = T^*$ be a Yang–Baxter–Hecke operator. Then for each $n \geq 1$*

$$\left(P_T^{(n)}\right)^2 = \underline{n}!\, P_T^{(n)} \geq 0, \qquad (*)$$

where $\underline{n} = 1 + q + \cdots + q^{n-1}$ and $\underline{n}! = \underline{1} \cdot \underline{2} \cdot \ldots \cdot \underline{n}$.

Moreover, for $q \geq -1$

$$\left\| P_T^{(n)} \right\| = \underline{n}! = \frac{\prod_{k=1}^{n} (1 - q^k)}{(1-q)^n}.$$

Remark. Proposition 1 solves the problem of L. Accardi: $P_T^{(n)}$ is similar to a projection if and only if T is a Hecke operator.

2.3. T-symmetric Fock Hilbert space

$$\mathcal{F}_T(\mathcal{H}) = \bigoplus_{n=0}^{\infty} \mathcal{H}^{\circledast n} \underline{n}! = \mathbb{C}\Omega \oplus \mathcal{H} \oplus \mathcal{H}^{\circledast 2} \oplus \cdots,$$

where

$$\mathcal{H}^{\circledast n} = P_T^{(n)}(\mathcal{H}^{\otimes n})$$

is the space of T-symmetric tensors. By Proposition 1 we have that for $f \in \mathcal{H}^{\circledast n}$, $P_T^{(n)}(f) = \underline{n}! f$. Therefore the Hilbert norm $\|f\|_T^2$ for $f = (f_0, f_1, f_2, \dots) \in \mathcal{F}_T(\mathcal{H})$ is defined as:

$$\|f\|_T^2 = \langle P_T(f) | f \rangle = \sum_{n=0}^{\infty} \langle P_T^{(n)}(f_n) | f_n \rangle = \sum_{n=0}^{\infty} \underline{n}!\, \|f_n\|^2 \leq \infty.$$

One can see that we have the following description of T-symmetric tensors:

Lemma 1. *For $n > 1$ we have*

$$\begin{aligned}\mathcal{H}^{\circledast n} &= \{f \in \mathcal{H}^{\otimes n} : T_j(f) = qf \text{ for } j \in \{1,2,\dots,n-1\}\} \\ &= \left\{f \in \mathcal{H}^{\otimes n} : \tilde{P}_T^{(n)}(f) = f\right\},\end{aligned}$$

where $\tilde{P}_T^{(n)} = \frac{1}{n!}P_T^{(n)}$.

Let us observe that the T-creation and T-annihilation operators on the T-Fock space can be defined as follows: for $f \in \mathcal{H}$

$$\begin{aligned}a_T^+(f) &= \tilde{P}\, l^+(f)\tilde{P} = \tilde{P}\, l^+(f), \\ a_T(f) &= \tilde{P}\, l(f)\tilde{P} \ = l(f)\, \tilde{P},\end{aligned}$$

where $\tilde{P} = \tilde{P}_T = \sum_{n=0}^{\infty} \frac{1}{n!}P_T^{(n)}$ is the orthogonal projection onto T-symmetric tensors and $l^+(f), l(f)$ are the free creation and free annihilation operators.

2.4. Boolean Fock spaces

The simplest among deformed T-symmetric Fock spaces is the Boolean Fock space $\mathcal{F}_{-1}(\mathcal{H}) = \mathbb{C}\Omega \oplus \mathcal{H}$ and the Yang–Baxter–Hecke operator $T = -I$, $P_T^{(n)} = 0$ for $n > 1$.

The Boolean creation and annihilation operators are following:

$$b^+(f)\xi = \begin{cases} 0 & \text{if } \xi \in \mathcal{H}, \\ f & \text{if } \xi = \Omega, \end{cases}$$

$$b(f)\xi = \begin{cases} \langle f|\xi\rangle & \text{if } \xi \in \mathcal{H}, \\ 0 & \text{if } \xi \in \mathbb{C}\Omega. \end{cases}$$

They satisfy the following relations: if $(e_1, e_2, \dots, e_N)$ is an orthonormal basis of $\mathcal{H}$ and $b_i^{\pm} := b^{\pm}(e_i)$ then

$$b_i b_j^+ = \delta_{i,j}\left(1 - \sum_{k=0}^{N} b_k^+ b_k\right) = \delta_{i,j} P_\Omega,$$

where P_Ω is the projection on the vacuum vector Ω.

For the Boolean Gaussian random variables $G^B(f) = b(f) + b^+(f)$, the following proposition is known to be true:

Proposition 2 ([9]). *For arbitrary operators $\alpha_i \in B(\mathcal{H})$ and $f_i \in \mathcal{H}_{\mathbb{R}}$, $\|f_i\| = 1$, we have*

$$\left\|\sum_{i=1}^{N} \alpha_i \otimes G^B(f_i)\right\| = \max\left\{\left\|\left(\sum_{i=1}^{N} \alpha_i\alpha_i^*\right)^{1/2}\right\|, \left\|\left(\sum_{i=1}^{N} \alpha_i^*\alpha_i\right)^{1/2}\right\|\right\}.$$

That means that Boolean Gaussian random variables span the operator space completely isometrically isomorphic to row and column operator space. Similar results were obtained for the free and q-Gaussian random variables and free generators (see Haagerup–Pisier, Bożejko–Speicher).

2.5. Monotone Fock spaces

Now we recall the definition of the monotone Fock space following the fundamental paper of Muraki [16], and we show that it is equal to special case of the T-symmetric Fock space for the Pusz–Woronowicz operator considered in the example (H_6) $T_0^{\mathrm{CAR}} = T^M$.

Let $\mathbb{N}$ be the set of all natural numbers. For $r \geq 1$ we define $I_r = \{(i_1, i_2, \ldots, i_r) : i_1 < i_2 < \cdots < i_r,\ i_j \in \mathbb{N}\}$ and for $r = 0$ we set $I_0 = \{\emptyset\}$, where $\emptyset$ denotes the null sequence.

We define $\mathrm{Inc}(\mathbb{N}) = \bigcup_r I_r$. Let $\mathcal{H}_r = l^2(I_r)$ be the r-particle Hilbert space and $\Phi = \bigoplus_{r=0}^{\infty} \mathcal{H}_r$ the monotone Fock space.

For an increasing sequence $\sigma = (i_1, i_2, \ldots, i_r) \in \mathrm{Inc}(\mathbb{N})$, denote by $[\sigma] = \{i_1, i_2, \ldots, i_r\}$ the associated set and by $\{e_\sigma\}$ the canonical basis vector in the monotone Fock space Φ.

We will write $[\sigma] < [\tau]$ if for each $i \in [\sigma]$ and $j \in [\tau]$ we have $i < j$. The monotone creation operator δ_i^+ and the annihilation operator δ_i^- are defined for each $i \in \mathbb{N}$ by:

$$\delta_i^+ e_{(i_1,\ldots,i_r)} = \begin{cases} e_{(i,i_1,\ldots,i_r)} & \text{if } \{i\} < \{i_1, \ldots, i_r\}, \\ 0 & \text{otherwise,} \end{cases}$$

$$\delta_i^- e_{(i_1,\ldots,i_r)} = \begin{cases} e_{(i_2,\ldots,i_r)} & \text{if } r \geq 1, i = i_1, \\ 0 & \text{otherwise.} \end{cases}$$

Let us observe that if $P = \bigoplus P^{(n)}$ is the orthogonal projection from the full Fock space onto the monotone Fock space, then $\delta_i^{\pm} = P l_i^{\pm} P$, where $l_i^{\pm}$ are the free creation and the free annihilation operators. Moreover, the following relations hold:

$$\begin{aligned} \delta_i^+ \delta_j^+ = \delta_j^- \delta_i^- &= 0 && \text{for } i \geq j, \\ \delta_i^- \delta_j^+ &= 0 && \text{for } i \neq j, \\ \delta_i^- \delta_i^+ &= 1 - \sum_{j \leq i} \delta_j^+ \delta_j^- && \text{for } i = j. \end{aligned}$$

Proposition 3. *If $T^M = T_0^{\mathrm{CAR}}$ is the Pusz–Woronowicz Yang–Baxter–Hecke operator defined as*

$$T^M(e_i \otimes e_j) = \begin{cases} 0 & \text{if } i < j, \\ -(e_i \otimes e_j) & \text{if } i \geq j \end{cases}$$

then the T-symmetric Fock space is exactly Muraki monotone Fock space and the corresponding creation and annihilation operators are the following:

$$a_i^+ = \delta_i^+, \qquad a_i = \delta_i^-.$$

Proposition 4. *Let $\alpha_i \in B(\mathcal{H})$ and $G_i = \delta_i^- + \delta_i^+$ be the monotone Gaussian operators. Then*

$$\left\|\sum_{i=1}^{N} \alpha_i \otimes \delta_i^-\right\| = \left\|\sum_{i=1}^{N} \alpha_i \alpha_i^*\right\|^{1/2}, \tag{4}$$

$$\left\|\sum_{i=1}^{N} \alpha_i \otimes \delta_i^+\right\| = \left\|\sum_{i=1}^{N} \alpha_i^* \alpha_i\right\|^{1/2}, \tag{5}$$

$$\|\delta_i^-\| = \|\delta_i^+\| = 1, \tag{6}$$

$$1 \leq \|G_i\| \leq 2, \tag{7}$$

$$\begin{aligned} &\max\left\{\left\|\left(\sum_{i=1}^{N} \alpha_i \alpha_i^*\right)^{1/2}\right\|, \left\|\left(\sum_{i=1}^{N} \alpha_i^* \alpha_i\right)^{1/2}\right\|\right\} \\ &\leq \left\|\sum_{i=1}^{N} \alpha_i \otimes G_i\right\| \leq 2\max\left\{\left\|\left(\sum_{i=1}^{N} \alpha_i \alpha_i^*\right)^{1/2}\right\|, \left\|\left(\sum_{i=1}^{N} \alpha_i^* \alpha_i\right)^{1/2}\right\|\right\}. \end{aligned} \tag{8}$$

3. Connections of Woronowicz–Pusz operators T_μ^{CAR}

Bose monotone Fock spaces

If we consider bosonic type of the operator Pusz–Woronowicz defined as

$$T^B(c_i \otimes c_j) = T_0^{\mathrm{CCR}}(c_i \otimes c_j) = \begin{cases} 0 & \text{if } i \leq j, \\ -(e_i \otimes e_j) & \text{if } i > j \end{cases}$$

then one can see that in that case the nth particle space of the corresponding T-symmetric Fock space $\mathcal{F}_{T^B}(\mathcal{H})$ is of the following form:

$$P_T^{(n)}(\mathcal{H}^{\otimes n}) = \mathcal{H}_T^{\otimes n} = \mathrm{Lin}\{e_{i_1} \otimes e_{i_2} \otimes \cdots \otimes e_{i_n} : i_1 \leq i_2 \leq \cdots \leq i_n\}.$$

The action of the creation and annihilation operators is following:

$$\Delta_j^+(e_{i_1} \otimes e_{i_2} \otimes \cdots \otimes e_{i_n}) = \begin{cases} e_j \otimes e_{i_j} \otimes e_{i_1} \otimes \cdots \otimes e_{i_n} & \text{if } j \leq i_1, \\ 0 & \text{otherwise.} \end{cases}$$

$$\Delta_j(e_{i_1} \otimes e_{i_2} \otimes \cdots \otimes e_{i_n}) = \begin{cases} e_{i_2} \otimes \cdots \otimes e_{i_n} & \text{if } j = i_1, \\ 0 & \text{otherwise,} \end{cases}$$

and they satisfy the following commutation relations:

$$\begin{aligned} \Delta_i \Delta_j^+ &= 0 && \text{for } i \neq j, \\ \Delta_i \Delta_i^+ &= 1 - \sum_{k<i} \Delta_k^+ \Delta_k && \text{for } i = j. \end{aligned}$$

From the last formulas we get $\Delta_1\Delta_1^+ = 1$ and $\|\Delta_1\| = 1$. Moreover, $\|\Delta_j^\pm\| \le 1$ and since $\|\Delta_j^+\Omega\| = 1$ we have $\|\Delta_j^\pm\| = 1$.

By the similar considerations like in the Fermi-monotone Muraki–Fock space we have the following:

Proposition 5. *Let $\alpha_i \in B(\mathcal{H})$ and $g_i = \Delta_i^- + \Delta_i^+$ be the monotone Bose Gaussian operators, then*

$$\max\left\{\left\|\left(\sum_{i=1}^N \alpha_i\alpha_i^*\right)^{1/2}\right\|, \left\|\left(\sum_{i=1}^N \alpha_i^*\alpha_i\right)^{1/2}\right\|\right\}$$
$$\le \left\|\sum_{i=1}^N \alpha_i \otimes g_i\right\| \le 2\max\left\{\left\|\left(\sum_{i=1}^N \alpha_i\alpha_i^*\right)^{1/2}\right\|, \left\|\left(\sum_{i=1}^N \alpha_i^*\alpha_i\right)^{1/2}\right\|\right\}.$$

If we take the vacuum state $\varepsilon(T) = \langle T\Omega, \Omega\rangle$, then one can show the following central limit theorem for the Bose-monotone Gaussian random variables $g_i = \Delta_i + \Delta_i^+$.

Proposition 6 (Central Limit Theorem, 2000). *If $S_N = \frac{1}{\sqrt{N}}\sum_{i=1}^N g_i$, then*

$$\lim_{N\to\infty} \varepsilon(S_N^{2n}) = \binom{2n}{n},$$

i.e., S_N weakly tends to arcsine law $\frac{1}{\pi}\frac{1}{\sqrt{1-x^2}}$.

In the case of the Fermi monotone case this same law was obtained by N. Muraki [16]. See also the paper of J. Wysoczanski (2011) for related generalization of the central limit theorems for the Boolean-monotonic case.

When we consider the Pusz–Woronowicz Hecke operator T_μ^{CCR} for $\mu = -1$, defined as

$$T^B(e_i \otimes e_j) = T_{-1}^{\mathrm{CCR}}(e_i \otimes e_j) = \begin{cases} (e_i \otimes e_j) & \text{if } i = j, \\ -(e_i \otimes e_j) & \text{if } i \ne j. \end{cases}$$

we get the model of mixed Bose–Fermi commutation relations:

$$b_i b_j^+ + b_j^+ b_i = 0 \text{ if } i \ne j,$$
$$b_i b_j^+ - b_j^+ b_i = 1 \text{ if } i = j.$$

These models correspond to so-called q_{ij}-CCR commutations relations of the form

$$A_i A_j^+ - q_{ij} A_j^+ A_i = \delta_i j 1,$$

where $q_{ij} = \bar{q}_{ji}$ and $|q_{ij}| \le 1$.

Such models were considered in many papers: Bożejko, Speicher, Jorgensen, Smith, Werner, Nou, Krolak, Yoshida, Hiai, Lust-Piquard, Śniady.

In our last case we have case of "anicommuting bosons" i.e.: $q_{ii} = 1$ and $q_{ij} = -1$ for $i \neq j$.

Similarly, if we consider the Pusz–Woronowicz–Hecke operator T_μ^{CAR}, for $\mu = -1$, we obtain again q_{ij}-CCR commutation relations of the type "commuting fermions", when $q_{ii} = -1$ and $q_{ij} = 1$ for $i \neq j$.

4. Non-commutative Levy process for generalized "ANYON"

A first rigorous interpolation between canonical commutation relations (CCR) and canonical anticommutation relations (CAR) was constructed in 1991 by Bożejko and Speicher. Given a Hilbert space $\mathcal{H}$, we constructed, for each $q \in (-1, 1)$, a deformation of the full Fock space over $\mathcal{H}$, denoted by $\mathcal{F}^q(\mathcal{H})$. For each $h \in \mathcal{H}$, one naturally defines a (bounded) creation operator, $a^+(h)$, in $\mathcal{F}^q(\mathcal{H})$. The corresponding annihilation operator, $a^-(h)$, is the adjoint of $a^+(h)$. These operators satisfy the q-commutation relations:

$$a^-(g)a^+(h) - qa^+(h)a^-(g) = (g, h)_{\mathcal{H}}, \quad g, h \in \mathcal{H}.$$

This is special case of Yang–Baxter deformation given by the $T_q(x \otimes y) = q(y \otimes x)$.

The limiting cases, $q = 1$ and $q = -1$, correspond to the Bose and Fermi statistics, respectively.

Another generalization of the CCR and CAR was proposed in 1995 by Ligouri and Mintchev. They fixed a *continuous* underlying space $X = \mathbb{R}$ and considered a function $Q : X^2 \to \mathbb{C}$ satisfying $Q(s, t) = \overline{Q(t, s)}$ and $|Q(s, t)| = 1$.

Setting $\mathcal{H}$ to be the complex space $L^2(R)$, one defines a bounded linear operator T acting on $\mathcal{H} \otimes \mathcal{H}$ by the formula

$$T(f \otimes g)(s, t) = Q(s, t)g(s)f(t), \quad f, g \in \mathcal{H}. \tag{9}$$

This operator is self-adjoint, its norm is equal to 1, and it satisfies the Yang–Baxter and Hecke relation.

One then defines corresponding creation and annihilation operators, $a^+(h)$ and $a^-(h)$, for $h \in \mathcal{H}$. By setting $a^+(h) = \int_T dt\, h(t)\partial_t^\dagger$ and $a^-(h) = \int_T dt\, \overline{h(t)}\partial_t$, one gets (at least informally) creation and annihilation operators, $\partial_t^\dagger$ and ∂_t, at point $t \in T$. These operators satisfy the Q-commutation relations

$$\partial_s\partial_t^\dagger - Q(s, t)\partial_t^\dagger\partial_s = \delta(s, t),$$
$$\partial_s\partial_t - Q(t, s)\partial_t\partial_s = 0, \quad \partial_s^\dagger\partial_t^\dagger - Q(t, s)\partial_t^\dagger\partial_s^\dagger = 0. \tag{10}$$

From the point of view of physics, the most important case of a generalized statistics (10) is the anyon statistics. For the anyon statistics, the function Q is given by

$$Q(s, t) = \begin{cases} q, & \text{if } s < t, \\ \bar{q}, & \text{if } s > t \end{cases}$$

for a fixed $q \in \mathbb{C}$ with $|q| = 1$. Hence, the commutation relations (10) become

$$\partial_s \partial_t^\dagger - q\partial_t^\dagger \partial_s = \delta(s,t),$$
$$\partial_s \partial_t - \bar{q}\partial_t \partial_s = 0, \quad \partial_s^\dagger \partial_t^\dagger - \bar{q}\partial_t^\dagger \partial_s^\dagger = 0, \tag{11}$$

for $s < t$.

The free Levy processes, i.e., case $Q(s,t) = q = 0$, was done in our paper with E. Lytvynov.

Having creation, neutral, and annihilation operators at our disposal, we define and study, a noncommutative stochastic process (white noise)

$$\omega(t) = \partial_t^\dagger + \partial_t + \lambda \partial_t^\dagger \partial_t, \quad t \in T.$$

Here $\lambda \in \mathbb{R}$ is a fixed parameter. The case $\lambda = 0$ corresponds to a Q-analog of Brownian motion, while the case $\lambda \neq 0$ (in particular, $\lambda = 1$) corresponds to a (centered) Q-Poisson process.

We identify corresponding Q-Hermite (Q-Charlier respectively) polynomials, denoted by $\omega(t_1)\cdots\omega(t_n)$, of infinitely many noncommutative variables $(\omega(t))_{t\in T}$.

Then we introduce the notion of independence for a generalized statistics, and to derive corresponding Lévy processes. We know from experience both in free probability and in q-deformed probability that a natural way to explain that certain noncommutative random variables are independent (relative to a given statistics/deformation of commutation relations) is to do this through corresponding deformed cumulants-like q-deformed cumulants $(-1 < q < 1)$.

Noncommutative Lévy processes have most actively been studied in the framework of free probability. Using q-deformed cumulants, Anshelevich constructed and studied noncommutative Lévy processes for q-commutation relations.

Acknowledgement

The author would like to thank professor Emma Previato for many discussions at the 2016 Bialowieża Workshop and for great help in the preparation of this paper.

References

[1] M. Bozejko and R. Speicher, *An example of a generalized Brownian motion*, Commun. Math. Phys. **137** (1991), 519–531.

[2] M. Bozejko and R. Speicher, *Completely positive maps on Coxeter groups, deformed commutation relations, and operator spaces*, Math. Ann. **300** (1994), 97–120.

[3] M. Bozejko and R. Speicher, *Interpolations between bosonic and fermionic relations given by generalized Brownian motions*, Math. Z. **22** (1996), 135–159.

[4] M. Bozejko, B. Kümmerer and R. Speicher, *d-Gaussian processes: non-commutative and classical aspects*, Comm. Math. Phys., **185**(1), (1997), 129–154.

[5] M. Bozejko and J. Wysoczanski, *Remarks on t-transformations of measures and convolutions*, Ann. Inst. Henri Poincaré-PR **37**(6) (2001), 737–761.

[6] M. Bozejko and M. Guta, *Functors of white noise associated to characters of the infinite symmetric group*, Comm. Math. Phys. **229** (2002), 209–227.

[7] M. Bozejko and R. Szwarc, *Asymptotic Combinatorics with Applications to Mathematical Physics*, Lecture Notes in Mathematics **1815**, (2003), 201–221.

[8] M. Bozejko and H. Yoshida, *Generalized d-deformed Gaussian random variables*, Banach Center Publ. **73** (2006), 127–140.

[9] M. Bozejko, A.D. Krystek, L.J. Wojakowski, *Remarks on the r and Δ convolution*, Math. Z. **253** (2006), pp. 177–196,

[10] M. Bożejko, E. Lytvynov: *Meixner class of noncommutative generalized stochastic processes with freely independent values. I. A characterization.* Comm. Math. Phys. 292, 99–129 (2009).

[11] M. Bożejko, E. Lytvynov, J. Wysoczanski: *Noncommutative Lévy Processes for Generalized (Particularly Anyon) Statistics*, Comm. Math. Phys. **313**, Issue 2 (2012), 535–569.

[12] M. Bożejko, Deformed Fock spaces, *Hecke operators and monotone Fock space of Muraki*, Demonstratio Math. **45** (2012) 399–413.

[13] M. Bozejko, W. Ejsmont and T. Hasebe, *Fock space associated to Coxeter groups of type B*, Journal of Functional Analysis J. Funct. Anal. **269**(6) (2015), 1769–1795.

[14] M. Bozejko, E. Lytvynov, I. Rodionova, *An extended Fock space and noncommutative Meixner-type orthogonal polynomials*, Uspiekhi Mat. Nauk, **70**(5), 2015, 47 pp., there is also a Russian version.

[15] Roe Goodman and Nolan R. Wallach, *Symmetry, Representations, and Invariants*, Springer, Dordrecht-Heidelberg-London-New York, 2009, ISBN 978-0-387-79851-6, e-ISBN 978-0-387-79852-3.

[16] Muraki, Naofumi. *Noncommutative Brownian motion in monotone Fock space.* Comm. Math. Phys. **183** (1997), no. 3, 557–570.
`http://projecteuclid.org/euclid.cmp/1158328657`.

Marek Bozejko
Institute of Mathematics (Wroclaw Branch)
Polish Academy of Sciences
Poland
e-mail: `bozejko@gmail.com`

Geometric Methods in Physics. XXXV Workshop 2016
Trends in Mathematics, 249–253

Differential and Integral Forms on Non-commutative Algebras

Tomasz Brzeziński

Mathematics Subject Classification (2010). Primary 58B30.
Keywords. Noncommutative geometry; differential forms; integral forms.

1. Introduction

These lectures describe an algebraic approach to differentiation and integration that is characteristic for non-commutative geometry. The material contained in Section 2 is standard and can be found in any text on non-commutative geometry, for example [4]. Items 5 and 6, which describe concepts introduced in [3], are exceptions. The bulk of Section 3 is based on [1] and [3], while the Berezin integral example is taken from [2].

2. Differential forms

1. Differential graded algebras. A *differential graded algebra* is a pair (Ω, d), where $\Omega = \oplus_{n\in\mathbb{Z}}\Omega^n$ is a graded algebra and $d : \Omega \to \Omega$ is a degree-one map that squares to zero and satisfies the graded Leibniz rule. Note that Ω^0 is an associative algebra and all the Ω^n are Ω^0-bimodules.

2. Differential calculus. Given an associative algebra A (over a field $\mathbb{K}$ of characteristic not 2), by a *differential calculus over* A we mean a differential graded algebra $(\Omega A, d)$, such that $\Omega^0 A = A$, ΩA is generated by $\Omega^1 A = Ad(A)$, and $\Omega^n A = 0$, for all $n < 0$. A calculus is said to be N*-dimensional*, if $\Omega^N A \neq 0$ and $\Omega^n A = 0$, for all $n > N$. The pair $(\Omega^1 A, d : A \to \Omega^1 A)$ is called a *first-order differential calculus.*

3. Universal differential calculus. Every algebra admits the *universal differential calculus*, defined as the tensor product algebra over the kernel of the multiplication map μ on A, with the exterior derivation $d : a \mapsto 1 \otimes a - a \otimes 1$, for all $a \in A$, and then extended to the whole of $T_A(\ker \mu)$ by the graded Leibniz rule.

Every first-order differential calculus $(\Omega^1 A, d)$ can by extended universally to the full calculus by defining ΩA as the quotient of the tensor algebra $T_A(\Omega^1 A)$ by the relations coming from the graded Leibniz rule and $d^2 = 0$.

4. Volume form. An N-dimensional calculus is said to *admit a volume form* if $\Omega^N A \cong A$ as a left and right A-module. Any free generator v of $\Omega^N A$ as a left and right A-module (if it exists) is called a *volume form.*

5. Skew multi-derivations. If $(\Omega A, d)$ is such that $\Omega^1 A$ is a finitely generated as a left A-module, then any left A-module basis $\{\omega_1, \ldots, \omega_n\}$ of $\Omega^1 A$ induces maps $\partial_i, \sigma_{ij} : A \to A$, $i, j = 1, \ldots, n$, by

$$da = \sum_i \partial_i(a)\omega_i, \qquad \omega_i a = \sum_j \sigma_{ij}(a)\omega_j. \tag{1}$$

These necessarily satisfy

$$\sigma_{ij}(1) = 1 \qquad \sigma_{ij}(ab) = \sum_k \sigma_{ik}(a)\sigma_{kj}(b), \tag{2a}$$

$$\partial_j(ab) = \sum_i \partial_i(a)\sigma_{ij}(b) + a\partial_j(b). \tag{2b}$$

A system $(\partial_j, \sigma_{ij})_{i,j=1}^n$ is called a *skew multi-derivation.* Any skew multi-derivation induces a calculus on A by formulae (1), provided there exist $a_i^{\alpha_i}, b_i^{\alpha_i} \in A$ such that $\sum_{\alpha_i} a_i^{\alpha_i}\partial_j(b_i^{\alpha_i}) = \delta_{ij}$. If such elements exist $(\partial_j, \sigma_{ij})_{i,j=1}^n$ is said to be *orthogonal.*

The conditions (2a) are equivalent to the statement that the map $\sigma : A \to M_n(A)$, $a \mapsto (\sigma_{ij}(a))_{i,j=1}^n$, where $M_n(A)$ denotes the ring of $n \times n$ matrices with entries from A, is an algebra homomorphism.

6. Free multi-derivations. A skew multi-derivation $(\partial_j, \sigma_{ij})_{i,j=1}^n$ is said to be *free,* provided there exist $\bar{\sigma}_{ij}, \hat{\sigma}_{ij} : A \to A$ that satisfy conditions (2a) and are such that, for all $a \in A$,

$$\sum_k \bar{\sigma}_{jk}(\sigma_{ik}(a)) = \sum_k \sigma_{kj}(\bar{\sigma}_{ki}(a)) = \sum_k \hat{\sigma}_{jk}(\bar{\sigma}_{ik}(a)) = \sum_k \bar{\sigma}_{kj}(\hat{\sigma}_{ki}(a)) = \delta_{ij}a.$$

If the matrix $\sigma = (\sigma_{ij})_{i,j=1}^n$ is triangular with invertible diagonal entries, then $(\partial_j, \sigma_{ij})_{i,j=1}^n$ is free.

7. Diagonal and skew q-derivations. If $\sigma = (\sigma_{ij})_{i,j=1}^n$ is diagonal, then relations (2b) separate into twisted Leibniz rules, $\partial_i(ab) = \partial_i(a)\sigma_{ii}(b) + a\partial_i(b)$, for all $i = 1, \ldots, n$. Furthermore, if σ_{ii} is an invertible map and there exists $q_i \in \mathbb{K}$ such that $\sigma_{ii}^{-1} \circ \partial_i \circ \sigma_{ii} = q_i\partial_i$, then $(\partial_i, \sigma_{ii})$ is called a *skew q_i-derivation.*

8. Calculus for q-polynomials. A *q-polynomial algebra* or the *quantum plane* is the algebra $A = \mathbb{K}_q[x, y]$ generated by x, y subject to the relation $xy = qyx$. The elements of A are finite combinations of monomials $x^r y^s$.

The algebra A admits a first-order calculus freely generated by one-forms dx, dy and relations:

$$dxx = pxdx, \quad dyx = pq^{-1}xdy, \quad dxy = qydx + (p-1)xdy, \quad dyy = pydy,$$

where p is a non-zero scalar. It is understood that $d : x \mapsto dx, y \mapsto dy$. The universal extension of this calculus necessarily yields $dxdy = -qp^{-1}dydx$, $(dx)^2 = (dy)^2 = 0$, and it is a two-dimensional calculus with a volume form, e.g., $v = dxdy$.

Setting $\omega_1 = dx$, $\omega_2 = dy$ one easily finds that the associated skew multi-derivation $(\partial_i, \sigma_{ij})_{i,j=1}^2$ is free with the upper-triangular matrix-valued endomorphism

$$\sigma(x^r y^s) = \begin{pmatrix} p^r q^s x^r y^s & p^r(p^s - 1)x^{r+1}y^{s-1} \\ 0 & p^{r+s}q^{-r}x^r y^s \end{pmatrix}.$$

9. Inner calculus. A calculus $(\Omega A, d)$ is said to be *inner* if there exists $\theta \in \Omega^1 A$, such that $d(\omega) = \theta\omega - (-1)^n\omega$, for all $\omega \in \Omega^n A$. Note that $d^2(\omega) = 0$ implies that θ^2 is central in ΩA. Also θ satisfies the Cartan–Maurer equations $d\theta = 2\theta^2$.

As an example, consider a one-dimensional calculus on the Laurent polynomial ring $A = \mathbb{K}[x, x^{-1}]$, given by Jackson's q-derivation

$$\partial_q(f) = \frac{f(qx) - f(x)}{(q-1)x}, \tag{3}$$

where taking the limit is understood in case $q = 1$. If $q \neq 1$, this calculus is inner with $\theta = \frac{1}{q-1}x^{-1}dx$, otherwise it is not inner.

3. Integral forms

10. Divergence. Let $(\Omega A, d)$ be a differential calculus on an algebra A. We will denote by $\mathfrak{I}_n A$, the Abelian group of all right A-linear maps $\Omega^n A \to A$. For all $n \geq m$, consider maps

$$\cdot : \mathfrak{I}_n A \otimes \Omega^m A \to \mathfrak{I}_{n-m} A, \qquad f \otimes \omega \mapsto f \cdot \omega, \quad (f \cdot \omega)(\omega') = f(\omega\omega').$$

In particular, $\cdot$ makes $\mathfrak{I}_n A$ into a right A-module.

A *divergence* is a linear map $\nabla_0 : \mathfrak{I}_1 A \to A$, such that, for all $a \in A$, $\nabla_0(f \cdot a) = \nabla_0(f)a + f(da)$. A divergence is extended to a family of maps $\nabla_n : \mathfrak{I}_{n+1} A \to \mathfrak{I}_n A$, by $\nabla_n(f)(\omega) = \nabla_0(f \cdot \omega) + (-1)^{n+1} f(d\omega)$, for all $\omega \in \Omega^n A$.

The cokernel map $\Lambda : A \to A/\mathrm{Im}\nabla_0$ is called the *integral* associated to ∇_0.

11. Integral forms. A divergence ∇_0 is said to be *flat*, provided $\nabla_0 \circ \nabla_1 = 0$. It is then the case that, for all n, $\nabla_n \circ \nabla_{n+1} = 0$, and hence there is a complex, $\cdots \xrightarrow{\nabla_2} \mathfrak{I}_2 \xrightarrow{\nabla_1} \mathfrak{I}_1 \xrightarrow{\nabla_0} A$, known as the complex of *integral forms*.

12. The inner case. If $(\Omega A, d)$ is an inner differential calculus with the exterior derivative given by a graded commutator with $\theta \in \Omega^1 A$, then $\nabla_0 : \mathfrak{I}_1 A \to A$, $f \mapsto -f(\theta)$ is a divergence. One easily finds that $\nabla_1(f)(\omega) = f(\theta\omega)$, and so this divergence is flat, provided $\theta^2 = 0$.

13. Divergences and multi-derivations. Let $(\partial_i, \sigma_{ij}; \bar{\sigma}_{ij}, \hat{\sigma}_{ij})_{i,j=1}^n$ be a free skew multi-derivation on A, and let $(\Omega^1 A, d)$ be the associated first-order calculus with generators $\omega_1, \ldots, \omega_n$. Let $\xi_i \in \mathfrak{I}_i$ be the dual basis to the ω_i, i.e., the ξ_i are given by $\xi_i(\omega_j) = \delta_{ij}$. Then

$$\nabla_0 : \mathfrak{I}_1 A \to A, \qquad f \mapsto \sum_{i,j,k} \bar{\sigma}_{kj} \left(\partial_i \left(f \left(\hat{\sigma}_{ki} \left(\omega_i\right)\right)\right)\right), \tag{4}$$

is a unique divergence such that $\nabla_0(\xi_i) = 0$, for all $i = 1, \ldots, n$.

In particular, if σ is diagonal and all the $(\partial_i, \sigma_{ii})$ are skew q_i-derivations, then

$$\nabla_0(f) = \sum_i q_i \partial_i \left(f\left(\omega_i\right)\right).$$

14. Cauchy's integral formula. Let $A = \mathbb{K}[x, x^{-1}]$ be the Laurent polynomial ring with the one-dimensional calculus $(\Omega A, d)$ given by Jackson's q-derivative (3). In this case ∂_q is twisted by the automorphism $\sigma(f(x)) = f(qx)$, and (∂_q, σ) is a skew q-derivation. $\Omega^1 A$ is generated by dx, and hence the corresponding divergence (4) is $\nabla_0(f) = q\partial_q\left(f\left(dx\right)\right)$. For all $f \in \mathfrak{I}_1 A$, define $f_x \in \mathbb{K}[x, x^{-1}]$ by $f_x(x) := f(dx)$. Then

$$\nabla_0(f) = q\frac{f_x(qx) - f_x(x)}{(q-1)x}.$$

The image of ∇_0 consists of all of $\mathbb{K}[x, x^{-1}]$ except the monomials αx^{-1}. Therefore, the integral is

$$\Lambda : \mathbb{K}[x, x^{-1}] \to \mathbb{K}, \qquad a \mapsto \mathrm{res}(a)\Lambda(x^{-1}).$$

In case $\mathbb{K} = \mathbb{C}$ we can normalise Λ as $\Lambda(x^{-1}) = 2\pi i$, and obtain the Cauchy integral formula.

15. Calculus on quantum groups. If A is a coordinate algebra of a compact quantum group (over $\mathbb{C}$), then every left-covariant differential calculus gives rise to a free multi-derivation [5], and hence there is a canonical divergence ∇_0 (4) and the corresponding integral Λ. Any right integral λ on A (the Haar measure) factors uniquely through Λ, i.e., there exists unique $\varphi : A/\mathrm{Im}\nabla_0 \to \mathbb{C}$ such that $\lambda = \varphi \circ \Lambda$.

16. Berezin's integral. Let A be a superalgebra of (integrable) real functions on the supercircle $S^{1|1}$. That is, A consists of $a(x, \vartheta) = a^0(x) + a^1(x)\vartheta$, where $a^i : [0, 1] \to \mathbb{R}$ are (integrable) functions such that $a^i(0) = a^i(1)$ and ϑ is a Grassmann variable, $\vartheta^2 = 0$. The differentiation on A is defined by

$$\partial_x a(x, \vartheta) := \frac{da^0(x)}{dx} + \frac{da^1(x)}{dx}\vartheta, \qquad \partial_\vartheta a(x, \vartheta) := a^1(x).$$

One easily checks that $\partial = (\partial_x, \partial_\vartheta)$ is a free skew multi-derivation with the twisting matrix-valued endomorphism $\sigma(a(x, \vartheta)) = \begin{pmatrix} a(x, \vartheta) & 0 \\ 0 & a(x, -\vartheta) \end{pmatrix}$. The calculus $\Omega^1 A$ is freely generated by dx and $d\vartheta$. The corresponding divergence in Item 13 comes out as

$$\nabla_0(f)(x, \vartheta) = \partial_x f_x(x, \vartheta) - \partial_\vartheta f_\vartheta(x, \vartheta) = \frac{df_x^0(x)}{dx} - f_\vartheta^1(x) + \frac{df_\vartheta^0(x)}{dx}\vartheta,$$

where $f_x(x,\vartheta) := f(dx)(x,\vartheta) = f_x^0(x) + f_x^1(x)\vartheta$ and $f_\vartheta(x,\vartheta) := f(d\vartheta)(x,\vartheta) = f_\vartheta^0(x) + f_\vartheta^1(x)\vartheta$.

If $a(x,\vartheta)$ is purely even, i.e., $a(x,\vartheta) = a(x)$, then setting $f_x(x,\vartheta) = 0$ and $f_\vartheta(x,\vartheta) = -a(x)\vartheta$ we obtain $a(x) = \nabla_0(f)$. Thus, the integral Λ vanishes on the even part of A. Since Λ is the cokernel map of ∇_0, for all $f \in \mathfrak{I}_1 A$,

$$0 = \Lambda \circ \nabla_0(f) = \Lambda\left(\frac{df_x^0(x)}{dx} - f_\vartheta^1(x) + \frac{df_\vartheta^0(x)}{dx}\vartheta\right) = \Lambda\left(\frac{d}{dx} f_\vartheta^0(x)\vartheta\right).$$

On the other hand, $f_\vartheta^0(0) = f_\vartheta^0(1)$, so $\int_0^1 \frac{d}{dx} f_\vartheta^0(x)dx = 0$. By the universality of Λ, $\Lambda(a^1(x)\vartheta) = \int_0^1 a^1(x)dx$. Therefore,

$$\Lambda(a^0(x) + a^1(x)\vartheta) = \int_0^1 a^1(x)dx,$$

i.e., Λ is the Berezin integral on the supercircle.

References

[1] T. Brzeziński, *Non-commutative connections of the second kind.* J. Algebra Appl. **7** (2008), 557–573.

[2] T. Brzeziński, *Divergences on projective modules and non-commutative integrals.* Int. J. Geom. Methods Mod. Phys. **8** (2011), 885–896.

[3] T. Brzeziński, L. El Kaoutit and C. Lomp, *Non-commutative integral forms and twisted multi-derivations.* J. Noncommut. Geom. **4** (2010), 281–312.

[4] M. Dubois-Violette, *Lectures on graded differential algebras and noncommutative geometry.* Noncommutative differential geometry and its applications to physics (Shonan, 1999), pp. 245–306, Math. Phys. Stud. **23**, Kluwer, Dordrecht (2001).

[5] S.L. Woronowicz, *Differential calculus on compact matrix pseudogroups (quantum groups).* Comm. Math. Phys. **122** (1989), 125–170.

Tomasz Brzeziński
Department of Mathematics
Swansea University
Swansea SA2 8PP, UK

and

Department of Mathematics
University of Białystok
K. Ciołkowskiego 1M
15-245 Białystok, Poland
e-mail: T.Brzezinski@swansea.ac.uk

Geometric Methods in Physics. XXXV Workshop 2016
Trends in Mathematics, 255–260

General Relativity Theory and Its Canonical Structure

Jerzy Kijowski

Mathematics Subject Classification (2010). 83C05, 83C10, 49S99.
Keywords. General relativity, symplectic structure, calculus of variations, canonical structure.

Overview of the lectures

Symplectic structures arising naturally in the calculus of variations were thoroughly discussed. Different variational principles for General Relativity were presented (metric, metric-affine, purely affine) and the relation between them deeply analyzed. Finally, Hamiltonian formulation of the field evolution were derived.

1. Gravity as a field of local refererence frames

Newton's first law: *there is a global inertial system.* In absence of any force a test body moves along a straight line with constant velocity:

$$\ddot{y}^\alpha = 0 \tag{1}$$

where (y^α) – linear coordinates in an *affine* spacetime $\mathcal{X}$. Inertial reference frame – not a coordinate system but an equivalence class

$$\left\{(y^\alpha) \sim (x^\lambda)\right\} \Leftrightarrow \left\{ \frac{\partial^2 y^\alpha}{\partial x^\mu x^\nu} = 0 \right\} . \tag{2}$$

Calculate equations of motion (1) in arbitrary (non-inertial, i.e., curvilinear) coordinates (x^λ):

$$\begin{aligned} \dot{y}^\alpha = \frac{\partial y^\alpha}{\partial x^\nu}\dot{x}^\nu, \quad \ddot{y}^\alpha = \frac{\partial y^\alpha}{\partial x^\nu}\ddot{x}^\nu + \frac{\partial^2 y^\alpha}{\partial x^\mu \partial x^\nu}\dot{x}^\mu \dot{x}^\nu = 0 \\ \frac{\partial x^\lambda}{\partial y^\alpha}\ddot{y}^\alpha = \delta^\lambda_\nu \ddot{x}^\nu + \frac{\partial x^\lambda}{\partial y^\alpha}\frac{\partial^2 y^\alpha}{\partial x^\mu \partial x^\nu}\dot{x}^\mu \dot{x}^\nu = \ddot{x}^\lambda + \Gamma^\lambda_{\mu\nu}\dot{x}^\mu \dot{x}^\nu = 0 \, , \end{aligned} \tag{3}$$

where the coefficients:

$$\Gamma^{\lambda}_{\mu\nu}(\mathbf{x}) := \frac{\partial x^{\lambda}}{\partial y^{\alpha}} \frac{\partial^2 y^{\alpha}}{\partial x^{\mu} \partial x^{\nu}}(\mathbf{x}) \,, \tag{4}$$

measure at each spacetime point $\mathbf{x} \in \mathcal{X}$ how much non-inertial is the system $\left(x^{\lambda}\right)$.

Equation of motion (3) can be written in a "Newtonian" form:

$$m\ddot{x}^{\lambda} = -m\Gamma^{\lambda}_{\mu\nu}\dot{x}^{\mu}\dot{x}^{\nu} =: F^{\lambda} \tag{5}$$

where m is the mass and F^{λ} are fictitious forces due to non-inertiality of the system. They can be *globally* eliminated if we use an inertial frame for which $\Gamma^{\lambda}_{\mu\nu} = 0$.

Einstein: gravity is nothing but the above fictitious force. It can *locally* can be eliminated (freely falling elevator!). Maybe, there is no *global* inertial system. But there is a *local* inertal system at each $\mathbf{x} \in \mathcal{X}$ separately.

Definition. A local reference frame at $\mathbf{x}$ is an equivalence class of local coordinate systems defined in a neighbourhood of $\mathbf{x}$, with respect to the local version $\sim_{\mathbf{x}}$ of the equivalence relation (2), namely:

$$\left\{ (y^{\alpha}) \sim_{\mathbf{x}} \left(x^{\lambda}\right) \right\} \Leftrightarrow \left\{ \frac{\partial^2 y^{\alpha}}{\partial x^{\mu} x^{\nu}}(\mathbf{x}) = 0 \right\} . \tag{6}$$

Collection of all reference frames is a fiber bundle $\mathbf{Ref}(\mathcal{X})$. Given a coordinate system (x^{μ}), a reference frame $\mathcal{R}$ at $\mathbf{x}$ is uniquely parameterized by coefficients (4), where (y^{α}) is any representative of $\mathcal{R}$, i.e., $[(y^{\alpha})]_{\mathbf{x}} = \mathcal{R}$. The system $\left(x^{\mu}, \Gamma^{\lambda}_{\mu\nu}\right)$ is compatible with the bundle structure of $\mathbf{Ref}(\mathcal{X})$. Its fibers carry a canonical affine structure.

Definition. Gravitational field in $\mathcal{X}$ is a section $s : \mathcal{X} \to \mathbf{Ref}(\mathcal{X})$ of the bundle of reference frames. Element $s(\mathbf{x})$ is called an *inertial frame* at $\mathbf{x}$.

Mathematical version of the Einstein's "free falling elevator" is given by the following

Lemma. *Any system of coordinates* (x^{μ}) *can be made inertial at* $\mathbf{x}$ *by a quadratic correction.*

Proof. Assume for simplicity that (x^{μ}) are *centered* at $\mathbf{x}$, i.e., $x^{\mu}(\mathbf{x}) = 0$. If (x^{μ}) is not inertial at $\mathbf{x}$, i.e., if $[(x^{\mu})]_{\mathbf{x}} \neq s(\mathbf{x})$, i.e., if $\Gamma^{\lambda}_{\mu\nu}(\mathbf{x}) \neq 0$, then

$$y^{\alpha} := x^{\alpha} + \frac{1}{2}\Gamma^{\lambda}_{\mu\nu}(\mathbf{x})x^{\mu}x^{\nu} \tag{7}$$

is already inertial, i.e., $[(y^{\alpha})]_{\mathbf{x}} = s(\mathbf{x})$. □

2. Curvature. Affine variational principle for gravitational field

Mathematically, gravitational field is a *symmetric* connection on spacetime. No metric tensor is necessary to formulate equations of motion (3) of test particles moving in a gravitational field:

$$\ddot{x}^\lambda + \Gamma^\lambda_{\mu\nu}\dot{x}^\mu\dot{x}^\nu = 0\,.$$

Metric is necessary for the remaining physics (electrodynamics and nuclear forces, see *Special Relativity Theory*).

Definition. A connection s is flat in a neighbourhood of $\mathbf{x} \in \mathcal{X}$ if there are coordinates which are *inertial* not only at $\mathbf{x} \in \mathcal{X}$ but also within its neighbourhood.

To check if this is the case, we first annihilate $\Gamma^\lambda_{\mu\nu}$ at $\mathbf{x}$ by (7) and then try to annihilate also its derivatives $\Gamma^\lambda_{\mu\nu\kappa} := \partial_\kappa \Gamma^\lambda_{\mu\nu}$. The only way to achieve this goal is to use cubic corrections in (7). Because (x^α) are already inertial, the quadratic correction must vanish. Hence, we put

$$y^\lambda := x^\lambda + \frac{1}{6} U^\lambda_{\mu\nu\kappa} x^\mu x^\nu x^\kappa\,, \tag{8}$$

where U is totally symmetric in $(\mu\nu\kappa)$ (in any case, only the totally symmetric part enters no-trivially in (8)). This correction of coordinates implies the following correction of $\Gamma^\lambda_{\mu\nu\kappa}$:

$$\Gamma^\lambda_{\mu\nu\kappa} \longrightarrow \Gamma^\lambda_{\mu\nu\kappa} + U^\lambda_{\mu\nu\kappa}\,. \tag{9}$$

Hence, we are able to annihilate only the totaly symmetric part $\Gamma^\lambda_{(\mu\nu\kappa)}$ of derivatives of Γ. What remains is invariant.

Definition. A curvature tensor $K^\lambda_{\mu\nu\kappa}$ of the connection s is the tensor $K^\lambda_{\mu\nu\kappa}$ whose components are given in inertial coordinates by:

$$K^\lambda_{\mu\nu\kappa} :\overset{*}{=} \Gamma^\lambda_{\mu\nu\kappa} - \Gamma^\lambda_{(\mu\nu\kappa)}\,. \tag{10}$$

Lemma. *Curvature tensor is symmetric in first two lower indices:* $K^\lambda_{\mu\nu\kappa} = K^\lambda_{\nu\mu\kappa}$. *Moreover, its totally symmetric part vanishes:* $K^\lambda_{(\mu\nu\kappa)}$ *(I-st type Bianchi identities).*

Corollary. *If* $K^\lambda_{\mu\nu\kappa} \neq 0$ *at* $\mathbf{x}$ *then the connection is not flat because there is no way to annihilate even its first derivatives.*

It is an easy exercise to recalculate the curvature in a generic (not necessarily inertial) coordinate system:

Lemma. *Formula* (10) *takes the following, universal form, valid in an arbitrary coordinate system:*

$$K^\lambda_{\mu\nu\kappa} = \Gamma^\lambda_{\mu\nu\kappa} - \Gamma^\lambda_{(\mu\nu\kappa)} + \Gamma^\lambda_{\gamma\kappa}\Gamma^\gamma_{\mu\nu} - \Gamma^\lambda_{\gamma(\kappa}\Gamma^\gamma_{\mu\nu)}\,. \tag{11}$$

Theorem. *Curvature tensor* $K^\lambda_{\mu\nu\kappa}$ *carries the same information as the Riemann tensor* $R^\lambda_{\mu\nu\kappa}$. *More precisely:*

$$K^\lambda_{\mu\nu\kappa} = -\frac{2}{3} R^\lambda_{(\mu\nu)\kappa} \quad ; \quad R^\lambda_{\mu\nu\kappa} = -2K^\lambda_{\mu[\nu\kappa]}\,. \tag{12}$$

In particular, the symmetric part of the Ricci tensor is given by its trace:

$$K_{\mu\nu} := R_{(\mu\nu)} = \frac{3}{2} K^{\lambda}_{\mu\nu\lambda} \tag{13}$$

(we do not know *a priori* that Ricci is symmetric because, *a priori*, Γ could be non-metric).

The simplest invariant Lagrangian manufactured from the gravitational field Γ is:

$$L = L(\Gamma, \partial\Gamma) = C \cdot \sqrt{\det K_{\mu\nu}} \tag{14}$$

(Remember that: 1) we have no metric tensor at our disposal to contract indices of $K^{\lambda}_{\mu\nu\kappa}$ and 2) L must be a scalar density!)

Theorem. *Variational principle* $\delta \int L = 0$ *is equivalent to the standard General Relativity Theory. More precisely, Euler–Lagrange equations derived from* (14) *can be written as:*

$$\pi^{\mu\nu} = \frac{\partial L}{\partial K_{\mu\nu}}, \tag{15}$$

$$\nabla_{\lambda} \pi^{\mu\nu} = 0. \tag{16}$$

Their equivalence with conventional GR is obtained if we interpret the momentum $\pi^{\mu\nu}$ as the contravariant density of the metric:

$$\pi^{\mu\nu} =: \frac{1}{16\pi} \sqrt{g}\, g^{\mu\nu}. \tag{17}$$

Then (16) implies that Γ is the metric (Levi-Civita) connection of g, whereas (15) are Einstein equations with cosmological constant $\Lambda = \frac{1}{4\pi C}$. In particular, $\Lambda = 0$ case corresponds to the constrained variations principle: $L \equiv 0$ on constraints $K_{\mu\nu} \equiv 0$.

3. Symplectic structure related to gravitational field

General Relativity Theory describes interaction of two different geometric structures: 1) the affine structure, described by the connection Γ, and 2) the metric structure, described by the metric tensor. In affine formulation the metric (or its contravariant density (17)) plays role of the momentum canonically conjugate to the field Γ. Any variational principle can be regarded as a symplectic control theory in an appropriate jet space equipped with a canonical symplectic structure.

Example. Consider a scalar field φ. Field equations derived from a Lagrangian density $L = L(\varphi, \partial\varphi)$ can be written as follows:

$$\delta L(\varphi, \varphi_{\mu}) = \partial_{\mu} \left(p^{\mu} \delta\varphi \right) = \left(\partial_{\mu} p^{\mu} \right) \delta\varphi + p^{\mu} \delta\varphi_{\mu}. \tag{18}$$

Here we use the jet-oriented notation: $\varphi_{\mu} := \partial_{\mu}\varphi$ and "δ" is just the standard exterior derivative within a fiber $J^1\mathcal{P}_{\mathbf{x}}$ of the first jet extension of a certain bundle over $\mathcal{X}$, describing the value of the field φ and the momenta p^μ. Indeed, (18) is equivalent to the Euler–Lagrange equation

$$\partial_\mu p^\mu = \frac{\partial L}{\partial \varphi} \tag{19}$$

together with the definition of momenta:

$$p^\mu = \frac{\partial L}{\partial \varphi_\mu} \, . \tag{20}$$

The symplectic interpretation is following: There is a canonical symplectic structure in each 10-dimensional space $J^1\mathcal{P}_{\mathbf{x}}$:

$$\omega := \partial_\mu \left(\delta p^\mu \wedge \delta\varphi \right) = \delta \left(\partial_\mu p^\mu \right) \wedge \delta\varphi + \delta p^\mu \wedge \delta\varphi_\mu \, . \tag{21}$$

Equation (18) is a definition of its 5-dimensional, Lagrangian (i.e., maximal isotropic), subspace of "physically admissible" jets and L is its generating function with respect to a specific control mode, where (φ, φ_μ) have been chosen as *control parameters*, whereas the remaining parameters $(\partial_\mu p^\mu, p^\mu)$ have been declared to be *response parameters.*

This formulation is analogous to the first law of thermodynamics:

$$dU(V, S) = -pdV + TdS,$$

where $\omega = dV \wedge dp + dT \wedge dS$ is the canonical symplectic structure in the four-dimensional phase space describing the pressure p, the volume V, the temperature T and the entropy S of a simple thermodynamical body. Here, the generating function U is the internal energy of the body.

In case of gravitation, the bundle $\mathcal{P}$ describes the field variables $\Gamma^\lambda_{\mu\nu}$ and the corresponding momentum $\pi_\lambda^{\mu\nu\kappa}$. Each fiber $J^1\mathcal{P}_{\mathbf{x}}$ of its first jet extension carries the canonical symplectic structure

$$\omega := \partial_\kappa \left(\delta\pi_\lambda^{\mu\nu\kappa} \wedge \delta\Gamma^\lambda_{\mu\nu} \right) = \delta \left(\partial_\kappa \pi_\lambda^{\mu\nu\kappa} \right) \wedge \delta\Gamma^\lambda_{\mu\nu} + \delta\pi_\lambda^{\mu\nu\kappa} \wedge \delta\Gamma^\lambda_{\mu\nu\kappa} \, . \tag{22}$$

The affine variational principle based on any $L = L(K_{\mu\nu})$ (e.g., (14)) is obtained if we choose the first jet of Γ, i.e., $(\Gamma^\lambda_{\mu\nu}, \Gamma^\lambda_{\mu\nu\kappa})$, as control parameters:

$$\delta L = \partial_\kappa \left(\pi_\lambda^{\mu\nu\kappa} \delta\Gamma^\lambda_{\mu\nu} \right) = \left(\partial_\kappa \pi_\lambda^{\mu\nu\kappa} \right) \delta\Gamma^\lambda_{\mu\nu} + \pi_\lambda^{\mu\nu\kappa} \delta\Gamma^\lambda_{\mu\nu\kappa} \, . \tag{23}$$

Because $\Gamma^\lambda_{\mu\nu\kappa}$ enters into L *via* the Ricci only, we have constraints:

$$\pi_\lambda^{\mu\nu\kappa} = \frac{\partial L}{\partial \Gamma^\lambda_{\mu\nu\kappa}} = \frac{\partial L}{\partial K_{\alpha\beta}} \frac{\partial K_{\alpha\beta}}{\partial \Gamma^\lambda_{\mu\nu\kappa}} = \pi^{\mu\nu} \delta^\kappa_\lambda - \pi^{\kappa(\mu} \delta^{\nu)}_\lambda \, , \tag{24}$$

where $\pi^{\mu\nu}$, given by (15), describes the metric tensor according to (16). Metric (Hilbert) variational principle is obtained if we perform Legendre transformation between Γ and the metric, i.e., we choose the jet of the metric metric $(\pi^{\mu\nu}, \partial_\lambda \pi^{\mu\nu})$ as control parameters and the first jet of the connection (i.e., also curvature) as response parameters.

The whole *canonical gravity* together with its Hamiltonian formulation is described by the symplectic structure (22) and its various control modes.

References

[1] W. Tulczyjew, *Hamiltonian systems, Lagrangian systems and the Legendre transformation*, Symposia Matematica **14** (1974) p. 247; M.R. Menzio and W.M. Tulczyjew, *Infinitesimal symplectic relations and generalized Hamiltonian dynamics*, Ann. Inst. Henri Poincaré, **28** (1978) p. 349

[2] J. Kijowski and W.M. Tulczyjew, *A Symplectic Framework for Field Theories*, Lecture Notes in Physics No. 107 (Springer-Verlag, Berlin, 1979)

[3] J. Kijowski, *On a new variational principle in general relativity and the energy of gravitational field*, GRG Journal **9** (1978), pp. 857–881

[4] J. Jezierski, J. Kijowski, *Thermo-hydrodynamics as a field theory*, in *Hamiltonian Thermodynamics*, editors S. Sieniutycz and P. Salamon, Taylor and Francis Publishing Company, New York-London, 1991

[5] J. Kijowski, *A finite-dimensional canonical formalism in classical field theory*, Comm. Math. Phys. **30** (1973) pp. 99–128

[6] P. Chruściel, J. Jezierski and J. Kijowski, *Hamiltonian Field Theory in the Radiating Regime* (together with P. Chruściel and J. Jezierski), monograph (174 pages), volume **70** of the series: Springer Lecture Notes in Physics, Monographs (2001)

[7] J. Kijowski, G. Moreno, *Symplectic structures related with higher order variational problems*, Int. J. of Geometric Methods in Modern Physics, **12** (2015) 1550084-1–1550084-45

[8] P. Chruściel, J. Jezierski, J. Kijowski, *Hamiltonian dynamics in the space of asymptotically Kerr–de Sitter spacetimes*, Phys. Rev. **D 92** (2015) 084030 (30 pages)

Jerzy Kijowski
Center for Theoretical Physics
Polish Academy of Sciences
Al. Lotników 32/46
02-668 Warsaw, Poland

Geometric Methods in Physics. XXXV Workshop 2016
Trends in Mathematics, 261–268

Exponential Formulae in Quantum Theories

Bogdan Mielnik

Mathematics Subject Classification (2010). 15A16, 81Q93.

Keywords. Baker–Campbell–Hausdorff formula, quantum dynamics, quantum control, evolution loops.

Due to the special form of the evolution equations in quantum theories, the exponential operators with constant or variable exponents $-iH\delta t$ (where H are Hamiltonians) are the most typical elements of quantum dynamics, even if sometimes implicit or difficult to perceive behind too many details of perturbative calculations.

The report below is aimed to present some simple laws permitting to use the algebras of exponential operators with non-commuting exponents.

The most elementary results concern just a pair of non-commuting exponents a, b (tentatively in an algebra of observables of some quantum system, $a = -iH_1\delta t_1$, $b = -iH_2\delta t_2$). Then in case when the commutator $[a, b]$ is just a number, $[a, b] = \alpha \in \mathbb{C}$, the law of multiplying two exponential operators e^a and e^b is simply:

$$e^a e^b = e^{a+b+\frac{1}{2}[a,b]}. \tag{1}$$

In general, if the commutators of some higher order of a, b do not vanish, the formula of Baker–Campbell–Hausdorff [1], leads to:

$$e^{\lambda a} e^{\lambda b} = e^{\lambda(a+b)+\frac{\lambda^2}{2}[a,b]+\frac{\lambda^3}{12}([a,[a,b]]+[b,[b,a]])+\cdots} \tag{2}$$

The inverse formula of Zassenhaus (see, e.g., Wilcox [2]) permits to decompose the operator $e^{\lambda(a+b)}$ into an infinite product of the exponential operators with exponents proportional to powers of λ which can facilitate the operations in the S-matrix formalism.

In what follows we shall be most interested in sequences of exponential operators $e^{-iH(t_n)\delta t_n} e^{-iH(t_{n-1})\delta t_{n-1}} \dots e^{-iH(t_1)\delta t_1}$. However, for symbolic simplicity, we shall consider slightly more general products:

$$U = e^{A(t_n)\delta t_n} e^{A(t_{n-1})\delta t_{n-1}} \dots e^{A(t_1)\delta t_1}, \tag{3}$$

where the algebraic properties of $A(t)$ are not *a priori* assumed. The limits for $\delta t_j \to 0$ and $n \to \infty$ are of special interest; they can approximate the continuous evolution processes described by the differential equations:

$$\frac{dU}{dt} = A(t)U(t, t_0). \tag{4}$$

The resulting operators can be equivalently expressed by the integral equation:

$$U(t, t_0) = 1 + \int_{t_0}^{t} A(t_1)U(t_1, t_0)dt_1. \tag{5}$$

Its iterations obtained by substituting the $U(t_1, t_0)$ under the integral on its own right-hand side offer the sequence of multi-integral expressions starting from:

$$U(t, t_0) = 1 + \int_{t_0}^{t} A(t_1)U(t_1, t_0)dt_1 + \int_{t_0}^{t}\int_{t_0}^{t_1} A(t_1)A(t_2)U(t_2, t_0)dt_2 dt_1. \tag{6}$$

leading in the limit to the solution: $U(t, t_0) = 1 + Z(t, t_0)$, where Z is expressed by the infinite formal series

$$\begin{aligned} Z = \int_{t_0}^{t} A(t_1)dt_1 &+ \int_{t_0}^{t}\int_{t_0}^{t} A(t_1)\theta_{1,2}A(t_2)dt_2 dt_1 \\ &+ \int_{t_0}^{t}\int_{t_0}^{t}\int_{t_0}^{t} A(t_1)\theta_{1,2}A(t_2)\theta_{2,3}A(t_3)dt_3 dt_2 dt_1 + \cdots, \end{aligned} \tag{7}$$

where the Heaviside functions $\theta_{j,k} = \theta(t_j - t_k)$ assure that the integrals over the subsequent parameters t_j run just from t_0 to t_{j-1}. The question thus arises, how to express $U(t, t_0)$ by the single exponential operator

$$U(t, t_0) = e^{\Omega(t, t_0)}. \tag{8}$$

The problem inspired a formal solution of the 'chronological product'

$$U(t, t_0) = T[e^{\int_{t_0}^{t} A(\tau)d\tau}], \tag{9}$$

which, however, contains a symbolic cheating since precisely $\Omega \neq \int_{t_0}^{t} A(\tau)d\tau$.

The problem inspired a sequence of papers using the non-linear algorithm, which permits to seek Ω as an infinite sum of the homogeneous contributions in form of multiple integrals of the A's, the most important of Magnus [3] and Wilcox [2]. However, the algorithm turned out overcomplicated. Magnus was able to calculate Ω up to the 3rd order in A's, and concluded that the rest is just a 'combinatorial mess'. Wilcox succeeded to obtain the 4th order, but no more. The problem seemed open for the future generations.

An authentic breakthrough, though, came from analyzing more carefully the structure of the series for Z given by (7) with the expression for $U = e^{\Omega} = 1 + Z$. It implies:

$$\Omega = \ln(1 + Z) = Z - \frac{1}{2}Z^2 + \frac{1}{3}Z^3 - \cdots \tag{10}$$

Then, one can notice that:

$$
\begin{aligned}
Z^2 = \cancel{\int_{t_0}^{t} A(t_1)dt_1} &+ \int_{t_0}^{t}\int_{t_0}^{t} A(t_1)\cancel{\theta}_{12}A(t_2)dt_2dt_1 \\
&+ \int_{t_0}^{t}\int_{t_0}^{t}\int_{t_0}^{t} A(t_1)\cancel{\theta}_{12}A(t_2)\theta_{23}A(t_3)dt_3dt_2dt_1 \\
&+ \int_{t_0}^{t}\int_{t_0}^{t}\int_{t_0}^{t} A(t_1)\theta_{12}A(t_2)\cancel{\theta}_{23}A(t_3)dt_3dt_2dt_1 + \cdots ,
\end{aligned}
\tag{11}
$$

obtained from Z by crossing some thetas. The first integral is removed altogether and in the remaining terms some θ functions are removed by the operation of crossing out $\cancel{\theta}$, which is indeed the formal differentiation of Z^2 by $\frac{d}{d\theta}$, so $Z^2 = \frac{d}{d\theta}Z$, in general $Z^{(n+1)} = \frac{1}{n!}\frac{d^n}{d\theta^n}Z$, implying:

$$
\Omega = \left[\frac{e^{-\frac{d}{d\theta}} - 1}{-\frac{d}{d\theta}}\right] Z. \tag{12}
$$

This leads to an explicit expression:

$$
\Omega = \sum_{n=1}^{\infty} \int_{t_0}^{t} \cdots \int_{t_0}^{t} L_n(t_1, \ldots, t_n), A(t_1) \cdots A(t_n)dt_n \cdots dt_1 \tag{13}
$$

where the kernels are:

$$
L_n(t_n, \ldots, t_1) = \frac{e^{-\frac{d}{d\theta}} - 1}{-\frac{d}{d\theta}}[\theta_{21} \ldots \theta_{n,n-1}]. \tag{14}
$$

Since all differentiations $\frac{d}{d\theta}$ act on finite θ products, all kernels L_n are explicitly known [4, 5] (cf. ample comments of Czyż [6] and Gelfand [7]).

Special applications. Some other aspects are also of interest. As can be shown, all terms of (12) are the Lie products (multiple commutators) of $A_1, \ldots, A_n$. Whenever they can be represented by generators of a finite-dimensional Lie algebra, some higher-order terms in (12) start to repeat and the problem of finding U leads to a finite-dimensional matrix equation. The simplest case occurs just in the Hilbert space $L^2(\mathbb{R})$ for the quantum motion of the 1-dimensional harmonic oscillator generated by the time-dependent Hamiltonians

$$
H(t) = \frac{p^2}{2} + \beta(t)\frac{q^2}{2}, \tag{15}
$$

where we put $m = \hbar = 1$ and $\beta(t)$ represents the time-dependent elastic force. The curious phenomenon of *classical-quantum* equivalence gives exactly the same differential equation for the evolution of either classical or quantum canonical variables q, p. Since the equations are linear, then the $q(t), p(t)$ pairs evolve linearly according to:

$$
\left\| \begin{matrix} q(t) \\ p(t) \end{matrix} \right\| = u(t, \tau) \left\| \begin{matrix} q(\tau) \\ p(\tau) \end{matrix} \right\|, \tag{16}
$$

where $u(t,\tau)$ is a simplectic, real 2×2 matrix, exactly the same for the classical and quantum canonical variables, given by the matrix equations:

$$\frac{du(t,\tau)}{dt} = \Lambda(t)u(t,\tau), \quad \frac{du(t,\tau)}{d\tau} = -u(t,\tau)\Lambda(\tau), \tag{17}$$

where

$$\Lambda(t) = \begin{Vmatrix} 0 & 1 \\ -\beta(t) & 0 \end{Vmatrix}. \tag{18}$$

An interesting property of the physical spinless particle in 1-space dimension is that if a pair of unitary operators U_1 and U_2 generate the same transformation of the canonical observables q, p, i.e., $U_1^\dagger q U_1 = U_2^\dagger q U_2$ and $U_1^\dagger p U_1 = U_2^\dagger p U_2$, then $U_1 U_2^\dagger$ commutes with both q and p and since the functions of q and p form an irreducible algebra in $L^2(R)$ then $U_1 U_2^\dagger = e^{i\phi}$ meaning that $U_2 = e^{i\phi} U_1$. We shall say that U_1 and U_2 are *equivalent*, $U_1 \equiv U_2$.

This means, however, that the time-dependent unitary operators $U(t,\tau)$ defining the evolution of quantum systems for the quadratic Hamiltonians (15) are defined with accuracy to the *c-number* phase factors by the corresponding evolution matrices $u(t,\tau)$. And since they are identical as in the classical theory, then the whole 'quantization problem' of the Hamiltonians (15) is basically solved by classical evolution – without an effort of using the chronological or normal products, Weyl's ordering of q, p-polynomials etc. The *only* problem is to solve the *c-number* matrix equation for the classical evolution trajectories with time-dependent $\beta(t)$, which may require some patient computer work.

Yet, in either quantum or classical cases there is a collection of exact data useful for the general solutions of the evolution problem. The simplest ones concern the fragments of free evolution interrupted by sudden δ-kicks of the oscillator potential. The free evolution operator in $L^2(\mathbb{R})$ in any interval $[t_1, t_2]$ where $\beta(t) = 0$ is given by $U_\tau = e^{-\tau \frac{p^2}{2}}$, where $\tau = t_2 - t_1$. In turn, the sudden kick of the oscillator force, $\beta(t) = a\delta(t)$ generates the evolution operator $U_a = e^{-a\frac{q^2}{2}}$. The corresponding evolution matrices are

$$u^\tau = \begin{Vmatrix} 1 & \tau \\ 0 & 1 \end{Vmatrix}, \quad u_a = \begin{Vmatrix} 1 & 0 \\ -a & 1 \end{Vmatrix}. \tag{19}$$

A curious effect which can happen by applying sequences of free evolution intervals, interrupted by the oscillator kicks are the *evolution loops* in which the sequences of evolution matrices give the identity: $u^{\tau_1} u_{a_1} \cdots u^{\tau_n} u_{a_n} = 1$ so that the quantum system returns to its initial state. The simplest case of this phenomenon in $L^2(\mathbb{R})$ is illustrated below in Figure 1, [8, 9].

The product of the 6 unitary operations represented here corresponds to the identity:

$$e^{-i\beta\Gamma\frac{q^2}{2}} e^{-i\gamma\frac{p^2}{2}} e^{-i\alpha\Gamma\frac{q^2}{2}} e^{-i\beta\frac{p^2}{2}} e^{-i\gamma\Gamma\frac{q^2}{2}} e^{-i\alpha\frac{p^2}{2}} \equiv 1. \tag{20}$$

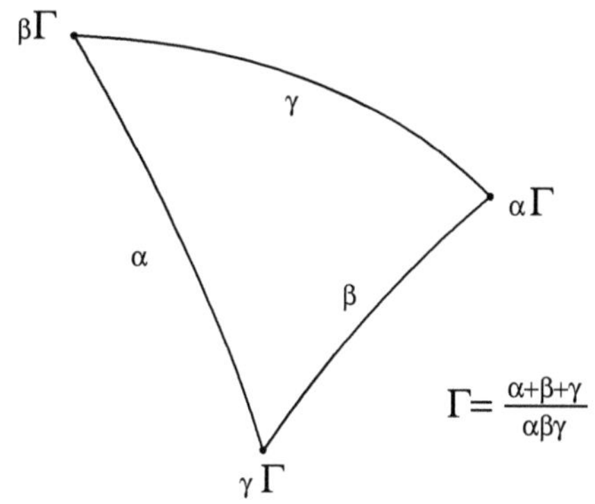

FIGURE 1. The evolution triangle. The free evolution intervals for $\tau = \alpha, \beta, \gamma$ (represented by the triangle sides), interrupted by three oscillator kicks $\alpha\Gamma, \beta\Gamma, \gamma\Gamma$ generate an *evolution loop* with the evolution matrices closing to 1 and the evolution operators (20) returning the system to its initial state.

Note that interesting operational effects. Indeed, any 5 subsequent operations of the triangle invert the 6th one, e.g.:

$$e^{i\alpha\frac{p^2}{2}} \equiv e^{-i\beta\Gamma\frac{q^2}{2}} e^{-i\gamma\frac{p^2}{2}} e^{-i\alpha\Gamma\frac{q^2}{2}} e^{-i\beta\frac{p^2}{2}} e^{-i\gamma\Gamma\frac{q^2}{2}} \tag{21}$$

producing an evolution incident inverting the free packet propagation during the time $\tau = \alpha$.

As subsequently found, the analogous incidents of the free evolution inverted can be generated for charged particle, under the influence of the homogeneous magnetic field pulses arriving subsequently from three orthogonal directions (non-relativistic approximation cf. [10]).

As interesting effects can occur also for charged particles in softly pulsating homogeneous magnetic fields $B(t) = B_1 \cos\omega t + B_2 \cos 2\omega t$ parallel to an axis of a solenoid [11]. A careful computer study of the stability limits for the charged particle in the dimensionless coordinates (A. Ramírez) succeeded to identify the map of the stability thresholds on which the trace of the evolution matrices becomes ± 1 (see Fig. 2).

Curiously, on all such thresholds the unitary evolution operators imitate either the free evolution incidents, but with the enlarged or slowed or even inverted evolution times (see Fig. 3). Alternatively, on a part of thresholds, one obtains the 'soft imitations' of the sharp oscillator kicks (non-relativistic, but it was found that the similar effects occur also in the special-relativistic approximation [12]).

The search for variable oscillator pulses, permitting to achieve some physically interesting result, in spite of its narrow subject, is still an open area. In particular, one might be interested to consult [13] (nonhermitian problems), also [14] (non-linear equations for higher-dimensional models), [15] (the exponential formula for

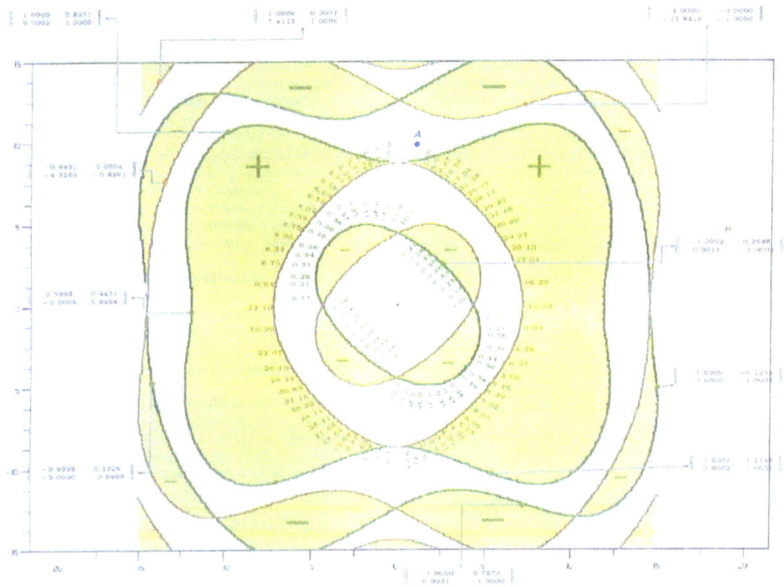

FIGURE 2. The Ramírez map for charged particle stability under the influence of double frequency pulses of a homogeneous magnetic field $B(t) = B_1 \cos \omega t + B_2 \cos 2\omega t$. The results expressed in dimensionless variables, $e = m = \hbar = \omega = 1$. On the horizontal and vertical axes β_1 and β_2 are the dimensionless equivalents of the amplitudes B_1 and B_2. The matrix B on the right side represents one of the effects of the inverted free evolution.

higher-dimensional matrices, though the physical applications are still an open problem).

References

[1] F. Hausdorff, Ber. Verhandl. Saechs. Akad. Wiss. Leipzig, Math. Naturwis. Kl., **58** (1906) 19.

[2] R.M. Wilcox, J. Math. Phys., **8** (1967) 962.

[3] W. Magnus, *On the exponential solution of differential equations for a linear operator*, Commun. Pure Appl. Math., **7** (1954) 649.

[4] I. Białynicki-Birula, B. Mielnik and J. Plebański, Ann. of Phys., **51** (1969) 187.

[5] B. Mielnik and J. Plebański, *Combinatorial approach to Baker–Campbell–Hausdorff exponents*, Ann. Inst. Henri Poincaré, Vol. **XII**, No. 3 (1970) pp. 215–254.

[6] J. Czyż, *Paradoxes of measures and dimensions originated in Felix Hausdorff ideas*, World Sci. Singapore, 1994.

[7] I.M. Gelfand, Adv. Math., **112** (1995) 218.

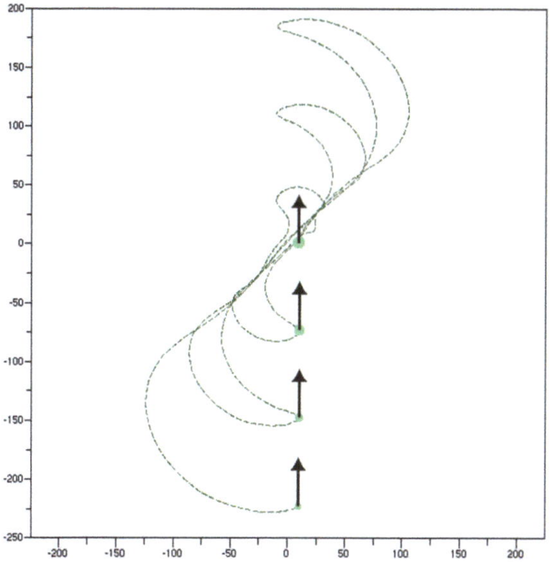

FIGURE 3. The trajectory for the center of a Gaussian packet evolving under the influence of the time inverting pulses of a threshold point of the map of Figure 2. After the ω period is over, the particle, recovers its initial velocity but does not respect its direction, instead it goes back, recovering its previous position. The involved trajectory illustrates the fact, that the free evolution inversion is not a continuous process but rather a special incident, caused by magnetic pulses. The incidents are repeated, while the pulses continue.

[8] B. Mielnik, J. Math. Phys., **27** (1986) 2290.

[9] For more information see
`http://wgmp.uwb.edu.pl/wgmp35/slides/Mielnik_school.pdf`

[10] D. Fernández C., *Bogdan Mielnik: Contributions to Quantum Control*, in: P. Kielanowski, Twareque Ali, A. Odzijewicz, M. Schlichenmaier and T. Voronov (eds.), *Proceedings of XXX Workshop on Geometric Methods in Physics*, Birkhäuser, Switzerland, 2013, pp. 135–156.

[11] B. Mielnik and A. Ramírez, *Magnetic operations: a little fuzzy mechanics?*, Phys. Scr., **84** (2011) 045008 (17 pp).

[12] L. Infeld and J. Plebański, *Motion and Relativity*, New York: Pergamon Press / Państwowe Wydawnictwo Naukowe, 1960, see EIH (Einstein, Infeld, Hoffman) approximation.

[13] H. Hernandez-Coronado, D. Krejčiřík, P. Siegl, *Perfect transmission scattering as a PT-symmetric spectral problem*, Phys. Lett. A, **375** (2011), 2149–2152.

[14] J. Wei and E. Norman, *On global representations of the solutions of linear differential equations as a product of exponentials*, Proc. Amer. Math. Soc. (4th ed.), **15** (1964) pp. 327–334.

[15] W.-X. Ma, B. Shekhtman, Do the chain rules for matrix functions hold without commutativity?, Linear and Multilinear Algebra, **58**, No. 1 (2010) 79087.

Bogdan Mielnik
Departamento de Física
Centro de Investigación y de Estudios Avanzados del I.P.N.
Av. Instituto Politécnico Nacional 2508
07360 Ciudad de México, Mexico
e-mail: bogdan@fis.cinvestav.mx

Geometric Methods in Physics. XXXV Workshop 2016
Trends in Mathematics, 269–280

Complex Algebraic Geometry Applied to Integrable Dynamics: Concrete Examples and Open Problems

E. Previato

Mathematics Subject Classification (2010). Primary 34M15; Secondary 14H70.
Keywords. Differential algebra; theta functions; vector bundles; integrable partial differential equations (PDEs).

WHAT IS…?

In September 2002, the *Notices* of the American Mathematical Society launched a new feature, published in each issue since then, with the following mission statement: "This is the inaugural installment of the "WHAT IS…?" column, which carries short (one- or two-page), nontechnical articles aimed at graduate students. Each article focuses on a single mathematical object, rather than a whole theory"[1].

This is a very popular feature of the *Notices*, and since the School's goal is to introduce an area of research, I tailored my three lectures after it. The original plan was to cover Elliptic and Hyperelliptic Theta Functions, and their generalization – Klein's higher-genus sigma function – specifically to construct solutions to integrable hierarchies such as the Toda Lattice [KMP]; introduce vector bundles over curves and their moduli, with applications to algebraically completely integrable Hamiltonian systems (ACIs) [Hi]; then bring the two topics together through classical theorems of projective geometry, in recent applications, for example, to random-matrix theory (Painlevé equations) [HaS]. As the lectures unfolded, more detail was required and the three lectures reorganized as follows: the first and second are concerned with aspects of elliptic/hyperelliptic curves in

This work was partly supported by the Conference Grant NSF DMS1609812. The author is sincerely grateful to the PI, Prof. Ekaterina Shemyakova, as well as the Organizers of Białowieża XXXV.

[1]A list can be found at `http://www.ams.org/publications/notices/whatis/noticesarchive`.

classical geometry, recently adapted to applications in integrability, in both the contexts of PDE hierarchies and of ACIs. The final lecture covered the Kleinian sigma function, concluding with Baker's striking interpretation in projective geometry of the PDEs that characterize it: this is a tool that brings vector bundles into integrability, but there was no time for specifics

1. Lecture I: What is an elliptic curve?

As Mumford says in [Mum1, Lect. I], "The beginning of the subject is the AMAZING SYNTHESIS, which surely overwhelmed each of us as graduate students", and which he illustrates by the three natures of curves: Algebra (finitely generated field extensions of transcendence degree one over $\mathbb{C}$); Geometry (subvarieties of projective space $\mathbb{P}^n$, locally defined by $n-1$ homogeneous polynomial equations with independent differentials); Analysis (compact Riemann surfaces). I started with the Analysis nature of the elliptic curve, the torus $E = \mathbb{C}/\{n + m\tau, n, m \in \mathbb{Z}\}$, which becomes Algebra by virtue of the ODE satisfied by the Weierstrass $\wp$ function, the doubly periodic meromorphic function whose poles occur at the vertices of the lattice with the smallest possible multiplicity, two. An introduction both accessible and comprehensive, including a proof that the field K of meromorphic functions on E is generated by $\wp$ and $\wp'$, can be found in [DuV].

Two remarks are relevant.

Remark 1. The role of the elliptic curve in integrability. The Korteweg–de Vries (KdV) equation,

$$u_t + \frac{3}{2}uu_x - \frac{1}{4}u_{xxx} = 0$$

was proposed in the 19th century to model waves in a shallow canal (the value of $u(x,t)$ represents the height of the wave, the coordinate x the position in the canal); it was therefore natural to make the 'one-wave ansatz', $u(z,t) = v(x-ct)$, c a constant, where the function $v(z)$ should satisfy the ordinary differential equation $-cv' + \frac{3}{2}vv' - \frac{1}{4}v''' = 0$. By integrating twice, it was originally observed that the general solution is then an elliptic function, $v(x) = 2\wp(z+\alpha)+a$ (α, a two additional constants introduced by integration; when a assumes special values so that the cubic polynomial defining the elliptic curve has repeated roots, the solution becomes an elementary function, given in terms of exponentials or trigonometric/hyperbolic functions.) A modern example arises in statistical mechanics, as a one-dimensional lattice with exponential (nearest-neighbor) interaction, the Toda (differential-difference) system:

$$\frac{d^2 r_n}{dt^2} = a[2\exp(-br_n) - \exp(-br_{n-1}) - \exp(-br_{n+1})],$$

where a, b are arbitrary constants and n is any integer. By the transformation:

$$r = -\frac{1}{b}\ln\left(1 + \frac{f}{a}\right),$$

$$\frac{d^2}{dt^2}\ln\left(1+\frac{f_n}{a}\right) = b(f_{n-1}+f_{n+1}-2f_n),$$

Toda [T] produced exact solutions, expressed algebraically in terms of (Jacobi) elliptic functions,

$$f_n = \frac{(2k\nu)^2}{b}\left[\mathrm{dn}^2\left(2K(\nu t - \frac{n}{\lambda})\right) - \frac{E}{K}\right],$$

where ν is the frequency, λ the wavelength, K and E are complete elliptic integrals of the first and second kind for the modulus k:

$$K = \int_0^{\pi/2} \frac{d\theta}{\sqrt{1-k^2\sin^2\theta}}, \quad K = \int_0^{\pi/2} \sqrt{1-k^2\sin^2\theta}d\theta;$$

the formula shows that the discrete evolution corresponds to the addition of a point on the elliptic curve: the addition law is arguably the reason for the "unreasonable effectiveness" of elliptic functions in dynamics.

Remark 2. Less famous than the Weierstrass equation, $(\wp')^2 = 4\wp^3 - g_2\wp - g_3$, two differential properties that characterize the $\wp$ function were forerunners of the theory of integrable PDEs. On the one hand, I mention $\wp'' = 6\wp^2 - g_2/2$, because the theory of the "higher-genus Kleinian function" σ [B, BEL], to which Lecture III is devoted, and which generalizes the genus-one Weierstrass sigma function, is centered on the search for a complete set of (partial, in higher genus) differential equations satisfied by σ; complete in the sense of differential algebra, for example, namely sets that are bases of differential ideals that define the differential rings of the algebraic varieties where σ is defined. On the other hand, Baker, as pointed out in [EE], wrote an equation for the σ function of a hyperelliptic curve of genus two, using the "bilinear operator" that Hirota rediscovered independently and yields the "Hirota form" of the Kadomtsev–Petviashvili (KP) equation $(u_t + 6uu_x + u_{xxx})_x + 3u_{yy} = 0$, namely

$$(D_xD_t + D_x^4 + 3D_y^2)\tau\cdot\tau = 0,$$

for $w(x,y,t) = 2\ln\partial_x^2\tau(x,y,t)$, where two differentiations D_uD_v applied to $\tau\cdot\tau$ signify

$$\frac{\partial}{\partial u}\frac{\partial}{\partial v}\big(\tau(u+v)\tau(u-v)\big)\,|_{u=v}$$

The Weierstrass equation thus provides the Algebra aspect of E, which Mumford (*loc. cit.*) describes as "field extensions $K \supset \mathbb{C}$, where K is finitely generated and of transcendence 1 over $\mathbb{C}$." I then gave three versions of the Geometry nature of an elliptic curve, all closely related to integrable systems, the third one less known. Briefly: The first, as a smooth cubic in the plane, in Weierstrass normal form, $y^2 = 4x^3 - g_2x - g_3$; The second, as the intersection of two quadrics in 3-dimensional projective space $\mathbb{P}^3$ – these embeddings are images under the divisor map for the linear series of 2∞ and 3∞ respectively, where ∞ is the point $[0,0,1]$ of the Weierstrass cubic in projective coordinates $[x_0,x_1,x_2]$ for which $x = x_1/x_0$, $y = x_2/x_0$. These two projective models can be brought together by the third geometric representation, namely, the incidence correspondence $I \subset C \times D^*$ with a choice of origin for the group law [BKOR]. Here the points of the curve are pairs (P, ℓ), with P a point in a fixed conic C and ℓ one of the two lines through P that are tangents to another fixed conic D; C and D must be in general position, and the limiting cases correspond to cubics that define rational curves. The model provides a beautiful proof of the classical "Poncelet's Porism Theorem", very much relevant to integrable dynamics, such as billiard or geodesic motion [P3].

2. Lecture II: Differential algebra

We now meet another, less known, nature of the elliptic curve.

2.1. Burchnall and Chaundy

A fourth nature of the elliptic curve emerged in the 20th century, in fact surprisingly early. In [BC], the authors pose the following question: what is the structure of a commutative subalgebra of the $\mathbb{C}$-algebra of Ordinary Differential Operators (ODOs) that is not of the form $\mathbb{C}[L]$, with L an ODO? We briefly recall the setting: we choose to work in the formal one of the algebra of Pseudo Differential Operators (ΨDOs), which is the most general, with the disadvantage that no convergence is addressed; for more restrictive (and precise) functional restrictions, cf. Sato's work, e.g., [SS].

Definitions. (i) The ring of formal pseudodifferential operators Ψ is the set

$$\Big\{ \sum_{j=-\infty}^{N} u_j(x)\partial^j,\ u_j \text{ a formal power series}\Big\}.$$

If we think of these symbols as acting on functions of x by multiplication and differentiation: $(u(x)\partial)f(x) = u\frac{d}{dx}f$, and formally integrate by parts: $\int(uf) = u\int f - \int(u'f)$, we can motivate the composition rules:

$$\partial^{-1}\partial = \partial\partial^{-1} = 1,\quad \partial u = u\partial + u',\quad \partial^{-1}u = u\partial^{-1} - u'\partial^{-2} + u''\partial^{-3} - \dots$$

and easily check an extended Leibniz rule for a function f and for $A, B \in \Psi$:

$$\partial^i f\cdot = \sum_{j=0}^{\infty}\binom{i}{j}(\partial^j f)\partial^{i-j}\cdot\,,\quad A\circ B = \sum_{i=0}^{\infty}\frac{1}{i!}\tilde{\partial}^i A * \partial^i B,$$

where $\tilde{\partial}$ is a partial differentiation w.r.t. the symbol ∂ and $*$ has the effect of bringing all functions to the left and powers of ∂ to the right.

(ii) Ψ contains the subring $\mathcal{D}$ of differential operators $A = \sum_0^N u_j\partial^j$ and we denote by $(\ )_+$ the projection $B_+ = \sum_0^N u_j\partial^j$ where $B = \sum_{-\infty}^N u_j\partial^j$. The much studied Weyl algebra in two generators, $\mathbb{C}[p,q]$ with multiplication rule defined by the commutator $[p,q] = 1$ can be viewed as a subring of $\mathcal{D}$, namely the operators with polynomial coefficients, by letting $p = \partial$ and $q = x$.

(iii) The Burchnall–Chaundy (hereafter BC for short) problem asks to find and classify all commutative subrings of $\mathcal{D}$. If we denote by $\mathcal{C}_{\mathcal{D}}(L)$ the centralizer in $\mathcal{D}$ of an element $L \in \mathcal{D}$, we see that the polynomial ring $\mathbb{C}[L]$ is always contained in $\mathcal{C}_{\mathcal{D}}(L)$. We also see that if L has order $n > 0$ then L can be brought to standard form:

$$L = \partial^n + u_{n-2}(x)\partial^{n-2} + u_{n-3}(x)\partial^{n-3} + \cdots + u_0(x)$$

by using change of variable and conjugation by a function, which are the only two automorphisms of $\mathcal{D}$; we shall always assume L to be in standard form, and define a BC solution to be such an L for which $\mathcal{C}_{\mathcal{D}}(L)$ is not a polynomial ring $\mathbb{C}[M]$, $M \in \mathcal{D}$. Notice that any translation in x: $x \mapsto x - a$, transforms a BC solution L into another solution L_a. We refer to this operation as the "x-flow".

(iv) The rank of a subset of $\mathcal{D}$ is the greatest common divisor of the orders of all the elements of $\mathcal{D}$.

Now we can give two new models for the elliptic curve (for references and more examples cf. [P2]):

The classical "Lamé operator" $L = \partial^2 - c\wp(x)$, where $c \in \mathbb{C}$ is a constant, is a BC solution iff $c = n(n+1)$ with n an integer greater than zero; if this is the case, the centralizer $\mathcal{C}_{\mathcal{D}}(L)$ is the affine ring of a hyperelliptic curve of genus n, given by an equation: $\mu^2 = \lambda^{2n+1}+$ lower order, or an elliptic curve when $n = 1$. A singular-cubic example is given by:

$$L = \partial^2 - \frac{2}{x^2}, \quad B = \partial^3 - \frac{3}{x^2}\partial + \frac{3}{x^3}$$

which satisfy $B^2 \equiv L^3$.

In the Weyl algebra, define $u = p^3 + q^2 + \alpha$, $v = \frac{1}{2}p$, $L = u^2 + 4v$, $B = u^3 + 3(uv - vu)$; then $\mathcal{C}(L) = \mathbb{C}[L, B]$ and $B^2 - L^3 = -\alpha$, as shown in [Di]. By the assignment $p = \partial$, $q = x$ we obtain $L, B \in \mathcal{D}$ of order 6,9, but notice that the automorphism $\partial \mapsto -x$, $x \mapsto \partial$ will turn the orders into 4,6. Again, $\mathcal{C}_{\mathcal{D}}(L) = \mathbb{C}[L, B]$, the affine ring of the curve $\mu^2 = \lambda^3 - \alpha$; in particular, L is a BC solution, and the rank of this algebra is three, two, respectively.

It can be shown that centralizers $\mathcal{C}_{\mathcal{D}}(L)$ are maximal-commutative subalgebras of $\mathcal{D}$. How large can they be? Not very: since their quotient fields are function fields of one variable (cf. Th. 3), they are affine rings of curves, and in a formal sense these are indeed spectral curves; the algebras that correspond to a fixed curve make up the (generalized) Jacobian of that curve, and the x-flow is a holomorphic vector field on it. We may (formally) view this as a "direct" spectral problem; the "inverse" spectral problem allows us to reconstruct the coefficients of the operators (in terms of theta functions) from the data of a point on the Jacobian. The x-flow is tangent to the Abel image of the curve in its Jacobian, at a specific point. The higher osculating flows form a sequence (essentially finite): $x = t_1, t_2, \ldots, t_s, \ldots$ and the corresponding operators depend on these parameters in such a way as to satisfy the KP hierarchy. The higher-rank algebras are still much of a mystery.

In Ψ any (normalized) L has a unique nth root, n =ord L, of the form $\mathcal{L} = \partial + u_{-1}(x)\partial^{-1} + u_{-2}(x)\partial^{-2} + \cdots$. By a dimension count based on the orders, I. Schur showed that

Theorem 3. $\mathcal{C}_{\mathcal{D}}(L) = \{\sum_{-\infty}^{N} c_j \mathcal{L}^j, c_j \in \mathbb{C}\} \cap \mathcal{D}$.

This shows that the quotient field of $\mathcal{C}_{\mathcal{D}}(L)$ is a function field of one variable; indeed, a B which commutes with L must satisfy an algebraic equation $f(L, B) = 0$ (identically in x). In the case that the algebra can be generated by two elements L, B, the curve has a plane model, where L, B can be viewed as affine coordinates x, y. I offered a little-known algorithm for computing the equation of the curve, the "differential resultant" ([BP, P1]). Since the algebra $\mathbb{C}[L, B]$ has no zero-divisors, it can be viewed as the affine ring $\mathbb{C}[X, Y]/(h)$ of a plane curve, with $h(X, Y)$ an irreducible polynomial. The BC curve $= \{(\lambda, \mu) \mid L, B$ have a joint eigenfunction $Ly = \lambda y$, $By = \mu y\}$ is included in the curve Spec $\mathbb{C}[L, B]$ and since the latter is irreducible, they must coincide; this shows in particular that the BC polynomial is some power of an irreducible polynomial h: $f(\lambda, \mu) = h^{r_1}$. In addition, each point of the spectral curve has a solution space: this gives a vector bundle over the curve. More precisely, let $r_2 = \text{rank}\, \mathbb{C}[L, B]$, and $r_3 = \dim V_{(\lambda,\mu)}$ where $V_{(\lambda,\mu)}$ is the vector space of common eigenfunctions at any smooth point (λ, μ) of the BC curve. Then $r_1 = r_2 = r_3$. Moreover, this integer is the order of $G = gcd(L - \lambda, B - \mu)$, the operator (found by the Euclidean algorithm) of highest order for which a factorization holds, $B - \mu = T_1 G$, $L - \lambda = T_2 G$.

In theory, higher-rank algebras are classified by vector bundles over curves [Mul], but there is no explicit dictionary between the vector bundles and the coefficients of the operators; a recent paper [BZ] completed the result in [PW], covering the genus-one case of the spectral curve.

Lastly, we introduce the KP deformations, following [SS].

Definitions. (i) In Ψ, it is possible to conjugate any $\mathcal{L} = \partial + u_{-1}(x)\partial^{-1} + \cdots$ into ∂ by a $K \in \Psi$, $K = 1 + v_{-1}(x)\partial^{-1} + \cdots$, determined up to elements of $\mathbb{C}[\partial] = \mathcal{C}_{\mathcal{D}}(\partial)$. From now on we assume that $K^{-1}\mathcal{L}K = \partial$.

(ii) We define a formal Baker function for $\mathcal{L}$ as the element of the module of formal eigenfunctions such that $\mathcal{L}\psi = z\psi$; notice that $\psi = Ke^{xz}$.

The KP hierarchy is determined by the Lax equations ($\partial_n = \partial/\partial t_n$),

$$\partial_n \mathcal{L} = [B_n, \mathcal{L}] := B_n \mathcal{L} - \mathcal{L} B_n,$$

where $B_n = (\mathcal{L}^n)_+$. Motivated by an algebraic conjugation,

$$\partial e^{x\lambda} = \lambda e^{x\lambda}, \quad \mathcal{L}\psi = \lambda\psi, \quad \psi = We^{x\lambda},$$

$$W\partial W^{-1}We^{x\lambda} = \lambda We^{x\lambda}, \quad W = 1 + \sum_{1}^{\infty} w_n \partial^{-n}$$

set: $\mathcal{L} = W\partial W^{-1}$, then the KP hierarchy is given by the Sato equations:

$$\partial_n W\, W^{-1} = -(\mathcal{L})^n_- = (\mathcal{L})^n_+ - (\mathcal{L})^n.$$

The "inverse spectral construction", which holds for any number of variables and yields explicit, exact solutions to the KP equations, is largely due to Krichever [Kr]:

Inverse spectral problem. The following choices: (i) A Riemann surface X of genus g; (ii) A point $\infty \in X$; (iii) A local parameter λ^{-1} near ∞; (iv) A generic divisor $P_1 + \cdots + P_g = D$ (the condition is that $h^0(P_1 + \cdots + P_g - \infty) = 0$, equivalently, there are no meromorphic functions on X with a zero at ∞ and poles bounded by $P_1 + \cdots + P_g$); determine uniquely a function $\psi(\underline{t}, P)$, the "Baker–Akhiezer (BA) function," such that near ∞, $\psi \sim \exp(\sum_{i\geq 1} t_i\lambda^i)(1 + \sum \xi_i(\underline{t})\lambda^{-i})$ and at finite points P of the curve, ψ has poles bounded by D and is analytic elsewhere.

For such a ψ there exist unique operators K_j such that $K_j\psi = \partial_{t_j}\psi$ and these operators are a solution to the KP hierarchy, in particular $\mathcal{L}\psi = \lambda\psi$ gives $\mathcal{L} \in \Psi$ as above. All statements are local in $\underline{t}$. Explicitly,

$$\psi(\underline{t}) = e^{(\sum_{i\geq 1} t_i(\int_{P_0}^{P} \eta_i - c_i))} \cdot \frac{\vartheta(A(P) + \sum_{i\geq 1} U_i t_i + \delta)\vartheta(A(\infty) + \delta - A(D))}{\vartheta(A(P) + \delta - A(D))\vartheta(A(\infty) + \sum_{i\geq 1} U_i t_i + \delta)},$$

where δ is Riemann's constant so that $\vartheta(A(P)+\delta-A(D))$ vanishes for $P = P_j, j = 1, \ldots, g$; η_i are suitably normalized meromorphic differentials; $U_i \in \mathbb{C}^g$ are suitable vectors that make ψ into a function of P independent of the path of integration; $c_i \in \mathbb{C}$ are constants that normalize ψ as above.

In conclusion, the general (algebro-geometric) solution of KP is:

$$u(\underline{t}) = 2\partial_x^2 \log \vartheta\left(\sum_{j\geq 1} t_j U_j + A(P) + \delta\right) + \text{const}.$$

Most strikingly, Novikov conjectured that a theta function which satisfies the KP hierarchy arises from a Jacobian, and this was shown to be true, thus settling the "Shottky Problem" [BD].

To conclude the lecture on differential algebra, I mentioned a second major still largely open problem: what is the answer to the Burchnall–Chaundy question if we consider the algebra of Partial Differential Operators (PDOs)? Is there an analog of the spectral curve, such as, in two variables, a surface, and are its equations given by a differential resultant? Much work has been done, but concrete results are scarce, and the answer to simple questions is not known; for example, given that the multivariate resultant, a multivariate polynomial, vanishes identically when evaluated on a set of PDOs that have a common eigenfunction (cf. [KP] for a precise formulation and references), is the differential resultant independent of the differential variables? This is what happens in the ODO case, where the resultant if the equation of the spectral curve.

3. Lecture III: The Kleinian sigma function

Why switch from theta, which yields exact KP solutions, to sigma? I offer three reasons, of which the previous leads to the next: Modular invariance, thus a stronger relationship with the Jacobian (briefly put, in the following sense: the symplectic group $Sp(2g, \mathbb{Z})$ acts in the standard way [Mum2, II.5, (5.3)] on the two variables of ϑ, the Abelian variable $z \in \mathbb{C}^g$ and the period lattice Λ; the action produces a multiplicative non-zero factor, whereas σ is invariant); an explicit transliteration between meromorphic and transcendental functions; More useful formulas for solutions of integrable equations, since meromorphic functions lend themselves more clearly to a qualitative analysis.

Recall the definition of the Weierstrass sigma function (genus one):

$$\wp(u) = -\frac{d^2}{du^2} \ln \sigma(u), \ \ (\wp')^2 = 4\wp^3 - g_2\wp - g_3$$

Recall σ is an odd function (ϑ is even), with expansion:

$$\sigma(u) = u - \frac{1}{240} g_2 u^5 - \frac{1}{840} g_3 u^7 - \cdots .$$

Klein defined σ for two variables, then for any hyperelliptic curve, and for a trigonal curve [KS].

The sigma function for a hyperelliptic curve X of genus $g \geq 2$ defined in the affine plane by:

$$y^2 = f(x) := x^{2g+1} + \lambda_{2g} x^{2g} + \cdots + \lambda_0$$

(where λ_j's are generic complex numbers so that X, completed by ∞ at infinity, is smooth), is easy to define, because there is an explicit basis of differentials of the first kind:

$$\omega_i := \frac{x^{i-1}dx}{2y} \qquad (i = 1, \ldots, g),$$

and differentials of the second kind,

$$\eta_i := \frac{1}{2y} \sum_{k=j}^{2g-j} (k+1-j)\lambda_{k+1+j} x^k dx, \quad (j = 1, \ldots, g)$$

so that when taking the periods around a symplectic homology basis $\{\alpha_i, \beta_j\}$, $1 \leq i, j \leq g$, the matrices $\omega = \begin{bmatrix} \omega' \\ \omega'' \end{bmatrix}$ where

$$\omega' = \frac{1}{2}\left[\oint_{\alpha_j} \omega_i\right], \quad \omega'' = \frac{1}{2}\left[\oint_{\beta_j} \omega_i\right],$$

$$\eta' = \frac{1}{2}\left[\oint_{\alpha_j} \eta_i\right], \quad \eta'' = \frac{1}{2}\left[\oint_{\beta_j} \eta_i\right],$$

satisfy the generalized Legendre relation

$$\mathfrak{M} \begin{pmatrix} 0 & -1_g \\ 1_g & 0 \end{pmatrix} \mathfrak{M}^T = \frac{\imath\pi}{2} \begin{pmatrix} 0 & -1_g \\ 1_g & 0 \end{pmatrix}, \tag{1}$$

where $\mathfrak{M} = \begin{pmatrix} \omega' & \omega'' \\ \eta' & \eta'' \end{pmatrix}$. We let Λ be the lattice in $\mathbb{C}^g$ spanned by the column vectors of $2\omega'$ and $2\omega''$. The Jacobian variety of X is identified with $\mathbb{C}^g/\Lambda$. We let κ be the projection $\mathbb{C}^g \to \mathbb{C}^g/\Lambda$. For a non-negative integer k, we define the Abel map from the kth symmetric product $\mathrm{Sym}^k X$ of the curve X to $\mathbb{C}^g$ by first choosing any (suitable) path of integration[2]:

$$w : \mathrm{Sym}^k X \to \mathbb{C}^g, \quad w((x_1, y_1), \ldots, (x_k, y_k)) = \sum_{i=1}^{k} \int_{\infty}^{(x_i, y_i)} \begin{pmatrix} \omega_1 \\ \vdots \\ \omega_g \end{pmatrix}.$$

We denote the image by W_k. Let $\mathbb{T} = {\omega'}^{-1}\omega''$. The theta function on $\mathbb{C}^g$ with "modulus" $\mathbb{T}$ and characteristics $\mathbb{T}a + b$ for $a, b \in \mathbb{C}^g$ is given by

$$\vartheta \begin{bmatrix} a \\ b \end{bmatrix} (z; \mathbb{T}) = \sum_{n \in \mathbb{Z}^g} \exp \left[2\pi i \left\{ \frac{1}{2}\, {}^t(n+a)\mathbb{T}(n+a) + {}^t(n+a)(z+b) \right\} \right].$$

The σ-function, an analytic function on the space $\mathbb{C}^g$ and a theta-series having modular invariance of a given weight with respect to $\mathfrak{M}$, is given by the formula

$$\sigma(u) = \gamma_0 \exp\left\{ -\frac{1}{2}\, {}^t u \eta' {\omega'}^{-1} u \right\} \vartheta \begin{bmatrix} \delta'' \\ \delta' \end{bmatrix} \left(\frac{1}{2} {\omega'}^{-1} u;\, \mathbb{T} \right),$$

where δ' and δ'' are half-integer characteristics giving the vector of Riemann constants with basepoint at ∞ and γ_0 is a non-zero constant. Computing γ_0 is again possible because the curve is hyperelliptic: the result is based on a normalization, thus it is achieved by expanding the function at ∞ as a power series in the Abelian variables:

$$\gamma_0 = \frac{\epsilon_4}{\vartheta(0; \mathbb{T})} \prod_{r=1}^{g} \frac{\sqrt{P'(a_r)}}{\sqrt[4]{f'(a_r)}} \frac{1}{\prod_{k<l} \sqrt{e_k - e_l}}.$$

Since this constant plays no role in this paper, we have retained the slightly different notation of [BEL], where the curve is written as

$$y^2 = \sum_{i=0}^{2g+1} \lambda_i x^i = \lambda_{2g+1} \prod_{k=1}^{2g+1} (x - e_k) = 4P(x)Q(x)$$

with:

$$P(x) = \prod_{i=1}^{g} (x - a_i), \quad Q(x) = (x - b) \prod_{i=1}^{g} (x - b_i),$$

for the homology basis whose loops correspond to the branch cuts beginning at a_i and ending at b_i, with an additional one beginning at $a = \infty$ and ending at b. The fourth root of unity ϵ_4 is difficult to compute, but clearly does not depend on the moduli of the curve, since it is a discrete parameter and σ depends holomorphically on the moduli parameters. In genus one, the formula reduces to Weierstrass' σ,

[2]The results presented are independent of the particular choice.

and in that case this root of unity is related to the eight root of unity appearing in the functional equation of ϑ under the action of the congruence subgroup

$$\Gamma := \left(\begin{bmatrix} a & b \\ c & d \end{bmatrix} | ad - bc = 1, \ cd \ \text{ even} \right),$$

which is calculated in [Mum2, Vol. I, II.5] and involves the Jacobi symbol [Mum2, Vol. I, I.7, Th. 7.1].

The σ-function vanishes to the first order on $\kappa^{-1}(W_{g-1})$. The Kleinian $\wp$ and ζ functions are defined by

$$\wp_{ij} = -\frac{\partial^2}{\partial u_i \partial u_j} \log \sigma(u), \quad \zeta_i = \frac{\partial}{\partial u_i} \log \sigma(u).$$

I concluded returning full circle to Mumford's vision, but now for surfaces: indeed, Baker generalized Weierstrass' equation to cut out the Kummer surface in $\mathbb{P}^3$, the linear series of $|2\Theta|$, where Θ is the canonical theta divisor for a curve of genus two (hence also, up to translation, the zero locus of σ). This is Analysis turning into Geometry; Algebra is the field of meromorphic functions of the Kummer surface, but for surfaces the perfect synthesis no longer holds, since two surfaces may have isomorphic fields without being isomorphic, such as $\mathbb{P}^2$ and $\mathbb{P}^1 \times \mathbb{P}^1$.

The Kummer surface is the image of the $|2\Theta|$-divisor map $\mathrm{Jac}(X) \to \mathbb{P}^3$, using the basis $1, \wp_{11}, \wp_{12}, \wp_{22}$, and a quartic in these coordinates:

$$\det \begin{bmatrix} -\lambda_0 & \frac{1}{2}\lambda_1 & 2\wp_{11} & -2\wp_{12} \\ \frac{1}{2}\lambda_1 & -(\lambda_2 + 4\wp_{11}) & \frac{1}{2}\lambda_3 + 2\wp_{12} & 2\wp_{22} \\ 2\wp_{11} & \frac{1}{2}\lambda_3 + 2\wp_{12} & -(\lambda_4 + 4\wp_{22}) & 2 \\ -2\wp_{12} & 2\wp_{22} & 2 & 0 \end{bmatrix} = 0$$

is an algebraic differential equation that holds identically exactly on the Kummer surface.

This was generalized to all hyperelliptic Kummer varieties in [BÉ], and to trigonal Kummer varieties in [BLÉ]. For non-hyperelliptic curves, the Kummer variety is the singular locus of a projective model for the moduli space of rank-two, trivial-determinant vector bundles over X, a key ingredient in the construction of Hitchin-type ACIs [vGP]. This is one more area of intense study centered on the role of σ function in integrability.

References

[B] H.F. Baker, *On a system of differential equations leading to periodic functions.* Acta Math. **27** (1903), 135–156.

[BD] A. Beauville and O. Debarre, *Une relation entre deux approches du problème de Schottky.* Invent. Math. **86** (1986), no. 1, 195–207.

[BKOR] H.J.M. Bos, C. Kers, F. Oort and D.W. Raven, *Poncelet's closure theorem.* Expositiones Math. **5** (1987), 238–364.

[BP] J.-L. Brylinski and E. Previato, *Koszul complexes, differential operators, and the Weil–Tate reciprocity law.* J. Algebra **230** (2000), no. 1, 89–100.

[BÉ] V.M. Bukhshtaber and V.Z. Ènol'skiĭ, *Explicit algebraic description of hyperelliptic Jacobians based on Klein's σ-function.* Funktsional. Anal. i Prilozhen. **30** (1996), no. 1, 57–60; translation in Funct. Anal. Appl. **30** (1996), no. 1, 44–47

[BEL] V.M. Buchstaber, V.Z. Ènol'skiĭ and D.V. Leĭkin, *Hyperelliptic Kleinian functions and applications.* Solitons, geometry, and topology: on the crossroad, 1–33, Amer. Math. Soc. Transl. Ser. 2, 179, Adv. Math. Sci., 33, Amer. Math. Soc., Providence, RI, 1997.

[BLÉ] V.M. Bukhshtaber, D.V. Leĭkin, and V.Z. Ènol'skiĭ, *Matrix realization of hyperelliptic Kummer manifolds.* Uspekhi Mat. Nauk **51** (1996), no. 2(308), 147–148; translation in Russian Math. Surveys **51** (1996), no. 2, 319–320

[BC] J.L. Burchnall and T.W. Chaundy, *Commutative ordinary differential operators,* Proc. Roy. Soc. London Ser. A **118** (1928), 557–583.

[BZ] Igor Burban and Alexander Zheglov, *Fourier–Mukai transform on Weierstrass cubics and commuting differential operators.* Preprint arXiv:1602.08694

[Di] J. Dixmier, *Sur les algèbres de Weyl.* Bull. Soc. Math. France **96** (1968), 209–242.

[DuV] P. Du Val, Elliptic functions and elliptic curves. London Mathematical Society Lecture Note Series, No. 9. Cambridge University Press, London-New York, 1973.

[EE] J.C. Eilbeck and V.Z. Enolskii, *Bilinear operators and the power series for the Weierstrass σ function.* J. Phys. A **33** (2000), no. 4, 791–794

[vGP] B. van Geemen and E. Previato, *On the Hitchin system.* Duke Math. J. **85** (1996), no. 3, 659–683.

[HaS] L. Haine and J.-P. Semengue, *The Jacobi polynomial ensemble and the Painlevé VI equation.* J. Math. Phys. **40** (1999), no. 4, 2117–2134.

[Hi] N. Hitchin, *Stable bundles and integrable systems.* Duke Math. J. **54** (1987), no. 1, 91–114.

[KP] A. Kasman and E. Previato, *Factorization and resultants of partial differential operators.* Math. Comput. Sci. **4** (2010), no. 2-3, 169–184.

[KMP] Y. Kodama, S. Matsutani and E. Previato, *Quasi-periodic and periodic solutions of the Toda lattice via the hyperelliptic sigma function.* Ann. Inst. Fourier **63** (2013), no. 2, 655–688.

[KS] D. Korotkin and V. Shramchenko, *On higher genus Weierstrass sigma-function.* Physica D **241** (2012), 2086–2094.

[Kr] I.M. Krichever, *The integration of non-linear equations by methods of algebraic geometry.* Functional Anal. Appl. **12** (1977), 12–26.

[Mul] M. Mulase, *Geometric classification of commutative algebras of ordinary differential operators.* Differential geometric methods in theoretical physics (Davis, CA, 1988), 13–27, NATO Adv. Sci. Inst. Ser. B Phys., **245**, Plenum, New York, 1990.

[Mum1] D. Mumford, Curves and their Jacobians. The University of Michigan Press, Ann Arbor, Mich., 1975.

[Mum2] D. Mumford, Tata Lectures on Theta, I-II, Birkhäuser, Boston 1983–1984.

[P1] E. Previato, *Another algebraic proof of Weil's reciprocity.* Atti Accad. Naz. Lincei Cl. Sci. Fis. Mat. Natur. Rend. Lincei (9) Mat. Appl. **2** (1991), no. 2, 167–171.

[P2] E. Previato, *Seventy years of spectral curves:* 1923–1993. Integrable systems and quantum groups (Montecatini Terme, 1993), 419–481, Lecture Notes in Math., **1620**, Fond. CIME/CIME Found. Subser., Springer, Berlin, 1996.

[P3] E. Previato, *Poncelet's theorem in space.* Proc. Amer. Math. Soc. **127** (1999), no. 9, 2547–2556.

[PW] E. Previato and G. Wilson, *Differential operators and rank* 2 *bundles over elliptic curves.* Compositio Math. **81** (1992), no. 1, 107–119.

[SS] M. Sato and Y. Sato, *Soliton equations as dynamical systems on infinite-dimensional Grassmann manifold.* Nonlinear partial differential equations in applied science, 259–271, North-Holland Math. Stud. no. **81**, North-Holland, Amsterdam, 1983.

[T] M. Toda, *Vibration of a Chain with Nonlinear Interaction.* J. Phys. Soc. Jpn. **22**, pp. 431–436 (1967)

E. Previato
Department of Mathematics and Statistics
Boston University
Boston, MA 02215, USA
e-mail: `ep@bu.edu`

Geometric Methods in Physics. XXXV Workshop 2016
Trends in Mathematics, 281–282

Volumes of Classical Supermanifolds

Theodore Voronov

Mathematics Subject Classification (2010). 58A50, 32C30, 46L51.

Keywords. Classical supermanifolds, anticommuting variables, integration on supermanifolds.

This is only a sketch of the contents.

Lecture I covered the notion of volume for a compact manifold, starting from elements of volume induced by Riemannian, symplectic, and Kähler structures. Several analytic and geometric concepts were highlighted (e.g., fibre bundles that are Riemannian submersions and factorization of volume elements for them) and examples worked out (e.g., the sphere, whose volume was connected with the Gaussian integral in $\mathbb{R}^n$, the complex projective space with Fubini-Study metric, and Grassmannians).

Lecture II introduced supergeometry and supermanifolds, with emphasis on the Berezinian determinant [1], the key ingredient for the definition of volumes. Concrete examples were again given, harkening back to Lecture I, such as the supersphere and the projective superspace, and a Gaussian integral over superspace [2, 3].

Lecture III presented volumes of classical supermanifolds and recent results [3], the main result being the following property of certain supermanifolds (e.g., the supersphere, complex projective superspace, and Stiefel and Grassmann supermanifolds),

Theorem. *Up to a universal normalization, the volume of the supermanifold can be obtained as an analytic continuation of the volume of the corresponding ordinary manifold.*

References

[1] Berezin, F.A. Introduction to algebra and analysis with anticommuting variables. V.P. Palamodov, ed., Moscow State University Press, Moscow, 1983. Expanded transl.

into English: *Introduction to superanalysis.* A.A. Kirillov, ed., D. Reidel, Dordrecht, 1987.

[2] Voronov, Th. Geometric integration theory on supermanifolds. Soviet Scientific Reviews, Section C: Mathematical Physics Reviews, 9, Part 1. Harwood Academic Publishers, Chur, 1991. Second expanded edition: Cambridge Scientific Publ., 2014. (Vol. 3 in the series: Classic Reviews in Mathematics and Mathematical Physics, I.M. Krichever, ed.)

[3] Voronov, Th. *On volumes of classical supermanifolds.* Sbornik: Mathematics **207** (2016) no. 11, 1512–1536.

Theodore Voronov
School of Mathematics
University of Manchester
Oxford Road
Manchester, M13 9PL, UK
e-mail: theodore.voronov@manchester.ac.uk

MIX
Papier aus verantwortungsvollen Quellen
Paper from responsible sources
FSC® C105338

If you have any concerns about our products,
you can contact us on
ProductSafety@springernature.com

In case Publisher is established outside the EU,
the EU authorized representative is:
Springer Nature Customer Service Center GmbH
Europaplatz 3, 69115 Heidelberg, Germany

Printed by Libri Plureos GmbH
in Hamburg, Germany